Intelligent Control and Learning Systems

Volume 4

Series Editor

Dong Shen, School of Mathematics, Renmin University of China, Beijing, Beijing, China

The Springer book series Intelligent Control and Learning Systems addresses the emerging advances in intelligent control and learning systems from both mathematical theory and engineering application perspectives. It is a series of monographs and contributed volumes focusing on the in-depth exploration of learning theory in control such as iterative learning, machine learning, deep learning, and others sharing the learning concept, and their corresponding intelligent system frameworks in engineering applications. This series is featured by the comprehensive understanding and practical application of learning mechanisms. This book series involves applications in industrial engineering, control engineering, and material engineering, etc.

The Intelligent Control and Learning System book series promotes the exchange of emerging theory and technology of intelligent control and learning systems between academia and industry. It aims to provide a timely reflection of the advances in intelligent control and learning systems. This book series is distinguished by the combination of the system theory and emerging topics such as machine learning, artificial intelligence, and big data. As a collection, this book series provides valuable resources to a wide audience in academia, the engineering research community, industry and anyone else looking to expand their knowledge in intelligent control and learning systems.

Ziye Zhang · Zhen Wang · Jian Chen · Chong Lin

Complex-Valued Neural Networks Systems with Time Delay

Stability Analysis and (Anti-)Synchronization Control

Springer

Ziye Zhang
College of Mathematics and Systems Science
Shandong University of Science and Technology
Qingdao, Shandong, China

Zhen Wang
College of Mathematics and Systems Science
Shandong University of Science and Technology
Qingdao, Shandong, China

Jian Chen
School of Information and Control Engineering
Qingdao University of Technology
Qingdao, Shandong, China

Chong Lin
Institute of Complexity Science, College of Automation
Qingdao University
Qingdao, Shandong, China

ISSN 2662-5458 ISSN 2662-5466 (electronic)
Intelligent Control and Learning Systems
ISBN 978-981-19-5449-8 ISBN 978-981-19-5450-4 (eBook)
https://doi.org/10.1007/978-981-19-5450-4

This Springer imprint is published by the registered company Springer Nature Singapore Pte Ltd.
The registered company address is: 152 Beach Road, #21-01/04 Gateway East, Singapore 189721, Singapore

Preface

In recent years, with the emergence of new scientific and technological methods, complex signals have appeared in radar, imaging and other fields, which makes the state variables of the controlled system extended from the real number domain to the complex number one. As a result, the complex-valued neural networks (CVNNs) model has entered scholars' vision. It is a kind of network that processes information with complex parameters and variables in a complex field. Compared with real-valued neural networks, on one hand, the neuron state, activation function and weight of complex-valued ones are complex-valued, which can directly process two-dimensional data. On the other hand, CVNNs model is more efficient, faster and has better generalization ability. In particular, it can deal with many problems that cannot be solved by real-valued neural networks. So far, CVNNs have been widely used in many biological and engineering fields, such as image processing, optoelectronics, pattern recognition, signal processing and so on. Moreover, time delay is inevitable in the actual system, which has a very significant impact on the system. It often leads to the delay in information transmission, affects the performance of the system and even causes instability in the system. Therefore, it is of great significance to make a profound study of CVNNs model with time delay.

From the point of view of dynamic system, CVNNs with time delay model, as a complex nonlinear dynamic system, its wide applications in various fields largely depend on its dynamic behaviors. In the past ten or more years, a large number of outstanding achievements on dynamic behaviors for this model have been published, involving global stability, exponential stability, Lagrange stability, input-state stability, synchronization, stabilization, dissipativity, passivity, state estimation and so on. However, in this direction, there is a lack of a monograph that provides comprehensive reports on this new growth area.

Along this line, this book aims to provide the latest developments and advances in the dynamics analysis for complex-valued neural network systems with time delay. It is not only a collection of results in the existing references, but also covers a complete interpretation and establishes a systematic statement of the existing analysis and methods. Specifically, the book is mainly concentrated on the recent research development carried on by the authors and their co-workers in this domain, which includes asymptotic stability, Hopf bifurcation, finite-time stability, Lagrange exponential stability, synchronization control, anti-synchronization control, fixed-time synchronization, fixed-time pinning synchronization and adaptive synchronization. It gives an update on recent developments for analysis techniques and synthesis methods, put forward a systematic statement, brings out further insights and covers a complete topic on dynamical behaviors for complex-valued neural networks systems with time delay.

The book is suitable for postgraduates, senior undergraduates, graduate students and senior researchers interested in or working with the system theory of complex-valued neural networks, or for those relevant scientific and technical workers with research in stability and control theory in some areas containing engineering disciplines, applied mathematics, control engineering, system analysis and integration, automation, nonlinear science and computer science. The prerequisites for this book are modest, and it is assumed that the readers of this book have some background in related issues involving neural networks, system control theory and matrix theory.

We are deeply grateful to Dr. Mengchu Huang, Senior Editor at Springer, for his constant guidance and support during the preparation and production of this book. We would like to thank Dr. Dharaneeswaran Sundaramurthy in Springer, for his help in the project coordination and supervision during the production process of this. Both authors appreciate the love and support from their families during the writing of this book. Finally, we also acknowledge valuable support from the National Natural Science Foundation of China (Grant Nos. 62173214, 61803220, 61673227, 61503222, 61573008, 61174033), the Natural Science Foundation of Shandong Province of China (Grant Nos. ZR2021MF100, ZR2011FM006), the Research Fund for the Taishan Scholar Project of Shandong Province of China, the Science and Technology Support Plan for Youth Innovation of Colleges and Universities of Shandong Province of China (Grant Nos. 2019KJI005), and the SDUST Research Fund (Grant Nos. 2019TDJH102).

Qingdao, China
June 2022

Ziye Zhang
Zhen Wang
Jian Chen
Chong Lin

Contents

Symbols

$\mathbb{R}^+$	Set of positive real numbers
$\mathbb{R}$	Set of real numbers
$\mathbb{C}$	Set of complex numbers
$\mathbb{R}^n$	Set of n-dimensional real vectors
$\mathbb{C}^n$	Set of n-dimensional complex vectors
$\mathbb{R}^{n \times m}$	Set of $n \times m$ real matrices
$\mathbb{C}^{n \times m}$	Set of $n \times m$ complex matrices
i	Imaginary unit with $i = \sqrt{-1}$
I	Identity matrix with proper dimension
$\bar{z}$	Conjugate of complex number z
A^T	Transpose of real matrix A
A^*	Conjugate transpose of complex matrix A
$A > 0$	Symmetric positive definite as $A \in \mathbb{R}^{n \times n}$
$A \geq 0$	Symmetric positive semi-definite as $A \in \mathbb{R}^{n \times n}$
$A < 0$	Symmetric negative definite as $A \in \mathbb{R}^{n \times n}$
$A \leq 0$	Symmetric negative semi-definite as $A \in \mathbb{R}^{n \times n}$
$A > 0$	Hermitian positive definite as $A \in \mathbb{C}^{n \times n}$
$A \geq 0$	Hermitian positive semi-definite as $A \in \mathbb{C}^{n \times n}$
$A < 0$	Hermitian negative definite as $A \in \mathbb{C}^{n \times n}$
$A \leq 0$	Hermitian negative semi-definite as $A \in \mathbb{C}^{n \times n}$
$\text{diag}\{\ldots\}$	Diagonal matrix
$\lambda_{\min}(A)$	Minimum eigenvalue of matrix A
$\lambda_{\max}(A)$	Maximum eigenvalue of matrix A
$\forall$	For all
$\in$	Belong to
$\sum$	Sum
$\lvert . \rvert$	Absolute value (or modulus)
$\lVert . \rVert$	Norm
$\rightarrow$	Tend to or mapping to
sup	Supremum
inf	Infimum

D^+	Upper Dini-derivative
LMI	Linear matrix inequality

Chapter 1
Introduction

Abstract In this chapter, we will introduce the history of neural networks and complex-valued neural networks, and provide detailed recent progress and advances of complex-valued neural networks with emphasis on complex-valued neural networks with time delay, which include multiple dynamical behaviors achievements and analysis techniques and synthesis methods applied on them. Finally, we will state the purpose, scope, intended audience and organization of the book.

1.1 History of Neural Networks

Artificial neural networks (ANNs), usually simply called neural networks (NNs), are computing systems inspired by the biological neural networks that constitute animal brains. An ANN is based on a collection of connected units or nodes called artificial neurons, which loosely model the neurons in a biological brain. In most cases, NNs can change the internal structure on the basis of external information, which is a class of adaptive systems. Generally speaking, they own the function of learning.

The development of NNs has experienced the climax and trough, which is mainly divided into three stages. Firstly, in 1943, on the basis of mathematics and an algorithm called threshold logic, Mcculloch and Pitts opened the subject by creating a computational model for neural networks—MP model (Mcculloch and Pitts 1943), which divides the research for NNs into two different ideas. One mainly focuses on the biological processes in the brain, and the other centers on the application of artificial intelligence. In the late 1940s, Hebb created a learning hypothesis based on the mechanism of neural plasticity that became known as Hebbian learning (Hebb 1949), which lays the groundwork for the birth of NNs learning algorithm. In 1954, Farley and Clark first used computational machines, then called "calculators", to simulate a Hebbian network (Farley and Clark 1954) and Rochester et al. simulated the behavior of abstract NNs on an IBM 704 computer (Rochester et al. 1956) in 1956. Then, Rosenblatt put forward the perceptron neural networks model in 1958

Z. Zhang et al., *Complex-Valued Neural Networks Systems with Time Delay*,
Intelligent Control and Learning Systems 4,
https://doi.org/10.1007/978-981-19-5450-4_1

(Rosenblatt 1958), pushing the research of NNs to the first peak. Until 1969, Minsky and Papert discovered that the basic perceptrons were incapable of handling the exclusive-OR circuit and that computers lacked sufficient power to process useful NNs (Minsky and Papert 1969). Thereafter, the research has fallen into stagnation for many years.

Then, in 1980, Fukushima invented neocognitron (Fukushima 1980). Hopfield artificial neural networks model was established in the 1980s (Hopfield 1982, 1984). These opened the prelude to the revival of NNs. In 1986, Rumelhart, Hinton and Williams introduced multilayer perceptron into the backpropagation (BP) algorithm proposed by Werbos in 1975, enabling practical training of multi-layer networks (Werbos 1982; Rumelhart et al. 1986). At this moment, NNs became a hot spot again. BP algorithm is a very significant learning one in NNs, which still occupies an important position until now. In 1989, Lecun et al. proposed convolutional NNs (LeCun et al. 1989), but due to the long training time and the popularity of support vector machines (SVM) at that time, the research of NNs fell into a low tide again.

In 2006, Hinton et al. proposed deep belief networks (DBN) (Hinton et al. 2006), which used a greedy unsupervised training method to solve the problem and achieved good results. This training method reduced the difficulty of learning hidden layer parameters. With the development of distributed computing and GPU, the computing power is greatly improved. This allowed the use of larger networks, particularly in image and visual recognition problems, which became known as deep learning (DL) (Heaton 2016). NNs ushered in the third climax. The improvement of computing technology made many complex operations cheap. So, a large number of deep networks began to appear, which solved countless problems in the world. DL has become the most popular research direction in the field of artificial intelligence.

In the development process of NNs, many kinds of neural networks models were proposed, including bidirectional associative memory neural networks, cellular neural networks, inertial neural networks, Cohen-Grossberg neural networks, memristive neural networks, quantum neural networks, fractional-order neural networks and so on. In which, bidirectional associative memory neural networks (BAMNNs) were first proposed by Kosko in 1987 with two-layer hetero associative neurons (Kosko 1987, 1988). Because of iterations of forward and backward information flows between the two layers and imitating the thinking mode of human brain hetero associative, it is a kind of hetero associative model with information memory and association ability. The two layers of neurons are completely interconnected while the neurons in the same layer have no correlation. Because this kind of network has the characteristics of distributed information storage, it can associate a complete and clear pattern stored in memory from an incomplete or fuzzy one. Due to its special structure of connection weights, BAMNNs have been widely applied in many key fields, such as pattern recognition, image processing, associative memory, nuclear reactor and automatic control engineering.

Another important neural networks model, inertial neural networks (INNs), was put forward by Babcock and Westervelt in 1986 (Babcock and Westervelt 1986) via introducing inductor into artificial neural networks circuit. In fact, nonlinear oscillation resistance or inertial term can effectively improve the disordered search performance of the network, and it is also an important tool for the network to produce complex behaviors such as bifurcation and chaos (Babcock and Westervelt 1986; Wheeler and Schieve 1997). Moreover, the inertial term itself also has a strong biological background. For instance, the cell membrane of some organisms can be simulated by the circuit system including inductance, and the function of squid axon can also be realized by designing an inductance (Angelaki and Correia 1991).

1.2 History of Complex-Valued Neural Networks

It should be noted that the aforementioned neural network models are considered in the real domain, which are greatly limited in processing multidimensional data. In recent years, with the emergence of new scientific and technological methods, many industrial processes involve complex parameters and complex signals, which results in the variable range of the controlled system being extended to complex domain and attracts great attention from academic circles. This has caused a research upsurge for complex-valued neural networks (CVNNs) which can handle point data modeling in two-dimensional space as a whole. Compared with a real number, a complex one has great superiority since it can give a concise and natural way of representing problems.

Actually, the historical origin of CVNNs can be traced back to the middle of the twentieth century. The phase information in computation was first introduced by Eiichi Goto in his invention for parametron (Goto 1954, 1955). The parametron represents the values to be processed by discrete phase values of oscillation. The phase was used for a higher stability in the information representation in computing systems. In 1971, Aizenberg et al. reported an idea to extend binary output values of "0" and "1" to multiple values on the unit circle in the complex plane (Aizenberg et al. 1971). The multiple-valued output signals are placed on discrete points on the unit circle. Such a class of CVNNs is sometimes called the phasor neural networks where the amplitude is fixed at unity. In 1988, Noest proposed a multiple-valued associative memory network, which is the prototype of complex-valued Hopfield neural networks (Noest 1988a, b).

In 1991, Leung and Haykin discussed a kind of new CVNNs with complex variables and analyzed the dynamics by calculating partial derivatives of real and imaginary parts (Leung and Haykin 1991), respectively. In 1992, Benvenuto and Piazza considered a class of CVNNs with activation function expressed separately in real and imaginary parts as $f(z) = \text{sigmoid}(\text{Re}(z)) + i\,\text{sigmoid}(\text{Im}(z))$, namely, the real-imaginary-type activation function (Benvenuto and Piazza 1992), which is one of the simple extensions of a real sigmoid activation function. Its advantage is that we can also obtain a BP algorithm as a simple extension of that in the real-valued networks. Later, Nitta also considered this kind of CVNNs with the real-imaginary-type activation function and analyzed the dynamical behaviors by handling the real and imaginary parts separately (Nitta 1997). Of course, there also exist other references considering CVNNs with the real-imaginary-type activation function and analyzing steepest descent methods and BP algorithms by using independent real and imaginary parts (Kim and Guest 1990; Birx and Pipenberg 1993).

After some theoretical analysis for real-imaginary-type in the beginning, some researchers began to discuss which kind of activation function is meaningful. In 1992, Georgiou and Kutsougeras proposed a function whose amplitude is saturating, whereas the phase is unchanged (Georgiou and Koutsougeras 1992), that is, $f(z) = \frac{z}{c+|z|/r}$, where c and r are parameters determining the saturation characteristic. In the same year, Hirose proposed an associative memory employing an amplitude-phase-type activation function, and again provided the steepest descent method and a BP algorithm for CVNNs with the amplitude-phase-type activation function expressed as $f(z) = \tanh(|z|)\exp(i\arg(z))$ in Hirose (1992a, b, 2012a), which is basically equivalent to that one in Georgiou and Koutsougeras (1992). They were derived based on partial derivatives in amplitude and phase directions, which are compatible with this activation function. After that, the research on the theory and application of CVNNs has been going on.

For CVNNs, the information on amplitude and phase can be widely used, and amplitude and phase form the main characteristics. For instance, using a single complex neuron to solve the XOR problem (Nitta 2003) and treating speed and direction as an entire complex vector in wind profile model (Goh et al. 2006) made full use of this feature. In contrast with two-dimensional real-valued NNs, the functions of phase rotation and amplitude adjustment can reduce the freedom degree of the network in learning and self-organization, which is suitable for phase rotation phenomenon or information. Therefore, CVNNs are most suitable for robot technology with rotating joints, wave signal processing of phase transmission main information and so on. So far, the application of CVNNs has involved many aspects of production and life, such as ultrasonic detection, sonar system, radar design, electromagnetic wave, antenna design, UWB communication pulse beam, fuzzy compensation image processing, frequency division multiplexing microwave signal processing, optical wave learning logic circuit, optical tweezers adaptive three-dimensional holographic film, associative memory, independent component analysis, social system and traffic power system, etc. There have been lots of references collecting the application of CVNNs (Hirose 1994, 2012a, b, 2013; Hirose and Eckmiller 1996; Zhang and Ma 1997; Suksmono and Hirose 2005; Yang and Bose 2005; Hirose et al. 2006a, b; Lee 2006; Aizenberg et al. 2008; Goh and Mandic 2007; Kawata and Hirose 2008; Li and Adali 2008; Nitta 2009; Aizenberg 2011; Hirose and Yoshida 2012).

Up to now, various CVNNs models, such as complex-valued bidirectional associative memory neural networks (CVBAMNNs), complex-valued inertial neural networks (CVINNs), complex-valued Cohen-Grossberg neural networks, complex-valued memristive neural networks, complex-valued fractional-order neural networks and so on, have been successfully used in a number of real-world applications. Meanwhile, to support practical applications, numerous theoretical achievements have been reported. For instance, ensuring that the considered models are stable is a basic task of the first importance before putting them into application. Thus, it is very essential and significant to study the dynamics of various CVNNs models. So far, multiple dynamical behaviors containing stability, synchronization control, anti-synchronization control, bifurcation analysis and so on have been analyzed, and a large number of corresponding methodologies have been proposed one after another.

1.3 Recent Progress of Complex-Valued Neural Networks

From the proposal of the CVNNs model to the development of various dynamic behaviors, the issue for this model has always been a research hot point. It greatly promotes the research and development of system analysis and control in complex domain. For CVNNs, activation functions are complex-valued. Generally, for the research of real-valued neural networks, activation functions are generally assumed to be smooth and bounded, such as sigmoid function. If the assumption about smoothness and boundedness is directly extended to complex-valued activation functions, that is, they are analytic and bounded. According to the Liouville theorem (Rudin 1987), such functions are constant. Obviously, this is not suitable. In other words, complex-valued activation functions cannot be both bounded and analytic. Therefore, there always exists a great challenge and difficulty in choosing activation functions. This fact makes theoretical research for CVNNs more difficult.

At present, from the theoretical development, most results are developed based on two cases. One is that activation functions can be expressed by separating their real and imaginary parts and the other is that they can't be explicitly separated into their real and imaginary parts. In the following, we will give detailed reviews on the development of dynamical behaviors of three salient models involving CVNNs, CVBAMNNs and CVINNs on the basis of the aforementioned two cases.

1.3.1 Survey of Dynamics of CVNNs

A class of the generalized CVNNs is of the following form:

$$\dot{z}_j(t) = -d_j z_j(t) + \sum_{p=1}^{n} a_{jp} f_p(z_p(t)) + I_j(t), \tag{1.1}$$

where $j = 1, 2, \ldots, n$, $z_j(t) \in \mathbb{C}$ is the state of the jth neuron at time t; $d_j > 0$ is the self-feedback connection weight; $a_{jp} \in \mathbb{C}$ is the connection weight; $f_p(\cdot)$ is the complex-valued activation function; $I_j(t) \in \mathbb{C}$ is the external input of the jth neuron at time t. Equation (1.1) matches continuous-time CVNNs. For discrete-time ones, the expression form is difference equation, which is omitted here. For activation function expressed by separating their real and imaginary parts, there are several common types, such as Type A: $f(z) = f^R(z^R, z^I) + i f^I(z^R, z^I)$, Type

B: $f(z) = f^R(z^R) + if^I(z^I)$ (the special case of Type A), and Type C: $f(z) = \max(0, z^R) + i\min(0, z^I)$; for real and imaginary parts which cannot be explicitly separated, for example, $f(z) = \frac{1-e^{-\bar{z}}}{1+e^{-\bar{z}}}$, to name but a few.

(1) Achievements based on separation method The separation method, that is, activation functions are real-imaginary-type ones, then the original system is rewritten as two equivalent real-valued ones. From methodology, the basic theories and methods in real number field can be used directly. So, it is less difficult to achieve the purpose of theoretical analysis. From the presentation form of the theoretical results, it becomes complex and calculation pressure enlarges because of the increase of variables. Therefore, this separation method has both advantages and disadvantages. When we adopt it, in order to meet the actual needs, we need to find a balance to achieve our purpose.

In the early twenty-first century, for several classes of complex-valued activation functions including the above-mentioned Type A and Type B, and $f(z) = \psi(|z|)e^{i\phi(\arg(z))}$, via energy function method, their properties and how to find complex-valued functions to meet these properties were discussed in Kuroe et al. (2002, 2003). Then, for a class of discrete-time complex-valued Hopfield neural networks, Rao and Murthy (2008) provided a sufficient condition of global exponential stability for this system with activation functions which are bounded and satisfy the Lipschitz condition in complex domain and Liu et al. (2009) analyzed the synthesis problem with Type B. In the following ten years, the study of dynamics entered a peak period.

On one hand, numerous theoretical results have emerged on stability, dissipativity, passivity, bifurcation and state estimation. In 2012, Hu and Wang considered activation function of Type A, supposed the existence, continuity and boundedness of the partial derivatives for the real and imaginary parts, and derived sufficient conditions to ascertain the existence, global asymptotic stability and exponential stability of unique equilibrium for delayed CVNNs by applying homeomorphism theory and Lyapunov function theory (Hu and Wang 2012). Then, based on the same assumption in Hu and Wang (2012), global robust stability for delayed system under parameter uncertainties, exponential stability of system with mixed delays, global μ-stability for system with unbounded time-varying delays, Hopf bifurcation for delayed system, exponential stability for system with asynchronous time delays and exponential stability of system with mixed time delays and impulsive effect based on the vector Lyapunov function method are studied in Zhang and Li (2014), Li et al. (2018), Xu et al. (2014, 2017), Velmurugan et al. (2015), Dong et al. (2015, 2019), Liu and Chen (2016), Ji et al. (2018), respectively. In fact, this assumption for the existence, continuity and boundedness of the partial derivatives is a strong constraint. The references Rakkiyappan et al. (2015b), Gong et al. (2016, 2017a, 2019), Hu and Wang (2015), Wang et al. (2018), Hu et al. (2020), Xu et al. (2019), Li et al. (2021a), Li et al. (2021) removed this restriction and investigated multiple μ-stability, stability analysis based on nonlinear measure approach, exponential periodicity of discrete-time delayed system, state estimation for delayed system with norm-bounded parameter uncertainties, generalized stability and global exponential stability for systems with discontinuous activation functions via differential inclusions, mean square exponential stability and robust state estimation for stochastic delayed systems with Brownian motion, robust state estimation problem for semi-Markovian switching system with quantization effects and linear fractional uncertainties, and global exponential stability for system with proportional delays and inhibitory factors. Moreover, for discrete-time CVNNs with time-varying delays and distributed delays, Liu et al. (2022) considered the design problem of a stubborn state estimator.

Moreover, for Type A satisfying the Lipschitz condition in complex domain, when supposing again that activation functions are bounded, Xie and Jiang (2016) and Du (2018) studied existence and global exponential stability of periodic solution for delayed system by using Mawhin's continuation theorem. Ramasamy and Nagamani (2017) investigated dissipativity and passivity problems for discrete-time system with leakage delay and probabilistic time-varying delays. Considering the fact that the boundedness of activation functions can lead to limitations in choosing them, the authors of this book removed this restriction and derived delay-independent and delay-dependent stability criteria (Zhang et al. 2014, 2017b). Later, Cao et al. (2020) and Sriraman et al. (2020) discussed robust stability of stochastic system with additive time-varying delays and asymptotic stability of stochastic system with probabilistic time-varying delays, respectively. Rajchakit and Sriraman (2021) investigated the robust passivity and stability analysis for impulsive models with the uncertainty of norm-bounded parameters. Further, in Zhang et al. (2017), no assumption was made on activation functions of Type A, and equivariant bifurcation in coupled complex-valued neural network rings was discussed.

For activation function of Type B, global stability of system with both leakage time delay and discrete-time delay on time scales was discussed in Chen and Song (2013). Global exponential stability was analyzed based on the matrix measure method and the Halanay inequality instead of constructing Lyapunov functions in Gong et al. (2015a). Multistability, global asymptotic stability, and global stability of almost periodic solution were tackled for system with continuous or discontinuous activation functions in Huang et al. (2014, 2015), Chen et al. (2017), Gong et al. (2017b), Liang et al. (2016), Xu and Tan (2018), Zhang et al. (2021), Wang et al. (2017), Yan et al. (2018a, b). By using the matrix measure approach, global exponential stability problems for the impulsive Markovian switching system and deviating argument system were investigated in Li et al. (2019) and Zhou et al. (2020), respectively. For activation function of Type C—linear threshold function—the boundedness and complete stability were studied via the energy minimization method (Zhou and Song 2013), complete stability of delayed impulsive system (Rakkiyappan et al. 2015a) was analyzed and global attracting sets for a class of non-autonomous system (Li et al. 2016) were discussed.

On the other hand, as a typical dynamic behavior, synchronization was originally proposed by Pecora and Carroll (1990). It means that the trajectories of the master-slave systems can tend to be consistent with the change of time. Similarly, anti-synchronization means that the trajectories tend to be opposite. In fact, synchronization and anti-synchronization phenomena are ubiquitous in our lives and also exist in many practical systems and engineering fields. For instance, birds gathering, fireflies glowing synchronously, regulation of human heartbeat and secure communication system, etc. For CVNNs, many asymptotic (anti-)synchronization phenomena have been considered, such as common (anti-)synchronization, exponential (anti-)synchronization, quasi-synchronization, combination synchronization, projective synchronization, module-phase synchronization and so on (Zhang et al. 2016, 2017, 2020; Zhang and Wang 2017; Liu et al. 2019a, 2020; Xie et al. 2019; Kan et al. 2019; Duan et al. 2019; Li et al. 2019, 2021; Zhou et al. 2020; Wei et al. 2020; Guo et al. 2020; Kumar et al. 2021a, b; Chen et al. 2021; Kumar et al. 2021; Nian and Li 2021). Here, it should be noticed that the above (anti-)synchronization issues can be reached in infinite time. However, to meet the actual requirement, sometimes the (anti-)synchronization is required to be realized in a faster or limited time. So, various (anti-)synchronization problems in finite time have garnered a lot of attention in recent years (Zhou et al. 2017; Zhang et al. 2018a, d; Wu et al. 2018; Liu et al. 2019, 2020; Wang et al. 2019; Aouiti et al. 2020; Liu and Li 2020; Song et al. 2021; Pan and Zhang 2021; Hui et al. 2021; Aouiti and Bessifi 2021; Duan et al. 2021). Actually, as we know, in finite-time control theory, the settling time is estimated by the initial values which must be accurately given. When the initial values may be unavailable or not accurate enough owing to disturbance and obliviousness, the finite-time control will be inapplicable. At this moment, a special finite-time control scheme, named the fixed-time control, is often concerned. It was raised by Polyakov (2012). Its outstanding advantage is that the settling time can be bounded by a fixed upper bound independent of the initial values. Hence, it is more applicable to actual requirement. Further, for CVNNs, various (anti-)synchronization issues in fixed time have become popular (Ding et al. 2017; Aouiti et al. 2020; Mi et al. 2020; Song et al. 2021; Aouiti and Bessifi 2021; Duan et al. 2021).

For asymptotic (anti-)synchronization in infinite time, the earlier results were published in Zhang et al. (2016, 2017), Zhang and Wang (2017). In which, Zhang et al. (2016, 2017) considered activation function of Type A without any assumption and studied the module-phase synchronization by designing sliding mode controllers and adaptive feedback controllers, respectively. Zhang and Wang (2017) discussed the projective synchronization with a complex projective factor for system with activation function of Type B. Then, in nearly 3 years, for activation function of Type A without the assumption for the existence, continuity and boundedness of the partial derivatives, the references Kan et al. (2019), Chen et al. (2021), Liu et al. (2020), Kumar et al. (2021), Duan et al. (2019), Li et al. (2021), Wei et al. (2020), Guo et al. (2020), Nian and Li (2021) studied exponential synchronization under impulsive control strategies for systems with only discrete delays and with both discrete and distributed delays, exponential synchronization for networks with discontinuous neuron activations, matrix measure method to exponential synchronization for models with proportional delays, anti-synchronization for networks with leakage delay and time-varying delays, pth moment exponential anti-synchronization for stochastic system with Weiner process, and exponential module-phase synchronization of models with stochastic perturbation. Meanwhile, for activation function of Type A without the restriction of boundedness and satisfying the Lipschitz condition in complex domain, Liu et al. (2019a) focused on the synchronization problem of two delayed stochastic systems by using stochastic analysis technique and Xie et al. (2019) investigated global exponential synchronization based on matrix measure method. For the activation function of Type B, a special synchronization—combination synchronization, which can bring inspiration for multiple network-interactions in biological system—was considered via adaptive feedback control method in Zhang et al. (2020). Exponential synchronization problems were analyzed by applying matrix measure approach for some models involving impulsive Markovian switching system, deviating argument system, uncertain system, networks with interaction terms (Li et al. 2019; Zhou et al. 2020; Kumar et al. 2021a, b).

For finite-time synchronization of CVNNs, by designing a simple state feedback controller, several conditions were derived for the drive-response system realizing synchronization in finite-time in 2017 (Zhou et al. 2017). It is a relatively early result on this issue and the considered system owns the activation function of Type A without the assumption for the existence, continuity and boundedness of the partial derivatives. Later, for this type of function, many achievements have been produced in succession by designing an appropriate controller. For example, by estimating the derivative of Lyapunov function less than or equal to a negative number and using the definition of finite-time stability, Wu et al. (2018), Liu et al. (2019, 2020) provided several sufficient conditions for finite-time synchronization. By applying traditional finite-time stability theorems fully, for several models containing delayed system, stochastic system, discontinuous system and diffusive system, Zhang et al. (2018d), Hui et al. (2021), Aouiti et al. (2020), Aouiti and Bessifi (2021), Duan et al. (2021) discussed traditional synchronization, projective synchronization and synchronization based on sliding mode control scheme over finite-time interval. In reaching finite-time synchronization, besides these traditional methods, there have been some other ones, such as the integral inequality method in Zhang et al. (2018a) and the new finite-time convergence theorem proposed in Wang et al. (2019). Moreover, for activation function of Type A with the assumption for the existence and boundedness of the partial derivatives, the finite-time anti-synchronization was studied via two-phases-method in Liu and Li (2020). For activation function of Type B, by using inequality techniques and designing the exponential-type controllers of time variable, the finite-time synchronization for a class of delayed networks was concerned in Pan and Zhang (2021). Based on the traditional finite-time stability theorem, finite-time synchronization for nonlinear interconnected complex-valued neural networks with the Markovian jump parameters and reaction-diffusion terms was considered in Song et al. (2021) and the practicability of the corresponding result was verified via a secret communication example.

For fixed-time synchronization of CVNNs, via traditional fixed-time stability theorem, the synchronization and anti-synchronization for a class of systems with discontinuous activation functions and parameter uncertainties over fixed-time interval were first reported in Ding et al. (2017). Next, Aouiti et al. (2020), Aouiti and Bessifi (2021), Duan et al. (2021), Song et al. (2021) published fixed-time synchronization results for several models with activation functions of Type A and Type B. By establishing a new fixed-time stability theorem, the fixed-time synchronization issue was analyzed for a system with activation function of Type B (Mi et al. 2020) and a more accurate estimate for the settling time was given.

For CVNNs, based on the separation method, except that the above mentioned various dynamical behaviors were concerned, several stabilization problems were also addressed. For system with actuator failures and activation function of Type A satisfying the Lipschitz condition in complex domain, the fault-tolerant output-feedback stabilization issue was considered in Zhou et al. (2019). For piecewise-homogeneous Markovian switching system with activation function of Type A without the assumption for the existence, continuity and boundedness of the partial derivatives, stabilization criteria are presented by designing certain mode-dependent state feedback controller in Li et al. (2020). To reduce the frequency of data transmission, the exponential stabilization via aperiodic sampled-data control (Yao et al. 2020) and the global stabilization via event-triggered scheme (Wang et al. 2020) for delayed systems with activation functions of Type B were tackled, respectively.

(2) Achievements based on nonseparation method The nonseparation method, that is, the original system is treated as an entirety to investigate its dynamical behaviors. At this time, activation functions which cannot be explicitly expressed by real and imaginary parts are usually involved, but sometimes some real-imaginary separated ones are also considered. Either way, the addressed activation functions are mostly assumed to satisfy the Lipschitz condition in the complex domain. Then, by using the basic theories in the complex domain and developing some techniques in the complex domain, the original system is analyzed to establish a series of achievements.

On one hand, for stability, dissipativity, passivity, and state estimation, some results have been reported. The earlier ones mainly involve the references Zhou and Zurada (2009) and Duan and Song (2010), which studied the boundedness, global attractivity, complete stability and exponential stability for discrete-time CVNNs with linear threshold functions of Type C, respectively. Later, Rakkiyappan et al. (2015a) also focused on this type and finished the complete stability analysis for the system with impulsive effects. In fact, except for the above results related to linear threshold functions, in the published achievements, the forms of activation functions such as real-imaginary separated form or not are no longer noticed, but just assume that they satisfy the Lipschitz condition in complex domain. In 2012, Hu and Wang supposed that activation functions are bounded and satisfy the Lipschitz condition in complex domain, and proposed sufficient criteria of globally asymptotical stability by some inequalities of modulus of complex number (Hu and Wang 2012). Then, in 2013, Chen and Song extended Jensen-s inequality (Gu 2000) to the complex domain and provided stability criteria in forms of complex-valued linear matrix inequality (CVLMI) for system with both leakage delay and discrete-time delay on time scales (Chen and Song 2013). Meanwhile, the references Zhang and Li (2014), Xie and Jiang (2016), Song and Zhao (2016) discussed robust stability for networks with parameter uncertainties, exponential stability of periodic solution for networks with impulses, and exponential stability of system with leakage delay and time-varying delays, respectively. Based on the extended Jensen's inequality in complex domain in Chen and Song (2013), various inequalities dealing with integral terms containing tighter inequalities, inequality of double integral and inequality of triple integral, were extended to the complex domain and the relative conditions were expressed in terms of CVLMIs for multiple systems, such as system with probabilistic time-varying delays, switched system, neutral-type system and so on (Song et al. 2015a; Samidurai et al. 2018, 2019; Gunasekaran and Zhai 2019; Sriraman and Samidurai 2019). Moreover, Ramasamy and Nagamani (2017) extended two finite sum inequalities to the complex domain and investigated dissipativity and passivity problems for discrete-time delayed CVNNs.

As previously mentioned, the boundedness of activation functions will lead to the limitation of theoretical results in applications. Hence, many scholars removed the boundedness, relaxed the selection range of activation functions and showed the existence and uniqueness of equilibrium point via using the proposed conditions instead of the boundedness. The earlier result was introduced in Fang and Sun (2014), which gave less conservative than the Lipschitz condition and derived a new criterion. By applying M-matrix theory, Hu and Wang (2015), Song et al. (2015b) studied global exponential periodicity and exponential stability for discrete-time models, and Pan et al. (2015), Song et al. (2016a), Xu et al. (2018a), Yang and Liao (2019) analyzed exponential stability, stochastic robust exponential stability, and the attracting and invariant sets for continuous-time ones. Further, Velmurugan et al. (2015), Gong et al. (2015b), Chen et al. (2016) considered the problem of global μ-stability for continuous-time systems and discrete-time ones, respectively. For system with impulsive effects, system with proportional delays, system with interval parameter uncertainties and system with two additive time-varying delays, Song et al. (2016b, 2017, 2018a, b), Zhang et al. (2018), Xu et al. (2018b), Wang and Liu (2019), Liang et al. (2018), Zhang and Zhou (2022), Gunasekaran and Zhai (2020) presented the exponential stability, Lagrange stability, robust stability, state estimation, sampled-data state estimation and polynomial stability. In Zhang et al. (2017b), the authors of this book given new delay-dependent stability criterion that will be introduced in detail in Chap. 3. Zhang et al. (2017a), Zhang and Hao (2018) developed new complex-valued algebraic inequalities and derived new CVLMIs to ensure the asymptotic stability. Jian and Wan (2018) builded impulsive CVNNs model based on Takagi-Sugeno (T-S) fuzzy logic and discussed the global exponential convergence problem. For stochastic CVNNs models, Song et al. (2022), Liu et al. (2019b), Li and Liang (2020a), Li et al. (2021b) addressed input-to-state stability, robust dissipativity and H_∞ estimation.

On the other hand, for other dynamical behaviors, such as multiple synchronization control phenomena, based on the nonseparation method, there have already existed some results. For activation functions satisfying the Lipschitz condition in complex domain, asymptotic synchronization such as synchronization for two stochastic systems (Liu et al. 2019a), μ-synchronization for impulsive system (Hu et al. 2018), synchronization for two nonidentical models via integral sliding mode control approach (Wang et al. 2019), synchronization for system with mixed two additive time-varying delays (Yuan et al. 2019), exponential synchronization for impulsive system via matrix measure method (Li et al. 2019; Li and Mu 2019), anti-synchronization for system with leakage delay (Wei et al. 2020), and event-triggered synchronization via pinning impulsive control (Shen and Liu 2022) were explored deeply. For finite/fixed-time synchronization, in order to achieve the desired goal, Feng et al. (2021) proposed a complex-valued sign function, builded its properties, designed two discontinuous controllers under the quadratic norm and a new norm based on absolute values of real and imaginary parts, and then derived several finite/fixed-time synchronization criteria and provided the estimates of the settling time.

For CVNNs, based on the nonseparation method, some synthesis problems were also considered. For instance, by means of the newly developed reciprocal convex matrix inequality and summation inequalities in complex domain, the filtering problem for delayed discrete-time CVNNs was concentrated on in Ganesan and Gnaneswaran (2020). The exponential stabilization problem of delayed system under quantized sampled-data control scheme was addressed in Wang et al. (2021). For models with Markov jump, improved stabilization methods and sampling-based event-triggered control strategy were developed in depth in Li and Liang (2020b) and Wang et al. (2021), respectively.

1.3.2 Survey of Dynamics of CVBAMNNs

A class of the generalized CVBAMNNs is of the following form:

$$\begin{cases} \dot{u}_j(t) = -d_{1j}u_j(t) + \sum_{k=1}^{m} a_{jk} f_k(v_k(t)) + I_j(t), \\ \dot{v}_k(t) = -d_{2k}v_k(t) + \sum_{j=1}^{n} \tilde{a}_{kj} g_j(u_j(t)) + \tilde{I}_k(t), \end{cases} \tag{1.2}$$

where $j = 1, 2, \ldots, n$ and $k = 1, 2, \ldots, m$, $u_j(t)$ and $v_k(t)$ are complex-valued state variables of the jth neuron and kth neuron at time t; $d_{1j} > 0$ and $d_{2k} > 0$ are constants; a_{jk} and $\tilde{a}_{kj}$ are complex-valued connection weights; $f_k(\cdot)$ and $g_j(\cdot)$ are complex-valued activation functions; $I_j(t)$ and $\tilde{I}_k(t)$ are the external inputs of the jth neuron and kth neuron at time t.

For this class of model with or without multiple types of time delays including constant delay, time-varying delays, leakage delay, neutral-type delay and so on, on the basis of the separation approach and the nonseparation approach, some achievements on stability, (anti)-synchronization and stabilization have been published. In 2016, Wang and Huang proposed a CVBAMNNs model with constant delay and studied asymptotic stability problem via the separation approach (Wang and Huang 2016). Here, activation functions were assumed to satisfy the Lipschitz condition in complex domain. Later, under this assumption and separation technique, for neutral-type CVBAMNNs, asymptotic stability and finite-time state estimation were considered in Xu and Tan (2017), Guo et al. (2019). For CVBAMNNs with constant delay, a finite-time stability criterion was derived (Zhang et al. 2018c) and asymptotic periodic synchronization was concerned via vector-valued inequality in

Zhang et al. (2018e). The criteria derived in the above references were given in terms of LMIs. Then, for activation function of Type A without the assumption of the existence, continuity and boundedness of the partial derivatives, under the decomposition approach, Zhang et al. (2020, 2021), Wei et al. (2019), Cao et al. (2021) discussed the Lagrange exponential stability, anti-synchronization, fixed-time synchronization and finite-time stabilization for several kinds of delayed models. For delayed system with activation function of Type B, Rajivganthi et al. (2019), Wang et al. (2021) studied matrix measure method to dissipativity analysis and pinning fixed-time synchronization. The sufficient criteria derived in the above references (Zhang et al. 2020, 2021; Wei et al. 2019; Cao et al. 2021; Rajivganthi et al. 2019; Wang et al. 2021) were given in algebraic forms.

However, there have been few results based on the nonseparation method for CVBAMNNs model. For activation functions satisfying the Lipschitz condition in complex domain, Zhang et al. (2020) provided algebraic conditions and the CVLMI condition, and analyzed their superiority and inferiority. Popa (2020) developed complex-valued Wirtinger-based integral inequality and investigated global μ-stability for neutral-type system with leakage delay by using some inequalities techniques.

1.3.3 Survey of Dynamics of CVINNs

A class of the generalized CVINNs is of the following form:

$$\ddot{z}_j(t) = -a_j\dot{z}_j(t) - b_j z_j(t) + \sum_{p=1}^{n} c_{jp} f_p(z_p(t)) + I_j(t), \tag{1.3}$$

where $j = 1, 2, \ldots, n$, $z_j(t) \in \mathbb{C}$ is the state of the jth neuron at time t; the second derivative term is called an inertial term of (1.3); $a_j, b_j \in \mathbb{R}$ are positive constants; $c_{jp} \in \mathbb{C}$ is the connection weight; $f_p(\cdot)$ is the complex-valued activation function; $I_j(t) \in \mathbb{C}$ is the external input of the jth neuron at time t.

As we see, the CVINNs model is expressed by the second-order differential equation. So, there are two ways of dealing with this model. One is a reduced-order method, that is, via variable substitution the second-order original system is transformed into the first-order one and then the latter is focused on by using those methods on CVNNs or CVBAMNNs. The other is a nonreduced-order method, which means that the original second-order system is considered and analyzed directly. Based on reduced-order approach, Tang and Jian (2018) first studied CVINNs model with two kinds of activation functions including Type A without the assumption for the existence, continuity and boundedness of the partial derivatives and Type B, and established algebraic conditions to ensure the global exponential stabilization in line with the separation approach. Tang and Jian (2019) considered the exponential convergence for impulsive system with activation function satisfying the Lipschitz condition in complex domain and for Type B via decomposition approach and nondecomposition one, respectively. Then, for the activation function of Type B, by decomposition approach, Long et al. (2021) addressed p-norm fixed-time synchronization under reduced-order approach and Xiao et al. (2020), Li and Huang (2021), Yu et al. (2022) tackled the exponential stability for discrete-time model, adaptive synchronization for fuzzy system and fixed-time pinning synchronization under the nonreduced-order approach. Guo et al. (2021a) deeply studied fixed-time synchronization problems under reduced-order method for delayed system with two kinds of activation functions including Type A without the assumption of the existence, continuity and boundedness of the partial derivatives and the ones satisfying the Lipschitz condition in the complex domain by the separation method and the direct one, respectively, while Guo et al. (2021b) investigated them under the nonreduced-order method. Simultaneously, only considering activation function satisfying the Lipschitz condition in complex domain and applying the nonseparation approach, Yu et al. (2020) dealt with the exponential and adaptive synchronization without using the standard reduced-order method and Wei et al. (2021), Long et al. (2022) referred to the synchronization, anti-synchronization and finite-time stabilization issues under standard reduced-order method.

In fact, for the research on dynamical behaviors and synthesis problems of CVBAMNNs and CVINNs models, there are still many areas that have not been exploited, which are worthy of further exploration. Moreover, in order to meet the actual requirement, many other models such as memristor-based ones, fractional-order ones, Cohen-Grossberg ones and so on have also been widely concerned, see some related references (Zhang et al. 2018b, 2019; Gan et al. 2021; Ding et al. 2021; Ali et al. 2020; Arslan et al. 2020; Guo et al. 2017, 2018; Subramanian and Muthukumar 2018; Li et al. 2020; Iswarya et al. 2021). Here, we will no longer state them in detail.

1.4 This Book

With the development of information technology and the involvement of complex signals in many industrial production areas, the research for complex-valued neural networks has been always a rapidly growing issue. As complex nonlinear dynamical systems, their wide application in various fields largely depends on their dynamical behaviors. Meanwhile, considering the inevitability of time delay in practical systems, dynamical behaviors analysis of delayed complex-valued neural networks systems has been widely concerned and extensive results have been presented up to now. However, there lacks a monograph in this direction to make necessary insight.

This book aims to provide the latest developments in the dynamics analysis for complex-valued neural network systems with time delay. It serves as a comprehensive monograph on complex-valued neural network systems with time delay. It intends to present a fairly comprehensive, up-to-date and detailed treatment of dynamical behaviors including stability analysis and (anti-)synchronization control. The materials included in the book are mainly based on the recent research work carried on by the authors and their co-workers in this domain, see Li et al. (2018), Zhang et al. (2014, 2017b, 2018c, 2020, 2021), Wei et al. (2019, 2020, 2021), Yu et al. (2022). The topics of this book cover a wide range including asymptotic stability, Hopf bifurcation, finite-time stability, Lagrange exponential stability, synchronization control, anti-synchronization control, fixed-time synchronization, fixed-time pinning synchronization and adaptive synchronization from the point of saving cost and avoiding resource waste. It involves many interesting issues, gives the latest developments and advances in techniques and methods, and brings out further insights. It covers a complete interpretation of dynamical behaviors for complex-valued neural networks systems with time delay.

As a detailed treatment on dynamics analysis of complex-valued neural networks systems with time delay, this book is a useful source of reference for postgraduates, senior undergraduates, graduate students, senior researchers interested in or working with control theory, applied mathematics, system analysis and integration, automation, nonlinear science, computer and other related fields, especially those relevant scientific and technical workers in the research of complex-valued neural network systems, dynamic systems, and intelligent control theory.

This book is organized into 11 relevant but independent chapters.

This chapter introduces the history of neural networks and complex-valued neural networks, and the recent progress of complex-valued neural networks.

Chapter 2 studies the asymptotic stability of complex-valued neural networks with constant delay. It focuses on the delay-independent stability analysis based on the separable method and provides a sufficient condition in terms of linear matrix inequality (LMI) to ascertain the existence, uniqueness and global asymptotical stability of the equilibrium point of the addressed system.

Chapter 3 continues to study the asymptotic stability problem of complex-valued neural networks with constant delay. Based on separable method and nonseparable method, new delay-dependent stability criteria to guarantee the existence, uniqueness and global asymptotical stability of the equilibrium point of the addressed system are presented, respectively.

Chapter 4 deals with the stability and Hopf bifurcation problems of complex-valued neural networks model with discrete and distributed delays. Based on separable method, the normal form theory and center manifold theorem, it provides some sufficient conditions which guarantee the asymptotical stability and determine the direction of the Hopf bifurcation and the stability of the bifurcating periodic solutions, respectively.

Chapter 5 focuses on the finite-time stability problem of complex-valued bidirectional associative memory neural networks with constant delay. It presents a condition to ensure the existence and uniqueness of the equilibrium point for this system by using a nonlinear measure approach. It also establishes a finite-time stability criterion of the equilibrium point by Lyapunov function approach.

Chapter 6 investigates the Lagrange exponential stability problem of complex-valued bidirectional associative memory neural networks with time-varying delays. Based on different assumption conditions for activation functions, by using some inequalities techniques, different algebraic conditions and the condition in terms of LMI are given to guarantee Lagrange exponential stability of the addressed system, respectively. Moreover, the estimations of different globally attractive sets named the convergence balls are also provided.

Chapter 7 deals with the anti-synchronization control of complex-valued bidirectional associative memory neural networks with time-varying delays. By using a suitable Lyapunov functional and some inequalities techniques, a sufficient criterion based on algebraic structure for the anti-synchronization control design is proposed.

Chapter 8 investigates the anti-synchronization control of complex-valued neural networks with leakage delay and time-varying delays. By applying some inequalities techniques and choosing the suitable controllers, two sufficient criteria for the anti-synchronization problem of the addressed system based on the separable method and nonseparable method are derived, respectively.

Chapter 9 deals with the synchronization and anti-synchronization problems of complex-valued inertial neural networks with time-varying delays. Via reducing order method, constructing the appropriate Lyapunov functional and using LMIs approach, some sufficient criteria to ensure the synchronization and anti-synchronization are presented.

Chapter 10 considers the fixed-time synchronization problem for complex-valued bidirectional associative memory neural networks with time delays. Based on the fixed-time stability theory, the Lyapunov function method and some inequalities techniques, a new criterion to guarantee that the addressed system achieve the synchronization in a fixed time is established by designing a new nonlinear delayed controller.

Chapter 11 concentrates on pinning synchronization and adaptive synchronization problems of complex-valued inertial neural networks with time-varying delays in fixed-time interval. Via reducing order method, separating the real and imaginary parts, and establishing a novel Lyapunov function, the fixed-time stability for the closed-loop error system is guaranteed via partial nodes controlled directly by a new pinning controller. Further, to save cost and avoid resources waste, a new pinning adaptive controller is developed and sufficient condition ensuring the adaptive fixed-time stability for the closed-loop error system is also derived.

References

Aizenberg I (2011) Complex-valued neural networks with multi-valued neurons. Springer, Berlin

Aizenberg I, Paliy DV, Zurada JM, Astola JT (2008) Blur identification by multilayer neural network based on multivalued neurons. IEEE Trans Neural Netw 19(5):883–898

Aizenberg NN, Ivaskiv YL, Pospelov DA (1971) A certain generalization of threshold functions. Dokl Akad Nauk SSSR 196:1287–1290

Ali MS, Narayanan G, Shekher V, Alsaedi A, Ahmad B (2020) Global Mittag-Leffler stability analysis of impulsive fractional-order complex-valued BAM neural networks with time varying delays. Commun Nonlinear Sci Numer Simul 83:105088

Angelaki D, Correia M (1991) Models of membrane resonance in pigeon semicircular canal type II hair cells. Biol Cybern 65(1):1–10
Aouiti C, Bessifi M (2021) Sliding mode control for finite-time and fixed-time synchronization of delayed complex-value d recurrent neural networks with discontinuous activation functions and nonidentical parameters. Eur J Control 59:109–122
Aouiti C, Bessifi M, Li X (2020) Finite-time and fixed-time synchronization of complex-valued recurrent neural networks with discontinuous activations and time-varying delays. Circuits Syst Signal Process 39:5406–5428
Arslan E, Narayanan G, Ali MS, Arik S, Saroha S (2020) Controller design for finite-time and fixed-time stabilization of fractional-order memristive complex-valued BAM neural networks with uncertain parameters and time-varying delays. Neural Netw 130:60–74
Babcock K, Westervelt R (1986) Stability and dynamics of simple electronic neural networks with added inertia. Phys D 23(1–3):464–469
Benvenuto N, Piazza F (1992) On the complex backpropagation algorithm. IEEE Trans Signal Process 40(4):967–969
Birx DL, Pipenberg SJ (1993) A complex mapping network for phase sensitive classification. IEEE Trans Neural Netw 4(1):127–135
Cao Y, Sriraman R, Shyamsundarraj N, Samidurai R (2020) Robust stability of uncertain stochastic complex-valued neural networks with additive time-varying delays. Math Comput Simul 171:207–220
Cao Y, Ramajayam S, Sriraman R, Samidurai R (2021) Leakage delay on stabilization of finite-time complex-valued BAM neural network: decomposition approach. Neurocomputing 463:505–513
Chen Q, Bin H, Huang Z (2021) Synchronization of CVNNs: a time-scale impulsive strategy. IEEE Access 9:31762–31772
Chen X, Song Q (2013) Global stability of complex-valued neural networks with both leakage time delay and discrete time delay on time scales. Neurocomputing 121:254–264
Chen X, Song Q, Zhao Z, Liu Y (2016) Global μ-stability analysis of discrete-time complex-valued neural networks with leakage delay and mixed delays. Neurocomputing 175:723–735
Chen X, Zhao Z, Song Q, Hu J (2017) Multistability of complex-valued neural networks with time-varying delays. Appl Math Comput 294:18–35
Ding X, Cao J, Alsaedi A, Alsaadi FE, Hayat T (2017) Robust fixed-time synchronization for uncertain complex-valued neural networks with discontinuous activation functions. Neural Netw 90:42–55
Ding Z, Zhang H, Zeng Z, Yang L, Li S (2021) Global dissipativity and quasi-Mittag-Leffler synchronization of fractional-order discontinuous complex-valued neural networks. IEEE Trans Neural Netw Learn Syst. https://doi.org/10.1109/TNNLS.2021.3119647
Dong T, Liao X, Wang A (2015) Stability and Hopf bifurcation of a complex-valued neural network with two time delays. Nonlinear Dyn 82:173–184
Dong T, Bai J, Yang L (2019) Bifurcation analysis of delayed complex-valued neural network with diffusions. Neural Process Lett 50:1019–1033
Du B (2018) Stability analysis of periodic solution for a complex-valued neural networks with bounded and unbounded delays. Asian J Control 20(2):881–892
Duan C, Song Q (2010) Boundedness and stability for discrete-time delayed neural network with complex-valued linear threshold neurons. Discret Dyn Nat Soc 368379
Duan L, Shi M, Wang Z, Huang L (2019) Global exponential synchronization of delayed complex-valued recurrent neural networks with discontinuous activations. Neural Process Lett 50:2183–2200
Duan L, Shi M, Huang C, Fang X (2021) Synchronization in finite-/fixed-time of delayed diffusive complex-valued neural networks with discontinuous activations. Chaos Solitons Fractals 142:110386
Fang T, Sun J (2014) Further investigate the stability of complex-valued recurrent neural networks with time-delays. IEEE Trans Neural Netw Learn Syst 25(9):1709–1713

Farley BG, Clark WA (1954) Simulation of self-organizing systems by digital computer. IRE Trans Inform Theory 4(4):76–84

Feng L, Yu J, Hu C, Yang C, Jiang H (2021) Nonseparation method-based finite/fixed-time synchronization of fully complex-valued discontinuous neural networks. IEEE Trans Cybern 51(6):3212–3223

Fukushima K (1980) Neocognitron: a self-organizing neural network model for a mechanism of pattern recognition unaffected by shift in position. Biol Cybern 36(4):93–202

Gan Q, Li L, Yang J, Qin Y, Meng M (2021) Improved results on fixed-/preassigned-time synchronization for memristive complex-valued neural networks. IEEE Trans Neural Netw Learn Syst. https://doi.org/10.1109/TNNLS.2021.3070966

Ganesan S, Gnaneswaran N (2020) State-feedback filtering for delayed discrete-time complex-valued neural networks. IEEE Trans Neural Netw Learn Syst 31(11):4726–4736

Georgiou GM, Koutsougeras C (1992) Complex domain back propagation. IEEE Trans Circuits Syst II 39(5):330–334

Goh SL, Mandic DP (2007) An augmented extended Kalman filter algorithm for complex-valued recurrent neural networks. Neural Comput 19(4):1–17

Goh SL, Chen M, Popovic DH, Aihara K, Obradovic D, Mandic DP (2006) Complex-valued forecasting of wind profile. Renew Energy 31(11):1733–1750

Gong W, Liang J, Cao J (2015a) Matrix measure method for global exponential stability of complex-valued recurrent neural networks with time-varying delays. Neural Netw 70:81–89

Gong W, Liang J, Cao J (2015b) Global μ-stability of complex-valued delayed neural networks with leakage delay. Neurocomputing 168:135–144

Gong W, Liang J, Zhang C, Cao J (2016) Nonlinear measure approach for the stability analysis of complex-valued neural networks. Neural Process Lett 44(2):539–554

Gong W, Liang J, Kan X, Nie X (2017a) Robust state estimation for delayed complex-valued neural networks. Neural Process Lett 46:1009–1029

Gong W, Liang J, Zhang C (2017b) Multistability of complex-valued neural networks with distributed delays. Neural Comput Appl 28(Suppl 1):S1–S14

Gong W, Liang J, Kan X, Wang L, Dobaie AM (2019) Robust state estimation for stochastic complex-valued neural networks with sampled-data. Neural Comput Appl 31(Suppl 1):S523–S542

Goto E (1954) The parametron-a new circuit element which utilizes nonlinear reactors (in Japanese). Paper of Technical Group of Electronic Computers and Nonlinear Theory. IECE

Goto E (1955) On the application of parametrically excited non-linear resonators (in Japanese). J Inst Electr Commun Eng Jpn (IECE) 38(10):2761

Gu KQ (2000) An integral inequality in the stability problem of time-delay systems. In: Proceedings of the IEEE CDC 2805–2810

Gunasekaran N, Zhai G (2019) Stability analysis for uncertain switched delayed complex-valued neural networks. Neurocomputing 367:198–206

Gunasekaran N, Zhai G (2020) Sampled-data state-estimation of delayed complex-valued neural networks. Int J Syst Sci 51(2):303–312

Guo R, Zhang Z, Liu X, Lin C (2017) Existence, uniqueness, and exponential stability analysis for complex-valued memristor-based BAM neural networks with time delays. Appl Math Comput 311:100–117

Guo R, Zhang Z, Liu X, Lin C, Wang H, Chen J (2018) Exponential input-to-state stability for complex-valued memristor-based BAM neural networks with multiple time-varying delays. Neurocomputing 275:2041–2054

Guo R, Zhang Z, Lin C, Chu Y, Li Y (2019) Finite time state estimation of complex-valued BAM neutral-type neural networks with time-varying delays. Int J Control Autom Syst 17(3):801–809

Guo R, Xu S, Lv W (2020) pth moment stochastic exponential anti-synchronization of delayed complex-valued neural networks. Nonlinear Dyn 100:1257–1274

Guo R, Lu J, Li Y, Lv W (2021a) Fixed-time synchronization of inertial complex-valued neural networks with time delays. Nonlinear Dyn 105:1643–1656

Guo R, Xu S, Ma Q, Zhang Z (2021b) Fixed-time synchronization of complex-valued inertial neural networks via nonreduced-order method. IEEE Syst J. https://doi.org/10.1109/JSYST.2021.3117342

Heaton J (2016) Ian Goodfellow, Yoshua Bengio, and Aaron Courville: deep learning. MIT Press, Cambridge

Hebb D (1949) The organization of behavior. Wiley, New York

Hinton GE, Osindero S, Teh Y (2006) A fast learning algorithm for deep belief nets. Neural Comput 18(7):1527–1554

Hirose A (1992a) Dynamics of fully complex-valued neural networks. Electron Lett 28(16):1492–1494

Hirose A (1992b) Continuous complex-valued back-propagation learning. Electron Lett 28(20):1854–1855

Hirose A (1994) Applications of complex-valued neural networks to coherent optical computing using phase-sensitive detection scheme. Inf Sci Appl 2(2):103–117

Hirose A (2012a) Complex-valued neural networks. Springer, Heidelberg

Hirose A (2012b) Advances in applications of complex-valued neural networks. J Soc Instrum Control Eng 51(4):351–357

Hirose A (2013) Complex-valued neural networks: theories and applications. World Scientific, Singapore

Hirose A, Eckmiller R (1996) Behavior control of coherent-type neural networks by carrier frequency modulation. IEEE Trans Neural Netw 7(4):1032–1034

Hirose A, Yoshida S (2012) Generalization characteristics of complex-valued feedforward neural networks in relation to signal coherence. IEEE Trans Neural Netw Learn Syst 23(4):541–551

Hirose A, Higo T, Tanizawa K (2006a) Efficient generation of holographic movies with frame interpolation using a coherent neural network. IEICE Electron Express 3(19):417–423

Hirose A, Asano Y, Hamano T (2006b) Developmental learning with behavioral mode tuning by carrier-frequency modulation in coherent neural networks. IEEE Trans Neural Netw 7(6):1532–1543

Hopfield J (1982) Neural networks and physical systems with emergent collective computational abilities. Proc Natl Acad Sci 79(8):2554–2558

Hopfield J (1984) Neurons with graded response have collective computational properties like those of two-state neurons. Proc Natl Acad Sci 81(10):3088–3092

Hu B, Song Q, Li K, Zhao Z, Liu Y, Alsaadi FE (2018) Global μ-synchronization of impulsive complex-valued neural networks with leakage delay and mixed time-varying delays. Neurocomputing 307:106–116

Hu J, Wang J (2012) Global stability of complex-valued recurrent neural networks with time-delays. IEEE Trans Neural Netw Learn Syst 23(6):853–865

Hu J, Wang J (2015) Global exponential periodicity and stability of discrete-time complex-valued recurrent neural networks with time-delays. Neural Netw 66:119–130

Hu J, Tan H, Zeng C (2020) Global exponential stability of delayed complex-valued neural networks with discontinuous activation functions. Neurocomputing 416:1–11

Huang Y, Zhang H, Wang Z (2014) Multistability of complex-valued recurrent neural networks with real-imaginary-type activation functions. Appl Math Comput 229:187–200

Huang Y, Zhang H, Wang Z (2015) Multistability of delayed complex-valued recurrent neural networks with discontinuous real-imaginary type activation functions. Chin Phys B 24(12):120701

Hui M, Zhang JH, Zhang J, Iu HHC, Yao R, Bai L (2021) Finite-time projective synchronization of stochastic complex-valued neural networks with probabilistic time-varying delays. IEEE Access 9:44784–44796

Iswarya M, Raja R, Cao J, Niezabitowski M, Alzabut J, Maharajan C (2021) New results on exponential input-to-state stability analysis of memristor based complex-valued inertial neural networks with proportional and distributed delays. Math Comput Simul. https://doi.org/10.1016/j.matcom.2021.01.020

Ji C, Qiao Y, Miao J, Duan L (2018) Stability and Hopf bifurcation analysis of a complex-valued Wilson-Cowan neural network with time delay. Chaos Solitons Fractals 115:45–61

Jian J, Wan P (2018) Global exponential convergence of fuzzy complex-valued neural networks with time-varying delays and impulsive effects. Fuzzy Sets Syst 338:23–39

Kan Y, Lu J, Qiu J, Kurths J (2019) Exponential synchronization of time-varying delayed complex-valued neural networks under hybrid impulsive controllers. Neural Netw 114:157–163

Kawata S, Hirose A (2008) Frequency-multiplexing ability of complex-valued Hebbian learning in logic gates. Int J Neural Syst 12(1):43–51

Kim MS, Guest CC (1990) Modification of backpropagation networks for complex-valued signal processing in frequency domain. In: International joint conference on neural networks (IJCNN), San Diego, pp 27–31

Kosko B (1987) Adaptive bidirectional associative memories. Appl Opt 26(23):4947–4960

Kosko B (1988) Bidirectional associative memories. IEEE Trans Syst Man Cybern 18(1):49–60

Kumar A, Das S, Rajeev, Yadav VK (2021a) Global exponential synchronization of complex-valued recurrent neural networks in presence of uncertainty along with time-varying bounded and unbounded delay terms. Int J Dyn Control. https://doi.org/10.1007/s40435-021-00838-9

Kumar A, Das S, Yadav VK, Rajeev (2021b) Global quasi-synchronization of complex-valued recurrent neural networks with time-varying delay and interaction terms. Chaos Solitons Fractals 152:111323

Kumar R, Kumar U, Das S, Qiu J, Lu J (2021) Effects of heterogeneous impulses on synchronization of complex-valued neural networks with mixed time-varying delays. Inf Sci 551:228–244

Kuroe Y, Hashimoto N, Mori T (2002) On energy function for complex-valued neural networks and its applications. In: Proceedings of the 9th international conference on neural information processing, pp 1079–1083

Kuroe Y, Yoshid M, Mori T (2003) On activation functions for complex-valued neural networks-existence of energy functions. In: Artificial neural networks and neural information processing, pp 174–175

LeCun Y, Boser B, Denker JS, Henderson D, Howard RE, Hubbard W, Jackel LD (1989) Back-propagation applied to handwritten zip code recognition. Neural Comput 1(4):541–551

Lee DL (2006) Improvements of complex-valued Hopfield associative memory by using generalized projection rules. IEEE Trans Neural Netw 17(5):1341–1347

Leung H, Haykin S (1991) The complex backpropagation algorithm. IEEE Trans Signal Process 39(9):2101–2104

Li H, Adali T (2008) A class of complex ICA algorithms based on the kurtosis cost function. IEEE Trans Neural Netw 19(3):408–420

Li L, Mu G (2019) Synchronization of coupled complex-valued impulsive neural networks with time delays. Neural Process Lett 50:2515–2527

Li L, Wang Z, Li Y, Shen H, Lu J (2018) Hopf bifurcation analysis of a complex-valued neural network model with discrete and distributed delays. Appl Math Comput 330:152–169

Li L, Shi X, Liang J (2019) Synchronization of impulsive coupled complex-valued neural networks with delay: the matrix measure method. Neural Netw 117:285–294

Li L, Chen W, Wu X (2021) Global exponential stability and synchronization for a novel complex-valued neural networks with proportional delays and inhibitory factors. IEEE Trans Cybern 51(4):2142–2152

Li Q, Liang J (2020a) Dissipativity of the stochastic Markovian switching CVNNs with randomly occurring uncertainties and general uncertain transition rates. Int J Syst Sci 51(6):1102–1118

Li Q, Liang J (2020b) Improved stabilization results for Markovian switching CVNNs with partly unknown transition rates. Neural Process Lett 52:1189–1205

Li Q, Liang J, Gong W (2019) Stability and synchronization for impulsive Markovian switching CVNNs: matrix measure approach. Commun Nonlinear Sci Numer Simul 77:126–140

Li Q, Liang J, Gong W (2020) Stabilization of piecewise-homogeneous Markovian switching CVNNs with mode-dependent delays and incomplete transition rates. Syst Sci Control Eng 8(1):206–221

Li Q, Liang J, Gong W (2021a) State estimation for semi-Markovian switching CVNNs with quantization effects and linear fractional uncertainties. J Frankl Inst 358:6326–6347
Li Q, Liang J, Qu H (2021b) H_∞ estimation for stochastic semi-Markovian switching CVNNs with missing measurements and mode-dependent delays. Neural Netw 141:281–293
Li X, Huang T (2021) Adaptive synchronization for fuzzy inertial complex-valued neural networks with state-dependent coefficients and mixed delays. Fuzzy Sets Syst 411:174–189
Li X, Fang J, Huang T (2020) Event-triggered exponential stabilization for state-based switched inertial complex-valued neural networks with multiple delays. IEEE Trans Cybern. https://doi.org/10.1109/TCYB.2020.3031379
Li Y, Liao X, Li H (2016) Global attracting sets of nonautonomous and complex-valued neural networks with time-varying delays. Neurocomputing 173(3):994–1000
Liang J, Gong W, Huang T (2016) Multistability of complex-valued neural networks with discontinuous activation functions. Neural Netw 84:125–142
Liang J, Li K, Song Q, Zhao Z, Liu Y, Alsaadi FE (2018) State estimation of complex-valued neural networks with two additive time-varying delays. Neurocomputing 309:54–61
Liu M, Wang X, Zhang Z, Wang Z (2019a) Global synchronization of complex-valued neural networks with stochastic disturbances and time-varying delay. IEEE Access 7:182600–182610
Liu M, Wang X, Zhang Z, Wang Z (2019b) Dissipativity analysis of complex-valued stochastic neural networks with time-varying delays. IEEE Access 7:165076–165087
Liu M, Li Z, Jiang H, Hu C, Yu Z (2020) Exponential synchronization of complex-valued neural networks via average impulsive interval strategy. Neural Process Lett 52:1377–1394
Liu X, Chen T (2016) Global exponential stability for complex-valued recurrent neural networks with asynchronous time delays. IEEE Trans Neural Netw Learn Syst 27(3):593–606
Liu X, Li Z (2020) Finite time anti-synchronization of complex-valued neural networks with bounded asynchronous time-varying delays. Neurocomputing 387:129–138
Liu X, Fang K, Liu B (2009) A synthesis method based on stability analysis for complex-valued Hopfield neural network. In: Proceeding of 7th Asian control conference, pp 1245–1250
Liu Y, Qin Y, Huang J, Huang T, Yang X (2019) Finite-time synchronization of complex-valued neural networks with multiple time-varying delays and infinite distributed delays. Neural Process Lett 50:1773–1787
Liu Y, Huang J, Qin Y, Yang X (2020) Finite-time synchronization of complex-valued neural networks with finite-time distributed delays. Neurocomputing 416:152–157
Liu Y, Shen B, Sun J (2022) Stubborn state estimation for complex-valued neural networks with mixed time delays: the discrete time case. Neural Comput Appl 34:5449–5464
Long C, Zhang G, Hu J (2021) Fixed-time synchronization for delayed inertial complex-valued neural networks. Appl Math Comput 405:126272
Long C, Zhang G, Zeng Z, Hu J (2022) Finite-time stabilization of complex-valued neural networks with proportional delays and inertial terms: a non-separation approach. Neural Netw 148:86–95
Mcculloch W, Pitts W (1943) A logical calculus of the ideas immanent in nervous activity. Bull Math Biophys 5(4):115–133
Mi L, Chen C, Qiu B, Xu L, Zhang L (2020) Fixed-time synchronization analysis for complex-valued neural networks via a new fixed-time stability theorem. IEEE Access 8:172799–172807
Minsky M, Papert S (1969) Perceptrons: an introduction to computational geometry. MIT Press, Cambridge
Nian F, Li J (2021) The module-phase synchronization of complex-valued neural networks with time-varying delay and stochastic perturbations. J Syst Sci Complex 34:2139–2154
Nitta T (1997) An extension of the back-propagation algorithm to complex numbers. Neural Netw 10(8):1391–1415
Nitta T (2003) Solving the XOR problem and the detection of symmetry using a single complex-valued neuron. Neural Netw 16(8):1101–1105
Nitta T (2009) Complex-valued neural networks: utilizing high-dimensional parameters. Information Science Reference, New York
Noest AJ (1988a) Discrete-state phasor neural networks. Phys Rev A 38(4):2196–2199

Noest AJ (1988b) Associative memory in sparse phasor neural networks. Europhys Lett 6(6):69–474
Pan J, Zhang Z (2021) Finite-time synchronization for delayed complex-valued neural networks via the exponential-type controllers of time variable. Chaos Solitons Fractals 146:110897
Pan J, Liu X, Xie W (2015) Exponential stability of a class of complex-valued neural networks with time-varying delays. Neurocomputing 164:293–299
Pecora LM, Carroll TL (1990) Synchronization in chaotic systems. Phys Rev Lett 64(8):821–824
Polyakov A (2012) Nonlinear feedback design for fixed-time stabilization of linear control systems. IEEE Trans Autom Control 57(8):2106–2110
Popa CA (2020) Global μ-stability of neutral-type impulsive complex-valued BAM neural networks with leakage delay and unbounded time-varying delays. Neurocomputing 376:73–94
Rajchakit G, Sriraman R (2021) Robust passivity and stability analysis of uncertain complex-valued impulsive neural networks with time-varying delays. Neural Process Lett 53:581–606
Rajivganthi C, Rihan FA, Lakshmanan S (2019) Dissipativity analysis of complex-valued BAM neural networks with time delay. Neural Comput Appl 31:127–137
Rakkiyappan R, Velmurugan G, Li X (2015a) Complete stability analysis of complex-valued neural networks with time delays and impulses. Neural Process Lett 41:435–468
Rakkiyappan R, Velmurugan G, Cao J (2015b) Multiple μ-stability analysis of complex-valued neural networks with unbounded time-varying delays. Neurocomputing 149:594–607
Ramasamy S, Nagamani G (2017) Dissipativity and passivity analysis for discrete-time complex-valued neural networks with leakage delay and probabilistic time-varying delays. Int J Adapt Control Signal Process 31:876–902
Rao VSH, Murthy GR (2008) Global dynamics of a class of complex valued neural networks. Int J Neural Syst 18(2):165–171
Rochester N, Holland J, Haibt L, Duda W (1956) Tests on a cell assembly theory of the action of the brain, using a large digital computer. IRE Trans Inform Theory 2(3):80–93
Rosenblatt F (1958) The perceptron: a probabilistic model for information storage and organization in the brain. Psychol Rev 65(6):386–408
Rudin W (1987) Real and complex analysis. McGraw-Hill, New York
Rumelhart DE, Hinton GE, Williams RJ (1986) Learning representations by back-propagating errors. Nature 323:533–536
Samidurai R, Sriraman R, Cao J, Tu Z (2018) Effects of leakage delay on global asymptotic stability of complex-valued neural networks with interval time-varying delays via new complex-valued Jensen's inequality. Int J Adapt Control Signal Process 32:1294–1312
Samidurai R, Sriraman R, Zhu S (2019) Leakage delay-dependent stability analysis for complex-valued neural networks with discrete and distributed time-varying delays. Neurocomputing 338:262–273
Shen Y, Liu X (2022) Event-based master-slave synchronization of complex-valued neural networks via pinning impulsive control. Neural Netw 145:374–385
Song Q, Zhao Z (2016) Stability criterion of complex-valued neural networks with both leakage delay and time-varying delays on time scales. Neurocomputing 171:179–184
Song Q, Zhao Z, Liu Y (2015a) Stability analysis of complex-valued neural networks with probabilistic time-varying delays. Neurocomputing 159:96–104
Song Q, Zhao Z, Liu Y (2015b) Impulsive effects on stability of discrete-time complex-valued neural networks with both discrete and distributed time-varying delays. Neurocomputing 168:1044–1050
Song Q, Yan H, Zhao Z, Liu Y (2016a) Global exponential stability of impulsive complex-valued neural networks with both asynchronous time-varying and continuously distributed delays. Neural Netw 81:1–10
Song Q, Yan H, Zhao Z, Liu Y (2016b) Global exponential stability of complex-valued neural networks with both time-varying delays and impulsive effects. Neural Netw 79:108–116
Song Q, Shu H, Zhao Z, Liu Y, Alsaadi FE (2017) Lagrange stability analysis for complex-valued neural networks with leakage delay and mixed time-varying delays. Neurocomputing 244:33–41

Song Q, Yu Q, Zhao Z, Liu Y, Alsaadi FE (2018a) Dynamics of complex-valued neural networks with variable coefficients and proportional delays. Neurocomputing 275:2762–2768
Song Q, Yu Q, Zhao Z, Liu Y, Alsaadi FE (2018b) Boundedness and global robust stability analysis of delayed complex-valued neural networks with interval parameter uncertainties. Neural Netw 103:55–62
Song Q, Zhao Z, Liu Y, Alsaadi FE (2022) Mean-square input-to-state stability for stochastic complex-valued neural networks with neutral delay. Neurocomputing 470:269–277
Song X, Man J, Ahn CK, Song S (2021) Synchronization in finite/fixed time for Markovian complex-valued nonlinear interconnected neural networks with reaction-diffusion terms. IEEE Trans Netw Sci Eng 8(4):3313–3324
Sriraman R, Samidurai R (2019) Global asymptotic stability analysis for neutral-type complex-valued neural networks with random time-varying delays. Int J Syst Sci 50(9):1742–1756
Sriraman R, Cao Y, Samidurai R (2020) Global asymptotic stability of stochastic complex-valued neural networks with probabilistic time-varying delays. Math Comput Simul 171:103–118
Subramanian K, Muthukumar P (2018) Existence, uniqueness, and global asymptotic stability analysis for delayed complex-valued Cohen-Grossberg BAM neural networks. Neural Comput Appl 29:565–584
Suksmono AB, Hirose A (2005) Beamforming of ultra-wideband pulses by a complex-valued spatio-temporal multilayer neural network. Int J Neural Syst 15(1):1–7
Tang Q, Jian J (2018) Matrix measure based exponential stabilization for complex-valued inertial neural networks with time-varying delays using impulsive control. Neurocomputing 273:251–259
Tang Q, Jian J (2019) Global exponential convergence for impulsive inertial complex-valued neural networks with time-varying delays. Math Comput Simul 159:39–56
Velmurugan G, Rakkiyappan R, Cao J (2015) Further analysis of global μ-stability of complex-valued neural networks with unbounded time-varying delays. Neural Netw 67:14–27
Wang L, Song Q, Zhao Z, Liu Y, Alsaadi FE (2019) Synchronization of two non-identical complex-valued neural networks with leakage delay and time-varying delays. Neurocomputing 356:52–59
Wang S, Zhang Z, Lin C, Chen J (2021) Fixed-time synchronization for complex-valued BAM neural networks with time-varying delays via pinning control and adaptive pinning control. Chaos Solitons Fractals 153:111583
Wang X, Wang Z, Song Q, Shen H, Huang X (2020) A waiting-time-based event-triggered scheme for stabilization of complex-valued neural networks. Neural Netw 121:329–338
Wang X, Wang Z, Xia J, Ma Q (2021) Quantized sampled-data control for exponential stabilization of delayed complex-valued neural networks. Neural Process Lett 53:983–1000
Wang Z, Huang L (2016) Global stability analysis for delayed complex-valued BAM neural networks. Neurocomputing 173:2083–2089
Wang Z, Liu X (2019) Exponential stability of impulsive complex-valued neural networks with time delay. Math Comput Simul 156:143–157
Wang Z, Guo Z, Huang L, Liu X (2017) Dynamical behavior of complex-valued Hopfield neural networks with discontinuous activation functions. Neural Process Lett 45:1039–1061
Wang Z, Cao J, Guo Z, Huang L (2018) Generalized stability for discontinuous complex-valued Hopfield neural networks via differential inclusions. Proc R Soc A 474:20180507
Wang Z, Cao J, Cai Z, Huang L (2019) Periodicity and finite-time periodic synchronization of discontinuous complex-valued neural networks. Neural Netw 119:249–260
Wang Z, Xiong L, Zhang H, Liu Y (2021) Sampling-based event-triggered control for neutral-type complex-valued neural networks with partly unknown Markov jump and time-varying delay. Complexity 5563888
Wei X, Zhang Z, Zhong M, Liu M, Wang Z (2019) Anti-synchronization for complex-valued bidirectional associative memory neural networks with time-varying delays. IEEE Access 7:97536–97548
Wei X, Zhang Z, Liu M, Wang Z, Chen J (2020) Anti-synchronization for complex-valued neural networks with leakage delay and time-varying delays. Neurocomputing 412:312–319

Wei X, Zhang Z, Lin C, Chen J (2021) Synchronization and anti-synchronization for complex-valued inertial neural networks with time-varying delays. Appl Math Comput 403:126194
Werbos P (1982) Applications of advances in nonlinear sensitivity analysis. System modeling and optimization, pp 762–770
Wheeler W, Schieve W (1997) Stability and chaos in an inertial two-neuron system. Phys B 105:267–284
Wu E, Yang X, Xu C, Alsaadi FE, Hayat T (2018) Finite-time synchronization of complex-valued delayed neural networks with discontinuous activations. Asian J Control 20(6):2237–2247
Xiao Q, Huang T, Zeng Z (2020) On exponential stability of delayed discrete-time complex-valued inertial neural networks. IEEE Trans Cybern. https://doi.org/10.1109/TCYB.2020.3009761
Xie D, Jiang Y (2016) Global exponential stability of periodic solution for delayed complex-valued neural networks with impulses. Neurocomputing 207:528–538
Xie D, Jiang Y, Han M (2019) Global exponential synchronization of complex-valued neural networks with time delays via matrix measure method. Neural Process Lett 49:187–201
Xu D, Tan M (2017) Delay-independent stability criteria for complex-valued BAM neutral-type neural networks with time delays. Nonlinear Dyn 89:819–832
Xu D, Tan M (2018) Multistability of delayed complex-valued competitive neural networks with discontinuous non-monotonic piecewise nonlinear activation functions. Commun Nonlinear Sci Numer Simul 62:352–377
Xu X, Zhang J, Shi J (2014) Exponential stability of complex-valued neural networks with mixed delays. Neurocomputing 128:483–490
Xu X, Zhang J, Shi J (2017) Dynamical behaviour analysis of delayed complex-valued neural networks with impulsive effect. Int J Syst Sci 48(4):686–694
Xu X, Xu Q, Peng Y, Zhang J, Xu Y (2018a) Stochastic exponential robust stability of delayed complex-valued neural networks with Markova jumping parameters. IEEE Access 6:839–849
Xu X, Xue H, Peng Y, Xu Q, Yang J (2018b) Robust exponential stability of switched complex-valued neural networks with interval parameter uncertainties and impulses. Complexity 4981812
Xu X, Yang J, Xu Y (2019) Mean square exponential stability of stochastic complex-valued neural networks with mixed delays. Complexity 3429326
Yan M, Qiu J, Chen XY, Chen X, Yang C, Zhang A (2018a) Almost periodic dynamics of the delayed complex-valued recurrent neural networks with discontinuous activation functions. Neural Comput Appl 30:3339–3352
Yan M, Qiu J, Chen XY, Chen X, Yang C, Zhang A, Alsaadi F (2018b) The global exponential stability of the delayed complex-valued neural networks with almost periodic coefficients and discontinuous activations. Neural Process Lett 48:577–601
Yang C, Bose N (2005) Landmine detection and classification with complex-valued hybrid neural network using scattering parameters dataset. IEEE Trans Neural Netw 16(3):743–753
Yang Z, Liao X (2019) Invariant and attracting sets of complex-valued neural networks with both time-varying and infinite distributed delays. Neural Process Lett 49:1201–1215
Yao L, Wang Z, Wang Q, Xia J, Shen H (2020) Exponential stabilization of delayed complex-valued neural networks with aperiodic sampling: a free-matrix-based time-dependent Lyapunov functional method. Int J Control Autom Syst 18:1894–1903
Yu J, Hu C, Jiang H, Wang L (2020) Exponential and adaptive synchronization of inertial complex-valued neural networks: a non-reduced order and non-separation approach. Neural Netw 124:50–59
Yu Y, Zhang Z, Zhong M, Wang Z (2022) Pinning synchronization and adaptive synchronization of complex-valued inertial neural networks with time-varying delays in fixed-time interval. J Frankl Inst 359(2):1434–1456
Yuan Y, Song Q, Liu Y, Alsaadi FE (2019) Synchronization of complex-valued neural networks with mixed two additive time-varying delays. Neurocomputing 332:149–158
Zhang C, Sui Z, Li H (2017) Equivariant bifurcation in a coupled complex-valued neural network rings. Chaos Solitons Fractals 98:22–30

Zhang D, Jiang H, Wang J, Yu Z (2018) Global stability of complex-valued recurrent neural networks with both mixed time delays and impulsive effect. Neurocomputing 282:157–166
Zhang F, Huang T, Feng D, Zeng Z (2021) Multistability and robustness of complex-valued neural networks with delays and input perturbation. Neurocomputing 447:319–328
Zhang H, Wang X (2017) Complex projective synchronization of complex-valued neural network with structure identification. J Frankl Inst 354:5011–5025
Zhang H, Wang X, Lin X (2016) Synchronization of complex-valued neural network with sliding mode control. J Frankl Inst 353:345–358
Zhang H, Wang X, Lin X (2017) Topology identification and module-phase synchronization of neural network with time delay. IEEE Trans Syst Man Cybern Syst 47(6):885–892
Zhang H, Wang X, Yan P, Sun Y (2020) Combination synchronization and stability analysis of time-varying complex-valued neural networks. Chaos Solitons Fractals 131:109485
Zhang W, Li C (2014) Global robust stability of complex-valued recurrent neural networks with time-delays and uncertainties. Int J Biomath 7(2):1450016
Zhang Y, Ma Y (1997) CGHA for principal component extraction in the complex domain. IEEE Trans Neural Netw 8(5):1031–1036
Zhang Y, Zhou L (2022) Novel global polynomial stability criteria of impulsive complex-valued neural networks with multi-proportional delays. Neural Comput Appl 34:2913–2924
Zhang Z, Hao D (2018) Global asymptotic stability for complex-valued neural networks with time-varying delays via new Lyapunov functionals and complex-valued inequalities. Neural Process Lett 48:995–1017
Zhang Z, Lin C, Chen B (2014) Global stability criterion for delayed complex-valued recurrent neural networks. IEEE Trans Neural Netw Learn Syst 25(9):1704–1708
Zhang Z, Hao D, Zhou D (2017a) Global asymptotic stability by complex-valued inequalities for complex-valued neural networks with delays on period time scales. Neurocomputing 219:494–501
Zhang Z, Liu X, Chen J, Guo R, Zhou S (2017b) Further stability analysis for delayed complex-valued recurrent neural networks. Neurocomputing 251:81–89
Zhang Z, Li A, Yu S (2018a) Finite-time synchronization for delayed complex-valued neural networks via integrating inequality method. Neurocomputing 318:248–260
Zhang Z, Liu X, Zhou D, Lin C, Chen J, Wang H (2018b) Finite-time stabilizability and instabilizability for complex-valued memristive neural networks with time delays. IEEE Trans Syst Man Cybern Syst 48(12):2371–2382
Zhang Z, Liu X, Guo R, Lin C (2018c) Finite-time stability for delayed complex-valued BAM neural networks. Neural Process Lett 48:179–193
Zhang Z, Liu X, Lin C, Chen B (2018d) Finite-time synchronization for complex-valued recurrent neural networks with time delays. Complexity 8456737
Zhang Z, Li A, Yang L (2018e) Global asymptotic periodic synchronization for delayed complex-valued BAM neural networks via vector-valued inequality techniques. Neural Process Lett 48:1019–1041
Zhang Z, Liu X, Lin C, Zhou S (2019) Exponential stability analysis for delayed complex-valued memristor-based recurrent neural networks. Neural Comput Appl 31:1893–1903
Zhang Z, Guo R, Liu X, Lin C (2020) Lagrange exponential stability of complex-valued BAM neural networks with time-varying delays. IEEE Trans Syst Man Cybern Syst 50(8):3072–3085
Zhang Z, Guo R, Liu X, Zhong M, Lin C, Chen B (2021) Fixed-time synchronization for complex-valued BAM neural networks with time delays. Asian J Control 23(1):298–314
Zhou B, Song Q (2013) Boundedness and complete stability of complex-valued neural networks with time delay. IEEE Trans Neural Netw Learn Syst 24(8):1227–1238
Zhou C, Zhang W, Yang X, Xu C, Feng J (2017) Finite-time synchronization of complex-valued neural networks with mixed delays and uncertain perturbations. Neural Process Lett 46:271–291
Zhou C, Fang M, Xia J, Zhou J, Wang Z (2019) Fault-tolerant output-feedback stabilization for complex-valued neural networks with time delay and actuator failures. AIP Adv 9:095106

Zhou W, Zurada JM (2009) Discrete-time recurrent neural networks with complex-valued linear threshold neurons. IEEE Trans Circuits Syst II Express Briefs 56(8):669–673

Zhou W, Li B, Zhang J (2020) Matrix measure approach for stability and synchronization of complex-valued neural networks with deviating argument. Math Probl Eng 8877129

Chapter 2
Stability Criterion for CVNNs with Constant Delay

Abstract This chapter studies the asymptotic stability of complex-valued neural networks (CVNNs) with constant delay. By separating complex-valued neural networks into real and imaginary parts, forming an equivalent real-valued system, and constructing appropriate Lyapunov functional, we will provide a delay-independent sufficient criterion to ascertain the existence, uniqueness, and globally asymptotical stability of the equilibrium point of the considered system, which is in terms of linear matrix inequality (LMI). Meanwhile, the errors in the recent work are pointed out, and even if the result therein is correct, it is shown that our result not only improves, but also generalizes in that work.

2.1 Introduction

As an extension of real-valued neural networks (NNs), complex-valued neural networks (CVNNs) with complex-valued state, output, connection weight and activation function become strongly desired because of their practical applications in physical systems dealing with electromagnetic, light, ultrasonic, and quantum waves. Especially, along with the widespread use of analytic signals, they exhibit a critical advantage in diverse fields of engineering, where signals are routinely analyzed and processed in time, frequency, and phase domains. This results in that much effort for CVNNs has been reported in recent years (Lee 2001a, b, 2006; Hirose 2003, 2006; Kuroe et al. 2002, 2003; Li et al. 2002; Goh and Mandic 2004, 2007; Liu et al. 2009; Nitta 2009; Rao and Murthy 2008; Muezzinoglu et al. 2003; Kobayashi 2010; Bohner et al. 2011; Zhou and Zurada 2009). In studying the dynamical behaviors of CVNNs, various methods have been developed, such as energy function approach, Lyapunov function approach and synthesis approach, etc., see Kuroe et al. (2002, 2003), Liu et al. (2009), Nitta (2009). However, compared with real-valued neural networks, research for complex-valued ones has achieved slow and little progress as there are more complicated properties inside such systems.

It is well-known that time delay often appears in engineering systems and it is a source of instability and oscillation. For CVNNs with time delay, there have also been some achievements, see Hu and Wang (2012), Duan and Song (2010).

Z. Zhang et al., *Complex-Valued Neural Networks Systems with Time Delay*,
Intelligent Control and Learning Systems 4,
https://doi.org/10.1007/978-981-19-5450-4_2

Among which, global stability conditions for delayed CVNNs with two assumptions of complex-valued activation functions are derived in Hu and Wang (2012). However, the assumption about the existence, continuity, and boundedness of the partial derivatives for the real and imaginary parts in Hu and Wang (2012) for activation functions is covered by that about the Lipschitz condition, and activation functions are chosen to be bounded under the Lipschitz condition. In light of Liouville's theorem (Rudin 1987), such activation functions are not analytic. This makes the applications of the results in Hu and Wang (2012) quite limited. Moreover, the proofs of Theorems 3 and 4 in Hu and Wang (2012) are erroneous, as pointed out later in Remark 2.4, and so the results in Hu and Wang (2012) are problematic.

In this chapter, inspired by Hu and Wang (2012), we will study the stability problem for delayed CVNNs. We only require the Lipschitz condition of Hu and Wang (2012), that is, activation functions are Lipschitz continuous and remove the constraint of bounded or analytic conditions. It is obvious that our hypothesis is more suitable for CVNNs as the importance of the analyticity. The basic idea of this work is to separate n dimensional delayed CVNNs into their real and imaginary parts and form an equivalent $2n$ dimensional delayed real-valued system. We will derive a sufficient condition to guarantee the existence, uniqueness, and global asymptotical stability of the equilibrium point.

2.2 Problem Formulation

Consider the following CVNNs with constant delay as in Hu and Wang (2012):

$$\dot{z}(t) = -Dz(t) + Af(z(t)) + Bg(z(t-\tau)) + u, \tag{2.1}$$

where $z(t) = [z_1(t), z_2(t), \ldots, z_n(t)]^T \in \mathbb{C}^n$ is the state vector; $f(z(t)) = [f_1(z_1(t)), f_2(z_2(t)), \ldots, f_n(z_n(t))]^T \in \mathbb{C}^n$ and $g(z(t-\tau)) = [g_1(z_1(t-\tau)), g_2(z_2(t-\tau)), \ldots, g_n(z_n(t-\tau))]^T \in \mathbb{C}^n$ are the vector-valued activation functions; $D = \text{diag}\{d_1, d_2, \ldots, d_n\} \in \mathbb{R}^{n\times n}$ with $d_j > 0$ $(j = 1, 2, \ldots, n)$ is the self-feedback connection weight matrix; $A = (a_{jk}) \in \mathbb{C}^{n\times n}$ and $B = (b_{jk}) \in \mathbb{C}^{n\times n}$ are the connection weight matrix and the delayed connection weight matrix, respectively; $u = [u_1, u_2, \ldots, u_n]^T \in \mathbb{C}^n$ is the external input vector. Note that the special case of $g = f$ is usually considered for the neuron activation functions in real-valued neural networks.

Let $z(t) = x(t) + iy(t)$, we will study the stability problem for system (2.1) with the activation functions $f(z(t)) = f^R(x(t), y(t)) + if^I(x(t), y(t))$ and $g(z(t-\tau)) = g^R(x(t-\tau), y(t-\tau)) + ig^I(x(t-\tau), y(t-\tau))$ satisfying the following Assumption 2.1.

Assumption 2.1 The neuron activation functions $f_j(.)$ and $g_j(.)$ satisfy the following Lipschitz condition:

$$\begin{aligned} |f_j(z) - f_j(z')| &\le l_j |z - z'|, \quad \forall z, z' \in \mathbb{C}, \\ |g_j(z) - g_j(z')| &\le \eta_j |z - z'|, \quad \forall z, z' \in \mathbb{C}, \end{aligned} \tag{2.2}$$

where $l_j > 0$ and $\eta_j > 0$ ($j = 1, 2, \ldots, n$) are constants.

In Hu and Wang (2012), authors investigated the stability of system (2.1) under Assumption 2.1 and the following assumption, respectively.

Assumption 2.2 The neuron activation functions $f_j(.)$ and $g_j(.)$ can be separated into real and imaginary parts as

$$f_j(z) = f_j^R(x, y) + i f_j^I(x, y), \; g_j(z) = g_j^R(x, y) + i g_j^I(x, y),$$

where $z = x + iy$ and the real-imaginary parts satisfy the following conditions:
(1) The partial derivatives of $f_j(.,.)$ and $g_j(.,.)$ with respect to the variables x and y exist and are continuous.
(2) All the partial derivatives are bounded, that is, there exist positive constant numbers $\lambda_j^{RR}, \lambda_j^{RI}, \lambda_j^{IR}, \lambda_j^{II}, \mu_j^{RR}, \mu_j^{RI}, \mu_j^{IR}, \mu_j^{II}$ such that

$$\begin{aligned} &|\partial f_j^R/\partial x| \le \lambda_j^{RR}, \; |\partial f_j^R/\partial y| \le \lambda_j^{RI}, \; |\partial f_j^I/\partial x| \le \lambda_j^{IR}, \; |\partial f_j^I/\partial y| \le \lambda_j^{II}, \\ &|\partial g_j^R/\partial x| \le \mu_j^{RR}, \; |\partial g_j^R/\partial y| \le \mu_j^{RI}, \; |\partial g_j^I/\partial x| \le \mu_j^{IR}, \; |\partial g_j^I/\partial y| \le \mu_j^{II}. \end{aligned}$$

Then, one can obtain that for any $x, x', y, y' \in \mathbb{R}$

$$\begin{aligned} |f_j^R(x, y) - f_j^R(x', y')| &\le \lambda_j^{RR}|x - x'| + \lambda_j^{RI}|y - y'|, \\ |f_j^I(x, y) - f_j^I(x', y')| &\le \lambda_j^{IR}|x - x'| + \lambda_j^{II}|y - y'|, \\ |g_j^R(x, y) - g_j^R(x', y')| &\le \mu_j^{RR}|x - x'| + \mu_j^{RI}|y - y'|, \\ |g_j^I(x, y) - g_j^I(x', y')| &\le \mu_j^{IR}|x - x'| + \mu_j^{II}|y - y'|. \end{aligned}$$

for all $j = 1, 2, \ldots, n$.

Remark 2.1 In Hu and Wang (2012), Assumptions 2.1 and 2.2 are made on the activation functions. In fact, it is easy to verify that Assumption 2.2 is a strong constraint and it is a special case of Assumption 2.1. Thus, in this chapter, we relax the assumptions in Hu and Wang (2012) to a moderate one that the complex-valued activation functions satisfy the Lipschitz condition as in the above Assumption 2.1.

Remark 2.2 It should be noted that, in this chapter, the boundedness of activation functions to guarantee the existence of the equilibrium points is not assumed. According to the Liouville's theorem (Rudin 1987), if $f(z)$ is analytic and bounded for all $z \in \mathbb{C}$, then $f(z)$ is a constant function. It is obvious that a bounded and analytic function is a trivial case and it is usually not suitable for complex-valued activation functions. However, Theorem 4 of Hu and Wang (2012) assumes that the activation functions are bounded beside the Lipschitz condition, and thus the activation functions are not analytic. This leads to limitations in choosing activation functions. So, compared with the existing work in Hu and Wang (2012), our study serves a wider class of complex-valued neural networks.

In order to study system (2.1), we separate it into its real and imaginary parts and transform it into a real-valued neural network. Let $A = A^R + iA^I$, $B = B^R + iB^I$, and $u = u^R + iu^I$, where i denotes the imaginary unit. For simplicity, we denote $x = x(t)$, $y = y(t)$, $x^\tau = x(t-\tau)$, and $y^\tau = y(t-\tau)$. So, system (2.1) can be separated into real and imaginary parts as

$$\begin{aligned}\dot{x} &= -Dx + A^R f^R(x, y) - A^I f^I(x, y) + B^R g^R(x^\tau, y^\tau) - B^I g^I(x^\tau, y^\tau) + u^R,\\ \dot{y} &= -Dy + A^I f^R(x, y) + A^R f^I(x, y) + B^I g^R(x^\tau, y^\tau) + B^R g^I(x^\tau, y^\tau) + u^I.\end{aligned} \tag{2.3}$$

Let

$$\omega = \begin{bmatrix} x \\ y \end{bmatrix}, \ \bar{u} = \begin{bmatrix} u^R \\ u^I \end{bmatrix}, \ \bar{D} = \begin{bmatrix} D & 0 \\ 0 & D \end{bmatrix}, \ \bar{A} = \begin{bmatrix} A^R & -A^I \\ A^I & A^R \end{bmatrix},$$
$$\bar{B} = \begin{bmatrix} B^R & -B^I \\ B^I & B^R \end{bmatrix}, \ \bar{f}(\omega) = \begin{bmatrix} f^R(x, y) \\ f^I(x, y) \end{bmatrix}, \ \bar{g}(\omega^\tau) = \begin{bmatrix} g^R(x^\tau, y^\tau) \\ g^I(x^\tau, y^\tau) \end{bmatrix}.$$

Then, Eq. (2.3) can be rewritten as

$$\dot{\omega} = -\bar{D}\omega + \bar{A}\bar{f}(\omega) + \bar{B}\bar{g}(\omega^\tau) + \bar{u}. \tag{2.4}$$

It is clear from (2.2) that

$$\begin{aligned}(f(z) - f(z'))^*(f(z) - f(z')) &\le (z - z')^* L^T L(z - z'),\\ (g(z) - g(z'))^*(g(z) - g(z')) &\le (z - z')^* U^T U(z - z'),\end{aligned} \tag{2.5}$$

where $L = \text{diag}\{l_1, l_2, \ldots, l_n\}$ and $U = \text{diag}\{\eta_1, \eta_2, \ldots, \eta_n\}$. These are expressed by real and imaginary parts as

$$\begin{aligned}(\bar{f}(\omega) - \bar{f}(\omega'))^T(\bar{f}(\omega) - \bar{f}(\omega')) &\le (\omega - \omega')^T \bar{L}(\omega - \omega'),\\ (\bar{g}(\omega) - \bar{g}(\omega'))^T(\bar{g}(\omega) - \bar{g}(\omega')) &\le (\omega - \omega')^T \bar{U}(\omega - \omega'),\end{aligned} \tag{2.6}$$

where

$$\bar{L} = \begin{bmatrix} L^T L & 0 \\ 0 & L^T L \end{bmatrix}, \ \bar{U} = \begin{bmatrix} U^T U & 0 \\ 0 & U^T U \end{bmatrix}.$$

Notice that the equilibrium point of system (2.1) is also the equilibrium point of system (2.4) and the stability for system (2.1) is equivalent to the stability for system (2.4). Therefore, in the sequel, we focus our study on the real-valued neural networks system (2.4).

Remark 2.3 We have transformed n dimensional delayed complex-valued system (2.1) into an equivalent $2n$ dimensional real-valued system (2.4) by introducing augmented vectors $\omega, \bar{u}$, augmented matrices $\bar{D}, \bar{A}, \bar{B}$, and augmented functions $\bar{f}(\omega), \bar{g}(\omega^{\tau})$. This idea is similar to but simplifies that of Hu and Wang (2012), whereas a complicated real-valued system is used by introducing four augmented weight matrices and four augmented functions except for ω, $\bar{u}$ and $\bar{D}$.

The following lemmas are useful in the derivation of the main result.

Lemma 2.1 (Forti and Tesi 1995) *If $H(x) : \mathbb{R}^n \to \mathbb{R}^n$ is a continuous map and satisfies the following conditions:*
(1) $H(x)$ is injective on $\mathbb{R}^n$;
(2) $\|H(x)\| \to \infty$ as $\|x\| \to \infty$;
then $H(x)$ is a homeomorphism of $\mathbb{R}^n$.

Lemma 2.2 *For any vectors $x, y \in \mathbb{R}^n$, positive definite matrix $P \in \mathbb{R}^{n\times n}$, and positive real constant ε, the following matrix inequality holds:*

$$2x^T y \leq \varepsilon x^T P x + \frac{1}{\varepsilon} y^T P^{-1} y. \tag{2.7}$$

Proof From Liao et al. (2002), it holds that $2X^T Y \leq X^T P X + Y^T P^{-1} Y$, for any vectors $X, Y \in \mathbb{R}^n$. Letting $X = \sqrt{\varepsilon} x$ and $Y = \frac{1}{\sqrt{\varepsilon}} y$, we immediately obtain (2.7). □

2.3 Main Result

In this section, we will present a LMI-based delay-independent sufficient condition for the existence, uniqueness, and globally asymptotical stability of the equilibrium point for system (2.4).

Theorem 2.1 *Under Assumption 2.1, system (2.4) has a unique equilibrium point and it is globally asymptotically stable if there exist scalars $\varepsilon_1 > 0$, $\varepsilon_2 > 0$ and a matrix $P > 0$ such that the following LMI holds*

$$\Omega = \begin{bmatrix} \Omega_{11} & P\bar{A} & P\bar{B} \\ * & -\varepsilon_1 I & 0 \\ * & * & -\varepsilon_2 I \end{bmatrix} < 0, \tag{2.8}$$
$$\Omega_{11} = -P\bar{D} - \bar{D}P + \varepsilon_1 \bar{L} + \varepsilon_2 \bar{U}.$$

Proof We first prove the existence and uniqueness of the equilibrium point. To this end, we define the following map associated with (2.4):

$$H(\omega) = -\bar{D}\omega + \bar{A}\bar{f}(\omega) + \bar{B}\bar{g}(\omega) + \bar{u}. \tag{2.9}$$

Firstly, we prove that the map $H(\omega)$ is injective under the given condition. We show this by contradiction. Suppose that there exist ω_1 and ω_2 with $\omega_1 \neq \omega_2$ such that $H(\omega_1) = H(\omega_2)$. By (2.9), we have

$$-\bar{D}(\omega_1 - \omega_2) + \bar{A}(\bar{f}(\omega_1) - \bar{f}(\omega_2)) + \bar{B}(\bar{g}(\omega_1) - \bar{g}(\omega_2)) = 0. \quad (2.10)$$

Left multiplying both sides of (2.10) by $2(\omega_1 - \omega_2)^T P$ results in

$$\begin{aligned} &2(\omega_1 - \omega_2)^T P(-\bar{D}(\omega_1 - \omega_2) + \bar{A}(\bar{f}(\omega_1) - \bar{f}(\omega_2)) \\ &+\bar{B}(\bar{g}(\omega_1) - \bar{g}(\omega_2))) = 0. \end{aligned} \quad (2.11)$$

By Lemma 2.2 and (2.6), for scalars $\varepsilon_1 > 0$ and $\varepsilon_2 > 0$, the left term in (2.11) can be bounded as

$$\begin{aligned} &(\omega_1 - \omega_2)^T(-P\bar{D} - \bar{D}P)(\omega_1 - \omega_2) + 2(\omega_1 - \omega_2)^T P\bar{A}(\bar{f}(\omega_1) - \bar{f}(\omega_2)) \\ &\quad + 2(\omega_1 - \omega_2)^T P\bar{B}(\bar{g}(\omega_1) - \bar{g}(\omega_2)) \\ &\leq (\omega_1 - \omega_2)^T(-P\bar{D} - \bar{D}P)(\omega_1 - \omega_2) + \varepsilon_1(\bar{f}(\omega_1) - \bar{f}(\omega_2))^T(\bar{f}(\omega_1) - \bar{f}(\omega_2)) \\ &+\frac{1}{\varepsilon_1}(\omega_1 - \omega_2)^T P\bar{A}\bar{A}^T P(\omega_1 - \omega_2) + \varepsilon_2(\bar{g}(\omega_1) - \bar{g}(\omega_2))^T(\bar{g}(\omega_1) - \bar{g}(\omega_2)) \\ &\quad + \frac{1}{\varepsilon_2}(\omega_1 - \omega_2)^T P\bar{B}\bar{B}^T P(\omega_1 - \omega_2) \\ &\leq (\omega_1 - \omega_2)^T(-P\bar{D} - \bar{D}P + \varepsilon_1\bar{L} + \varepsilon_2\bar{U} + \frac{1}{\varepsilon_1}P\bar{A}\bar{A}^T P \\ &\quad + \frac{1}{\varepsilon_2}P\bar{B}\bar{B}^T P)(\omega_1 - \omega_2). \end{aligned} \quad (2.12)$$

If (2.8) holds, by Schur Complement, we have

$$-P\bar{D} - \bar{D}P + \varepsilon_1\bar{L} + \varepsilon_2\bar{U} + \frac{1}{\varepsilon_1}P\bar{A}\bar{A}^T P + \frac{1}{\varepsilon_2}P\bar{B}\bar{B}^T P < 0, \quad (2.13)$$

and thus, $2(\omega_1 - \omega_2)^T P(-\bar{D}(\omega_1 - \omega_2) + \bar{A}(\bar{f}(\omega_1) - \bar{f}(\omega_2)) + \bar{B}(\bar{g}(\omega_1) - \bar{g}(\omega_2))) < 0$, which is a contradiction. So, the map $H(\omega)$ is injective.

Next, we prove that $\|H(\omega)\| \to \infty$ as $\|\omega\| \to \infty$. From (2.13) we have

$$-P\bar{D} - \bar{D}P + \varepsilon_1\bar{L} + \varepsilon_2\bar{U} + \frac{1}{\varepsilon_1}P\bar{A}\bar{A}^T P + \frac{1}{\varepsilon_2}P\bar{B}\bar{B}^T P < -\varepsilon I, \quad (2.14)$$

for sufficiently small $\varepsilon > 0$. Then, using a derivation procedure similar to that in the last paragraph, one can obtain

$$\begin{aligned} &2\omega^T P(H(\omega) - H(0)) \leq \omega^T(-P\bar{D} - \bar{D}P + \varepsilon_1\bar{L} + \varepsilon_2\bar{U} \\ &+\frac{1}{\varepsilon_1}P\bar{A}\bar{A}^T P + \frac{1}{\varepsilon_2}P\bar{B}\bar{B}^T P)\omega \leq -\varepsilon\|\omega\|^2. \end{aligned} \quad (2.15)$$

From (2.15) and using Schwartz inequality, we have

$$\varepsilon\|\omega\|^2 \leq 2\|\omega\|\|P\|(\|H(\omega)\| + \|H(0)\|),$$

which gives

$$\frac{\varepsilon\|\omega\|}{2\|P\|} \leq \|H(\omega)\| + \|H(0)\|.$$

Therefore, $\|H(\omega)\| \to \infty$ as $\|\omega\| \to \infty$. By Lemma 2.1, the map $H(\omega)$ is a homeomorphism of $\mathbb{R}^{2n}$. Hence there exists a unique point $\hat{\omega}$ such that $H(\hat{\omega}) = 0$, that is, system (2.4) has a unique equilibrium point $\hat{\omega}$.

Now we show that (2.8) also implies the globally asymptotical stability of the equilibrium point for system (2.4). Since there exists a unique equilibrium point $\hat{\omega}$ for system (2.4), by the transformation $\tilde{\omega} = \omega - \hat{\omega}$, we rewrite system (2.4) as

$$\dot{\tilde{\omega}} = -\bar{D}\tilde{\omega} + \bar{A}\tilde{f}(\tilde{\omega}) + \bar{B}\tilde{g}(\tilde{\omega}^{\tau}), \tag{2.16}$$

where $\tilde{f}(\tilde{\omega}) = \bar{f}(\tilde{\omega} + \hat{\omega}) - \bar{f}(\hat{\omega})$ and $\tilde{g}(\tilde{\omega}^{\tau}) = \bar{g}(\tilde{\omega}^{\tau} + \hat{\omega}) - \bar{g}(\hat{\omega})$. It is clear that the stability of the equilibrium point for system (2.4) is equivalent to that of the origin for system (2.16).

We construct the following Lyapunov functional:

$$V(\tilde{\omega}(t)) = \tilde{\omega}^T(t)P\tilde{\omega}(t) + \int_{t-\tau}^{t} \tilde{\omega}^T(s)Q\tilde{\omega}(s)ds, \tag{2.17}$$

where $Q > 0$ will not be used as seen later. The time derivative of $V(\tilde{\omega}(t))$ along the trajectory of (2.16) yields

$$\begin{aligned} \dot{V}(\tilde{\omega}) &= 2\tilde{\omega}^T P\dot{\tilde{\omega}} + \tilde{\omega}^T Q\tilde{\omega} - \tilde{\omega}^{\tau T} Q\tilde{\omega}^{\tau} \\ &= 2\tilde{\omega}^T P(-\bar{D}\tilde{\omega} + \bar{A}\tilde{f}(\tilde{\omega}) + \bar{B}\tilde{g}(\tilde{\omega}^{\tau})) + \tilde{\omega}^T Q\tilde{\omega} - \tilde{\omega}^{\tau T} Q\tilde{\omega}^{\tau}. \end{aligned} \tag{2.18}$$

On the other hand, (2.6) guarantees that the following is true for $\varepsilon_j > 0,\ j = 1, 2,$

$$\varepsilon_1\Big(\tilde{\omega}^T\bar{L}\tilde{\omega} - \tilde{f}^T(\tilde{\omega})\tilde{f}(\tilde{\omega})\Big) \geq 0, \tag{2.19}$$

$$\varepsilon_2\Big(\tilde{\omega}^{\tau T}\bar{U}\tilde{\omega}^{\tau} - \tilde{g}^T(\tilde{\omega}^{\tau})\tilde{g}(\tilde{\omega}^{\tau})\Big) \geq 0. \tag{2.20}$$

Hence, (2.18), together with (2.19) and (2.20), gives

$$\dot{V}(\tilde{\omega}) \leq \xi^T(t)\hat{\Omega}\xi(t), \tag{2.21}$$

where

$$\xi^T(t) = \begin{bmatrix} \tilde{\omega}^T & \tilde{\omega}^{\tau T} & \tilde{f}^T(\tilde{\omega}) & \tilde{g}^T(\tilde{\omega}^\tau) \end{bmatrix}^T,$$

$$\hat{\Omega} = \begin{bmatrix} \hat{\Omega}_{11} & 0 & P\bar{A} & P\bar{B} \\ * & -Q + \varepsilon_2 \bar{U} & 0 & 0 \\ * & * & -\varepsilon_1 I & 0 \\ * & * & * & -\varepsilon_2 I \end{bmatrix}, \tag{2.22}$$

$$\hat{\Omega}_{11} = -P\bar{D} - \bar{D}P + Q + \varepsilon_1 \bar{L}.$$

It is easy to see that $\hat{\Omega} < 0$ if and only if $\varepsilon_2 \bar{U} < Q$ and

$$\begin{bmatrix} \hat{\Omega}_{11} & P\bar{A} & P\bar{B} \\ * & -\varepsilon_1 I & 0 \\ * & * & -\varepsilon_2 I \end{bmatrix} < 0. \tag{2.23}$$

It is obvious that (2.8) is equivalent to that $\varepsilon_2 \bar{U} < Q$ and (2.23) hold for some matrix $Q > 0$. Thus, (2.8) implies $\dot{V}(\tilde{\omega}) < 0$, and so the origin of system (2.16), or equivalently the equilibrium point of system (2.4) is globally asymptotically stable. □

Remark 2.4 Theorem 2.1 provides a sufficient condition for the existence, uniqueness, and stability analysis of the equilibrium point of delayed CVNNs under Assumption 2.1. It should be pointed out that the proofs of Theorem 3 and Theorem 4 in Hu and Wang (2012) are erroneous. In the proof of Theorem 3 in Hu and Wang (2012), it is obviously incorrect that $DP + PD - 4P^TP - 2\varepsilon[(A^R)^T A^R + (A^I)^T A^I] - 2\theta[(B^R)^T B^R + (B^I)^T B^I] > 0$ implies $\lambda_{\min}(DP + PD) - 4\lambda_{\max}(P^TP) - 2\varepsilon\lambda_{\max}[(A^R)^T A^R + (A^I)^T A^I] - 2\theta\lambda_{\max}[(B^R)^T B^R + (B^I)^T B^I] > 0$. While, in the proof of Theorem 4 in Hu and Wang (2012), the errors lie in the step of taking modulus of complex-valued $\tilde{z}$ for $-\tilde{z}^*(DP + PD)\tilde{z}$ and in another step of taking modulus of $\tilde{z}^* PA\tilde{f}(\tilde{z})$ and $\tilde{z}^* PB\tilde{g}(\tilde{z}^\tau)$ in (22) of Hu and Wang (2012). Therefore, the results of Theorem 3 and Theorem 4 are problematic. However, even if the result of Theorem 4 in Hu and Wang (2012) is correct, it is easy to theoretically verify that the present condition is less conservative than that of Hu and Wang (2012) for some special cases, such as one-neuron systems and two-neuron systems with real or pure imaginary weight matrices. For the general case, large amounts of examples numerically show the less conservatism of the present method.

2.4 Illustrative Examples

In this section, we will use examples to show the effectiveness of our result, and also show the improvements over the recent result in Hu and Wang (2012) even if it is correct.

Example 2.1 Consider system (2.1) with

$$A = \begin{bmatrix} 1+i & 1-i \\ 2-i & 2+i \end{bmatrix}, \quad B = \begin{bmatrix} 1-i & 1+i \\ 2+2i & 2-2i \end{bmatrix},$$

$$D = \begin{bmatrix} 4 & 0 \\ 0 & 3 \end{bmatrix}, \quad L = \begin{bmatrix} \frac{1}{4} & 0 \\ 0 & \frac{1}{4} \end{bmatrix}, \quad M = \begin{bmatrix} \frac{1}{2} & 0 \\ 0 & \frac{1}{2} \end{bmatrix}.$$

For this system, the given LMI condition of Theorem 4 in Hu and Wang (2012) is infeasible. However, by our result in Theorem 2.1, it concludes that this system is asymptotically stable.

The above merely lists one numerical example to verify the obtained result in Theorem 2.1. In fact, we have checked many examples and all verify that our result in Theorem 2.1 is less conservative than that in Hu and Wang (2012) even if it is correct.

2.5 Conclusion

The existence, uniqueness, and the stability problems of the equilibrium point for CVNNs with time delay have been investigated in this chapter. By transforming the system into real-valued neural networks, and under a more general assumption, a delay-independent sufficient condition is given in terms of LMI. The present result is shown effective and better than the recent work in Hu and Wang (2012). In the next chapter, we will study the delay-dependent stability analysis for CVNNs with time delay.

References

Bohner M, Rao VSH, Sanyal S (2011) Global stability of complex-valued neural networks on time scales. Differ Equ Dyn Syst 19(1–2):3–11

Duan C, Song Q (2010) Boundedness and stability for discrete-time delayed neural network with complex-valued linear threshold neurons. Discret Dyn Nat Soc 368379

Forti M, Tesi A (1995) New conditions for global stability of neural networks with application to linear and quadratic programming problems. IEEE Trans Circuits Syst I Fundam Theory Appl 42(7):354–366

Goh SL, Mandic DP (2004) A complex-valued RTRL algorithm for recurrent neural networks. Neural Comput 16(12):2699–2713

Goh SL, Mandic DP (2007) An augmented extended Kalman filter algorithm for complex-valued recurrent neural networks. Neural Comput 19:1039–1055

Hirose A (2003) Complex-valued neural networks: theories and applications. World Scientific, Singapore

Hirose A (2006) Complex-valued neural networks. Springer, New York

Hu J, Wang J (2012) Global stability of complex-valued recurrent neural networks with time-delays. IEEE Trans Neural Netw Learn Syst 23(6):853–865

Kobayashi K (2010) Exceptional reducibility of complex-valued neural networks. IEEE Trans Neural Netw 21(7):1060–1072

Kuroe Y, Hashimoto N, Mori T (2002) On energy function for complex-valued neural networks and its applications. In: Proceedings of the 9th international conference on neural information processing, pp 1079–1083

Kuroe Y, Yoshid M, Mori T (2003) On activation functions for complex-valued neural networks-existence of energy functions. In: Artificial neural networks and neural information processing, pp 174–175

Lee DL (2001a) Improving the capacity of complex-valued neural networks with a modified gradient descent learning rule. IEEE Trans Neural Netw 12(2):439–443

Lee DL (2001b) Relaxation of the stability condition of the complex-valued neural networks. IEEE Trans Neural Netw 12(5):1260–1262

Lee DL (2006) Improvement of complex-valued Hopfield associative memory by using generalized projection rules. IEEE Trans Neural Netw 17(5):1341–1347

Li C, Liao X, Yu J (2002) Complex valued recurrent neural network with IIR neural model: training and applications. Circuits Syst Signal Process 21(5):461–471

Liao X, Chen G, Sanchez EN (2002) LMI-based approach for asymptotic stability analysis of delayed neural networks. IEEE Trans Circuits Syst I Fundam Theory Appl 49(7):1033–1039

Liu X, Fang K, Liu B (2009) A synthesis method based on stability analysis for complex-valued Hopfield neural network. In: Proceeding of 7th Asian control conference, pp 1245–1250

Muezzinoglu MK, Cuzelis C, Zurada JM (2003) A new design method for the complex-valued multistate Hopfield associative memory. IEEE Trans Neural Netw 14(4):891–899

Nitta T (2009) Complex-valued neural networks: utilizing high-dimensional parameters. Information Science Reference, New York

Rao VSH, Murthy GR (2008) Global dynamics of a class of complex valued neural networks. Int J Neural Syst 18(2):165–171

Rudin W (1987) Real and complex analysis. McGraw-Hill, New York

Zhou W, Zurada JM (2009) Discrete-time recurrent neural networks with complex-valued linear threshold neurons. IEEE Trans Circuits Syst II, Express Briefs 56(8):669–673

Chapter 3
Further Stability Analysis for CVNNs with Constant Delay

Abstract This chapter continues to study the asymptotic stability problem of CVNNs with constant delay. Based on the result of the previous chapter, we will develop deeply new delay-dependent stability criteria to guarantee the existence, uniqueness and globally asymptotical stability of the equilibrium point of the addressed systems by use of the separable method and nonseparable method, respectively.

3.1 Introduction

In the past few years, CVNNs have already become hot research fields and much attention from researchers has been paid to investigating them. In which, research on dynamical behaviors of neural networks is always an important issue since its foundation in the practical applications, even for CVNNs. However, CVNNs have more complicated properties than real-valued ones, which results in that it is more difficult in analyzing the dynamical behaviors. Anyway, there have been various methods developed, such as energy function approach, synthesis approach, matrix measure approach, Lyapunov function approach and so on Kuroe et al. (2002, 2003), Liu et al. (2009), Hu and Wang (2012), Zhang et al. (2014), Liu and Chen (2016), Gong et al. (2015b, 2016).

As everyone knows, when involving the dynamical behaviors of real-valued neural networks, activation functions are usually assumed to be smooth and bounded. If the assumption about smoothness and boundedness is directly extended to complex-valued activation functions, that is, they are analytic and bounded. In light of Liouville's theorem (Rudin 1987), such functions are constant. This is not suitable. Therefore, there exist the great challenge and difficulty in choosing activation functions of CVNNs. According to the application background, there are various kinds of complex-valued activation functions and different methods to study the relevant systems are needed for different kinds of functions. This fact reflects that the practical applications of CVNNs are more extensive, whereas the theoretical research is more difficult. Currently, two types of activation functions are usually considered in relevant publications. One is the case that the activation functions can be expressed by

Z. Zhang et al., *Complex-Valued Neural Networks Systems with Time Delay*,
Intelligent Control and Learning Systems 4,
https://doi.org/10.1007/978-981-19-5450-4_3

separating real and imaginary parts. For this case, by separating the original system into real and imaginary parts and forming an equivalent real-valued system, various types of dynamics are investigated in many references (Hu and Wang 2012; Zhang et al. 2014; Liu and Chen 2016; Gong et al. 2015a, b, 2016, 2017; Chen and Song 2013; Zhou and Song 2013; Xu et al. 2014, 2017; Song et al. 2014; Zhang et al. 2014; Rakkiyappan et al. 2015a, b; Velmurugan et al. 2015a, b; Zhang and Yu 2016; Dong et al. 2015; Wang et al. 2016, 2017; Huang et al. 2014; Li et al. 2016), for example, exponential stability, global μ-stability, complete stability and so on. The other is the case that the activation functions are not explicitly expressed by separating real and imaginary parts. For this case, the original system is treated as an entirety and the corresponding dynamical behaviors are developed in Hu and Wang (2012), Chen and Song (2013), Zhang et al. (2014), Rakkiyappan et al. (2015a), Velmurugan et al. (2015a, b), Zhang and Yu (2016), Gong et al. (2015a), Fang and Sun (2014), Pan et al. (2015), Song et al. (2015, 2016a, b), Song and Zhao (2016). In these existing papers, real-imaginary separate-type activation functions are generally required to satisfy some specific assumptions. For instance, the existence, continuity and boundedness of the partial derivatives of the real and imaginary parts, and the inequalities conditions similar to those in real domain for them. Activation functions which can be not explicitly separated are usually required to satisfy the globally Lipschitz continuity condition in the complex domain and in many references, they are also chosen to be bounded. In fact, it can be proved that the assumption of the former is covered by that of the latter, as stated in Zhang et al. (2014).

Motivated by the above discussions, in this chapter, we further analyze the stability problem for CVNNs with constant delay. The complex-valued activation functions are always supposed to satisfy the globally Lipschitz continuity condition in the complex domain whether they are explicitly expressed by separating real and imaginary parts or not. It is clear that our hypothesis expands the range of use of CVNNs. Then, by using the homeomorphism theory and Lyapunov functional approach, new delay-dependent sufficient conditions to guarantee the existence, uniqueness and globally asymptotical stability of the equilibrium point of the considered system with two types of activation functions are derived, respectively.

3.2 Preliminaries

In this chapter we will continue to regard system (2.1) as our research target, which satisfies the Assumption 2.1. Here, they are omitted.

It is worth mentioning that except for the Assumption 2.1 in the previous chapter, the following Assumption 3.1 is often made on the separation case (Chen and Song 2013; Zhou and Song 2013; Huang et al. 2014; Rakkiyappan et al. 2015b; Velmurugan et al. 2015a; Gong et al. 2015a, b, 2017; Li et al. 2016).

Assumption 3.1 The neuron activation functions $f_j(.)$ and $g_j(.)$ can be separated into real and imaginary parts of the complex number $z = x + iy$ as

$$f_j(z) = f_j^R(x) + i f_j^I(y), \; g_j(z) = g_j^R(x) + i g_j^I(y),$$

where $f_j^R(.)$, $f_j^I(.)$, $g_j^R(.)$, $g_j^I(.) : \mathbb{R} \to \mathbb{R}$. Then, for any $\omega_1, \omega_2, \nu_1, \nu_2 \in \mathbb{R}$ and $\omega_1 \neq \omega_2, \nu_1 \neq \nu_2$,

$$\xi_j^{R-} \leq \frac{f_j^R(\omega_1) - f_j^R(\omega_2)}{\omega_1 - \omega_2} \leq \xi_j^{R+}, \; \xi_j^{I-} \leq \frac{f_j^I(\omega_1) - f_j^I(\omega_2)}{\omega_1 - \omega_2} \leq \xi_j^{I+},$$
$$\zeta_j^{R-} \leq \frac{g_j^R(\mu_1) - g_j^R(\mu_2)}{\mu_1 - \mu_2} \leq \zeta_j^{R+}, \; \zeta_j^{I-} \leq \frac{g_j^I(\mu_1) - g_j^I(\mu_2)}{\mu_1 - \mu_2} \leq \zeta_j^{I+},$$

for all $j = 1, 2, \ldots, n$.

Remark 3.1 In fact, we can verify easily that Assumption 3.1 is a strong constraint and is also a special case of Assumption 2.1. For example, if the activation functions $f_j(z)$ $(j = 1, 2, \ldots, n)$ satisfy Assumption 3.1, that is, the following conditions hold:

$$|f_j^R(x_1) - f_j^R(x_2)| \leq \xi_j^R |x_1 - x_2|, \; |f_j^I(y_1) - f_j^I(y_2)| \leq \xi_j^I |y_1 - y_2|,$$

where $z_1 = x_1 + iy_1$, $z_2 = x_2 + iy_2$, $\xi_j^R = \max\{|\xi_j^{R+}|, |\xi_j^{R-}|\}$ and $\xi_j^I = \max\{|\xi_j^{I+}|, |\xi_j^{I-}|\}$, then, we can have that

$$\begin{aligned}
|f_j(z_1) - f_j(z_2)| &= |(f_j^R(x_1) + i f_j^I(y_1)) - (f_j^R(x_2) + i f_j^I(y_2))| \\
&\leq |f_j^R(x_1) - f_j^R(x_2)| + |f_j^I(y_1) - f_j^I(y_2)| \\
&\leq \xi_j^R |x_1 - x_2| + \xi_j^I |y_1 - y_2| \\
&\leq \sqrt{\xi_j^{R^2} + \xi_j^{I^2}} |z_1 - z_2|.
\end{aligned}$$

Similarly, we can also obtain that

$$|g_j(z_1) - g_j(z_2)| \leq \sqrt{\zeta_j^{R^2} + \zeta_j^{I^2}} |z_1 - z_2|,$$

where $\zeta_j^R = \max\{|\zeta_j^{R+}|, |\zeta_j^{R-}|\}$ and $\zeta_j^I = \max\{|\zeta_j^{I+}|, |\zeta_j^{I-}|\}$. Thus, the activation functions $f_j(z)$ and $g_j(z)$ $(j = 1, 2, \ldots, n)$ satisfy the Lipschitz condition of Assumption 2.1. Hence, Assumption 3.1 implies Assumption 2.1, that is, the condition of Assumption 2.1 is weaker than the one of Assumption 3.1.

Remark 3.2 According to the above discussions in Remark 3.1, in this chapter we always assume that the complex-valued activation functions satisfy the Lipschitz condition as stated in the Assumption 2.1 of the previous chapter whether or not they are separation case.

Remark 3.3 In fact, when the activation functions are expressed by separating real and imaginary parts, the expressed form $f_j(z) = f_j^R(x, y) + if_j^I(x, y)$ and $g_j(z) = g_j^R(x, y) + ig_j^I(x, y)$ include $f_j(z) = f_j^R(x) + if_j^I(y)$ and $g_j(z) = g_j^R(x) + ig_j^I(y)$ as a special case. Hence, in the next section we always denote $f_j(z)$ and $g_j(z)$ as the former for real-imaginary separate-type activation functions.

Next, we give the following lemmas which play a very important role in deriving the main results.

Lemma 3.1 (Fang and Sun 2014) *If $H(z) : \mathbb{C}^n \to \mathbb{C}^n$ is a continuous map and satisfies the following conditions:*
(1) $H(z)$ is injective on $\mathbb{C}^n$;
(2) $\|H(z)\| \to \infty$ as $\|z\| \to \infty$;
then $H(z)$ is a homeomorphism of $\mathbb{C}^n$.

Lemma 3.2 (Fang and Sun 2014) *For any vectors $X, Y \in \mathbb{C}^n$, a positive-definite Hermitian matrix $H \in \mathbb{C}^{n\times n}$ and positive constant ε, the following inequality holds:*

$$X^*HY + Y^*HX \leq \varepsilon X^*HX + \frac{1}{\varepsilon}Y^*HY.$$

Lemma 3.3 (Velmurugan et al. 2015a) *For any constant matrix $R \in \mathbb{R}^{n\times n}$ and $R > 0$, vector function $u(s) : [a, b] \to \mathbb{R}^n$ with scalars $a < b$ such that the following inequality holds*

$$\left(\int_a^b u(s)ds\right)^T R\left(\int_a^b u(s)ds\right) \leq (b-a)\int_a^b u^T(s)Ru(s)ds.$$

Lemma 3.4 (Velmurugan et al. 2015a) *For any constant Hermitian matrix $W \in \mathbb{C}^{n\times n}$ and $W > 0$, vector function $u(s) : [a, b] \to \mathbb{C}^n$ with scalars $a < b$ such that the following inequality holds*

$$\left(\int_a^b u(s)ds\right)^* W\left(\int_a^b u(s)ds\right) \leq (b-a)\int_a^b u^*(s)Wu(s)ds.$$

Lemma 3.5 (Zhou and Song 2013) *Given a Hermitian matrix Θ, then $\Theta < 0$ is equivalent to*

$$\begin{bmatrix} \Theta^R & -\Theta^I \\ \Theta^I & \Theta^R \end{bmatrix} < 0,$$

where $\Theta^R = \mathrm{Re}(\Theta)$ and $\Theta^I = \mathrm{Im}(\Theta)$.

3.3 Further Stability Analysis Based on Separable Method

In this section, to enrich the development of stability analysis for system (2.1), by using the homeomorphism theory and Lyapunov functional approach, we further propose a delay-dependent stability sufficient condition based on separable method. The main result is stated as follows.

Theorem 3.1 *Suppose that the activation functions are real-imaginary separate-type and satisfy Assumption 2.1. Given a scalar* $\tau > 0$*, system (2.1) has a unique equilibrium point and it is globally asymptotically stable if there exist positive-definite matrices* P*,* Q*,* R *and scalars* $\varepsilon_1 > 0$*,* $\varepsilon_2 > 0$ *such that the following LMI holds*

$$\Omega = \begin{bmatrix} \Omega_{11} & R & P\bar{A} & P\bar{B} & -\tau\bar{D}R \\ * & -Q-R+\varepsilon_2\bar{U} & 0 & 0 & 0 \\ * & * & -\varepsilon_1 I & 0 & \tau\bar{A}^T R \\ * & * & * & -\varepsilon_2 I & \tau\bar{B}^T R \\ * & * & * & * & -R \end{bmatrix} < 0, \tag{3.1}$$

$$\Omega_{11} = -P\bar{D} - \bar{D}P + Q - R + \varepsilon_1\bar{L}.$$

Proof Because the existence, uniqueness and global stability of the equilibrium point for system (2.1) is equivalent to those for (2.4), in the sequel we will study system (2.4) to obtain a delay-dependent sufficient condition. First, when proving the existence and uniqueness of the equilibrium point for system (2.4), let

$$\Gamma = \begin{bmatrix} I & I & 0 & 0 & 0 \\ 0 & 0 & I & 0 & 0 \\ 0 & 0 & 0 & I & 0 \end{bmatrix}.$$

If (3.1) holds, by multiplying Ω by Γ and Γ^T on its left and right sides, respectively, we get

$$\begin{bmatrix} -P\bar{D} - \bar{D}P + \varepsilon_1\bar{L} + \varepsilon_2\bar{U} & P\bar{A} & P\bar{B} \\ * & -\varepsilon_1 I & 0 \\ * & * & -\varepsilon_2 I \end{bmatrix} < 0. \tag{3.2}$$

By Schur Complement, one can have

$$-P\bar{D} - \bar{D}P + \varepsilon_1\bar{L} + \varepsilon_2\bar{U} + \frac{1}{\varepsilon_1}P\bar{A}\bar{A}^T P + \frac{1}{\varepsilon_2}P\bar{B}\bar{B}^T P < 0. \tag{3.3}$$

Then, we can show that system (2.4) has a unique equilibrium point by using the contradiction similar to the proof process of Theorem 2.1.

Second, to show the globally asymptotical stability of the equilibrium point for system (2.4), we construct the following Lyapunov functional:

$$V(t) = \tilde{\omega}^T(t)P\tilde{\omega}(t) + \int_{t-\tau}^{t} \tilde{\omega}^T(s)Q\tilde{\omega}(s)ds + \tau\int_{-\tau}^{0}\int_{t+\theta}^{t} \dot{\tilde{\omega}}^T(s)R\dot{\tilde{\omega}}(s)dsd\theta, \tag{3.4}$$

where $P > 0$, $Q > 0$ and $R > 0$. Then, based on Lemma 3.3, by Lyapunov functional approach, we can prove that the equilibrium point of system (2.4) is globally asymptotically stable. This proof process is also similar to that in Theorem 2.1. So, here, we omit them. □

Remark 3.4 Theorem 3.1 provides a delay-dependent sufficient condition for the existence, uniqueness and stability analysis of the equilibrium point for delayed CVNNs with real-imaginary separate-type activation functions satisfying Assumption 2.1. If real and imaginary parts can't be explicitly separated, for example, $f(z) = \frac{1-e^{-\bar{z}}}{1+e^{-\bar{z}}}$, system (2.1) needs to be regarded as an entirety so as to investigate the dynamical behaviors of the equilibrium point as stated in the following section.

3.4 Stability Analysis Based on Nonseparable Method

Theorem 3.2 *Suppose that the activation functions aren't explicitly expressed by separating real-imaginary parts and satisfy Assumption 2.1. Given a scalar $\tau > 0$, system (2.1) has a unique equilibrium point and it is globally asymptotically stable if there exist positive-definite Hermitian matrices M, S, W and scalars $\varepsilon_1 > 0$, $\varepsilon_2 > 0$ such that the following LMI holds*

$$\Theta = \begin{bmatrix} \Theta_{11} & W & MA & MB & -\tau DW \\ * & -S - W + \varepsilon_2 U^T U & 0 & 0 & 0 \\ * & * & -\varepsilon_1 I & 0 & \tau A^* W \\ * & * & * & -\varepsilon_2 I & \tau B^* W \\ * & * & * & * & -W \end{bmatrix} < 0, \tag{3.5}$$

$$\Theta_{11} = -MD - DM + S - W + \varepsilon_1 L^T L.$$

Proof First, we show the existence and uniqueness of the equilibrium point for system (2.1). Define the following map:

$$H(z) = -Dz + Af(z) + Bg(z) + u. \tag{3.6}$$

Next, we show that $H(z)$ is injective. Similarly, we apply contradiction to prove it. Suppose that there exist z_1 and z_2 with $z_1 \neq z_2$ such that $H(z_1) = H(z_2)$. Then one can get

$$-D(z_1 - z_2) + A(f(z_1) - f(z_2)) + B(g(z_1) - g(z_2)) = 0. \tag{3.7}$$

Left-multiplying both sides of (3.7) by $(z_1 - z_2)^* M$ results in

$$-(z_1 - z_2)^* MD(z_1 - z_2) + (z_1 - z_2)^* MA(f(z_1) - f(z_2))$$
$$+(z_1 - z_2)^* MB(g(z_1) - g(z_2)) = 0. \quad (3.8)$$

Next, on both sides of (3.8), we take the conjugate transpose and have

$$-(z_1 - z_2)^* DM(z_1 - z_2) + (f(z_1) - f(z_2))^* A^* M(z_1 - z_2)$$
$$+(g(z_1) - g(z_2))^* B^* M(z_1 - z_2) = 0. \quad (3.9)$$

By adding (3.8) to (3.9), based on Lemma 3.2 and (2.2), one can get

$$\begin{aligned}
&(z_1 - z_2)^*(-MD - DM)(z_1 - z_2) + (z_1 - z_2)^* MA(f(z_1) - f(z_2))\\
&+(f(z_1) - f(z_2))^* A^* M(z_1 - z_2) + (z_1 - z_2)^* MB(g(z_1) - g(z_2))\\
&+(g(z_1) - g(z_2))^* B^* M(z_1 - z_2) \le (z_1 - z_2)^*(-MD - DM)(z_1 - z_2)\\
&+\varepsilon_1 (f(z_1) - f(z_2))^*(f(z_1) - f(z_2)) + \frac{1}{\varepsilon_1}(z_1 - z_2)^* MAA^* M(z_1 - z_2)\\
&+\varepsilon_2 (g(z_1) - g(z_2))^*(g(z_1) - g(z_2)) + \frac{1}{\varepsilon_2}(z_1 - z_2)^* MBB^* M(z_1 - z_2)\\
&\le (z_1 - z_2)^*(-MD - DM + \varepsilon_1 L^T L + \varepsilon_2 U^T U\\
&+\frac{1}{\varepsilon_1} MAA^* M + \frac{1}{\varepsilon_2} MBB^* M)(z_1 - z_2),
\end{aligned} \quad (3.10)$$

for scalars $\varepsilon_1 > 0$ and $\varepsilon_2 > 0$. On the other hand, let

$$\Gamma = \begin{bmatrix} I & I & 0 & 0 & 0 \\ 0 & 0 & I & 0 & 0 \\ 0 & 0 & 0 & I & 0 \end{bmatrix}.$$

If (3.5) holds, by multiplying Θ by Γ and Γ^T on its left and right sides, respectively, we get

$$\begin{bmatrix} -MD - DM + \varepsilon_1 L^T L + \varepsilon_2 U^T U & MA & MB \\ * & -\varepsilon_1 I & 0 \\ * & * & -\varepsilon_2 I \end{bmatrix} < 0. \quad (3.11)$$

Then, by Schur Complement, we have

$$-MD - DM + \varepsilon_1 L^T L + \varepsilon_2 U^T U$$
$$+\frac{1}{\varepsilon_1} MAA^* M + \frac{1}{\varepsilon_2} MBB^* M < 0. \quad (3.12)$$

Hence, we can obtain

$$\begin{aligned}&(z_1-z_2)^*(-MD-DM)(z_1-z_2)+(z_1-z_2)^*MA(f(z_1)-f(z_2))\\&+(f(z_1)-f(z_2))^*A^*M(z_1-z_2)+(z_1-z_2)^*MB(g(z_1)-g(z_2))\\&+(g(z_1)-g(z_2))^*B^*M(z_1-z_2)<0.\end{aligned}\tag{3.13}$$

This contradicts with (3.8) and (3.9). So, the map $H(z)$ is injective.

Next, we show that $\|H(z)\|\to\infty$ as $\|z\|\to\infty$. By (3.12), we can get

$$-MD-DM+\varepsilon_1L^TL+\varepsilon_2U^TU+\frac{1}{\varepsilon_1}MAA^*M+\frac{1}{\varepsilon_2}MBB^*M<-\varepsilon I,$$

for sufficiently small $\varepsilon>0$. Then, by using a derivation procedure similar to that of (3.10), we can obtain

$$\begin{aligned}2z^*M(H(z)-H(0))&\le z^*(-MD-DM+\varepsilon_1L^TL+\varepsilon_2U^TU\\&+\frac{1}{\varepsilon_1}MAA^*M+\frac{1}{\varepsilon_2}MBB^*M)z<-\varepsilon\|z\|^2.\end{aligned}\tag{3.14}$$

It is easy to infer from (3.14) that

$$\varepsilon\|z\|\le 2\|M\|(\|H(z)\|+\|H(0)\|).\tag{3.15}$$

Thus, $\|H(z)\|\to\infty$ as $\|z\|\to\infty$. By Lemma 3.1, the map $H(z)$ is homeomorphic on $\mathbb{C}^n$. So there exists a unique point $\hat{z}$ such that $H(\hat{z})=0$, that is, system (2.1) has a unique equilibrium point $\hat{z}$.

Second, we prove the globally asymptotical stability of the equilibrium point for system (2.1). By the transformation $\tilde{z}=z-\hat{z}$, the equilibrium point for system (2.1) can be shifted to the origin of the following system

$$\dot{\tilde{z}}=-D\tilde{z}+A\tilde{f}(\tilde{z})+B\tilde{g}(\tilde{z}^\tau),\tag{3.16}$$

where $\tilde{f}(\tilde{z})=f(\tilde{z}+\hat{z})-f(\hat{z})$ and $\tilde{g}(\tilde{z}^\tau)=g(\tilde{z}^\tau+\hat{z})-g(\hat{z})$. Considering the following Lyapunov functional defined on complex domain:

$$\begin{aligned}V(t)&=\tilde{z}^*(t)M\tilde{z}(t)+\int_{t-\tau}^{t}\tilde{z}^*(s)S\tilde{z}(s)ds\\&+\tau\int_{-\tau}^{0}\int_{t+\theta}^{t}\dot{\tilde{z}}^*(s)W\dot{\tilde{z}}(s)dsd\theta.\end{aligned}\tag{3.17}$$

Then, based on Lemma 3.4 and along the solution of (3.16), the time derivative of $V(t)$ yields

$$\begin{aligned}\dot{V}(t) &= 2\tilde{z}^*M\dot{\tilde{z}} + \tilde{z}^*S\tilde{z} - \tilde{z}^{\tau*}S\tilde{z}^{\tau} + \tau^2\dot{\tilde{z}}^*W\dot{\tilde{z}} - \tau\int_{t-\tau}^{t}\dot{\tilde{z}}^*(s)W\dot{\tilde{z}}(s)ds \\ &\le 2\tilde{z}^*M(-D\tilde{z} + A\tilde{f}(\tilde{z}) + B\tilde{g}(\tilde{z}^{\tau})) + \tilde{z}^*S\tilde{z} - \tilde{z}^{\tau*}S\tilde{z}^{\tau} \\ &\quad + \tau^2(-D\tilde{z} + A\tilde{f}(\tilde{z}) + B\tilde{g}(\tilde{z}^{\tau}))^*W(-D\tilde{z} + A\tilde{f}(\tilde{z}) + B\tilde{g}(\tilde{z}^{\tau})) \\ &\quad - (\tilde{z} - \tilde{z}^{\tau})^*W(\tilde{z} - \tilde{z}^{\tau}).\end{aligned} \tag{3.18}$$

Moreover, from (2.2) it follows that

$$\begin{aligned}&\varepsilon_1\left(\tilde{z}^*L^TL\tilde{z} - \tilde{f}^*(\tilde{z})\tilde{f}(\tilde{z})\right) \ge 0, \\ &\varepsilon_2\left(\tilde{z}^{\tau*}U^TU\tilde{z}^{\tau} - \tilde{g}^*(\tilde{z}^{\tau})\tilde{g}(\tilde{z}^{\tau})\right) \ge 0,\end{aligned} \tag{3.19}$$

for $\varepsilon_1 > 0$ and $\varepsilon_2 > 0$. Then, combining with (3.18) and (3.19), one can have

$$\dot{V}(t) \le \varsigma^*(t)(\hat{\Theta} + \tau^2\Upsilon^*W\Upsilon)\varsigma(t), \tag{3.20}$$

where

$$\begin{aligned}\varsigma(t) &= \begin{bmatrix}\tilde{z}^* & \tilde{z}^{\tau*} & \tilde{f}^*(\tilde{z}) & \tilde{g}^*(\tilde{z}^{\tau})\end{bmatrix}^*, \\ \hat{\Theta} &= \begin{bmatrix}\Theta_{11} & W & MA & MB \\ * & -S - W + \varepsilon_2U^TU & 0 & 0 \\ * & * & -\varepsilon_1I & 0 \\ * & * & * & -\varepsilon_2I\end{bmatrix}, \\ \Upsilon &= \begin{bmatrix}-D & 0 & A & B\end{bmatrix}.\end{aligned}$$

By Schur Complement, it is obvious that $\Theta < 0$ in (3.5) is equivalent to that $\hat{\Theta} + \tau^2\Upsilon^*W\Upsilon < 0$. Thus, $\dot{V}(t) < 0$ if (3.5) holds. So the origin of system (3.16) is globally asymptotically stable, i.e., the equilibrium point of system (2.1) is globally asymptotically stable. □

Remark 3.5 Theorem 3.2 provides a new delay-dependent criterion for the existence, uniqueness, and stability analysis of the equilibrium point for delayed CVNNs with the activation functions which aren't explicitly expressed by separating real-imaginary parts and satisfy Assumption 2.1. However, this criterion is established by complex-valued LMI which cannot be solved by LMI Toolbox straightforwardly. Next, by means of Lemma 3.5, we convert the given complex-valued LMI into real-valued one presented in the following corollary.

Corollary 3.1 *Suppose that the activation functions aren't explicitly expressed by separating real-imaginary parts and satisfy Assumption 2.1. Given a scalar* $\tau > 0$, *system (2.1) has a unique equilibrium point and it is globally asymptotically stable if there exist positive-definite Hermitian matrices* $M = M_1 + iM_2$, $S = S_1 + iS_2$, $W = W_1 + iW_2$ *and scalars* $\varepsilon_1 > 0$, $\varepsilon_2 > 0$ *such that the following LMI holds*

$$\begin{bmatrix} \Theta^R & -\Theta^I \\ \Theta^I & \Theta^R \end{bmatrix} < 0, \tag{3.21}$$

where

$$\Theta^R = \begin{bmatrix} \Theta_{11}^R & W_1 & M_1A_1 - M_2A_2 & M_1B_1 - M_2B_2 & -\tau DW_1 \\ * & \Theta_{22}^R & 0 & 0 & 0 \\ * & * & -\varepsilon_1 I & 0 & \tau(A_1^TW_1 + A_2^TW_2) \\ * & * & * & -\varepsilon_2 I & \tau(B_1^TW_1 + B_2^TW_2) \\ * & * & * & * & -W_1 \end{bmatrix},$$

$$\Theta_{11}^R = -M_1D - DM_1 + S_1 - W_1 + \varepsilon_1 L^T L,$$
$$\Theta_{22}^R = -S_1 - W_1 + \varepsilon_2 U^T U,$$

$$\Theta^I = \begin{bmatrix} \Theta_{11}^I & W_2 & M_1A_2 + M_2A_1 & M_1B_2 + M_2B_1 & -\tau DW_2 \\ -W_2^T & -S_2 - W_2 & 0 & 0 & 0 \\ \Theta_{31}^I & 0 & 0 & 0 & \Theta_{35}^I \\ \Theta_{41}^I & 0 & 0 & 0 & \Theta_{45}^I \\ \tau W_2^T D & 0 & \tau(W_1^TA_2 - W_2^TA_1) & \tau(W_1^TB_2 - W_2^TB_1) & -W_2 \end{bmatrix},$$

$$\Theta_{11}^I = -M_2D - DM_2 + S_2 - W_2,$$
$$\Theta_{31}^I = -A_2^TM_1^T - A_1^TM_2^T, \ \Theta_{41}^I = -B_2^TM_1^T - B_1^TM_2^T,$$
$$\Theta_{35}^I = \tau(A_1^TW_2 - A_2^TW_1), \ \Theta_{45}^I = \tau(B_1^TW_2 - B_2^TW_1),$$
$$A_1 = Re(A), \ A_2 = Im(A), \ B_1 = Re(B), \ B_2 = Im(B).$$

3.5 Illustrative Examples

In this section, we provide several examples to show the effectiveness of our results.

Example 3.1 Consider system (2.1) with

$$f_j(z_j) = \frac{1 - e^{-2x_j - y_j}}{1 + e^{-2x_j - y_j}} + i\frac{1}{1 + e^{-x_j + 2y_j}}, \ j = 1, 2,$$
$$g_j(z_j) = \frac{1}{1 + e^{-x_j + 2y_j}} + i\frac{1 - e^{-2x_j - y_j}}{1 + e^{-2x_j - y_j}},$$
$$D = \begin{bmatrix} 10 & 0 \\ 0 & 12 \end{bmatrix}, \ u = (-3 + i, 2 + 4i)^T,$$

and the same A and B in Example 2.1. By simple calculation, we have

$$\bar{A} = \begin{bmatrix} 1 & 1 & -1 & 1 \\ 2 & 2 & 1 & -1 \\ 1 & -1 & 1 & 1 \\ -1 & 1 & 2 & 2 \end{bmatrix}, \quad \bar{B} = \begin{bmatrix} 1 & 1 & 1 & -1 \\ 2 & 2 & -2 & 2 \\ -1 & 1 & 1 & 1 \\ 2 & -2 & 2 & 2 \end{bmatrix},$$

$$\bar{D} = \begin{bmatrix} 10 & 0 & 0 & 0 \\ 0 & 12 & 0 & 0 \\ 0 & 0 & 10 & 0 \\ 0 & 0 & 0 & 12 \end{bmatrix}, \quad \bar{L} = \bar{U} = \begin{bmatrix} \frac{41}{16} & 0 & 0 & 0 \\ 0 & \frac{41}{16} & 0 & 0 \\ 0 & 0 & \frac{41}{16} & 0 \\ 0 & 0 & 0 & \frac{41}{16} \end{bmatrix}.$$

Let $\tau = 2$ and by applying Theorem 3.1, it is easy to conclude that this system is asymptotically stable. Moreover, by solving (3.1) we can obtain $\varepsilon_1 = 0.5314$, $\varepsilon_2 = 0.5855$ and

$$P = \begin{bmatrix} 0.2716 & 0.0031 & 0 & 0 \\ 0.0031 & 0.2410 & 0 & 0 \\ 0 & 0 & 0.2716 & 0.0031 \\ 0 & 0 & 0.0031 & 0.2410 \end{bmatrix}, \quad R = \begin{bmatrix} 0.0043 & 0.0002 & 0 & 0 \\ 0.0002 & 0.0041 & 0 & 0 \\ 0 & 0 & 0.0043 & 0.0002 \\ 0 & 0 & 0.0002 & 0.0041 \end{bmatrix},$$

$$Q = \begin{bmatrix} 2.0025 & -0.0681 & 0 & 0 \\ -0.0681 & 1.6095 & 0 & 0 \\ 0 & 0 & 2.0025 & -0.0681 \\ 0 & 0 & -0.0681 & 1.6095 \end{bmatrix}.$$

We assume that the initial conditions are $z_1(t) = -2 - 4i$ and $z_2(t) = 1 + 3i$ for $t \in [-2, 0]$. Time responses for the real-imaginary parts of the states z_1 and z_2 are depicted in Fig. 3.1.

Example 3.2 Consider system (2.1) with

$$A = \begin{bmatrix} -3+2i & 4-i \\ 2-i & 1+2i \end{bmatrix}, \quad B = \begin{bmatrix} 1-i & 3+4i \\ 2+3i & -3-2i \end{bmatrix}, \quad D = \begin{bmatrix} 5.5 & 0 \\ 0 & 5.3 \end{bmatrix},$$

$$f_j(z_j) = \frac{1}{1+e^{-\bar{z}_j}}, \quad g_j(z_j) = \frac{1-e^{-\bar{z}_j}}{1+e^{-\bar{z}_j}}, \quad j = 1, 2,$$
$$u = (2-3i, -4+i)^T.$$

We can calculate that

$$L = \begin{bmatrix} \frac{1}{3} & 0 \\ 0 & \frac{1}{3} \end{bmatrix}, \quad U = \begin{bmatrix} \frac{2}{3} & 0 \\ 0 & \frac{2}{3} \end{bmatrix}.$$

For this system, the complex-valued activation functions aren't explicitly expressed by separating real and imaginary parts. Thus, we use Corollary 3.1 to check the stability of this system. Let $\tau = 3$, it is clear to find that the given LMI condition

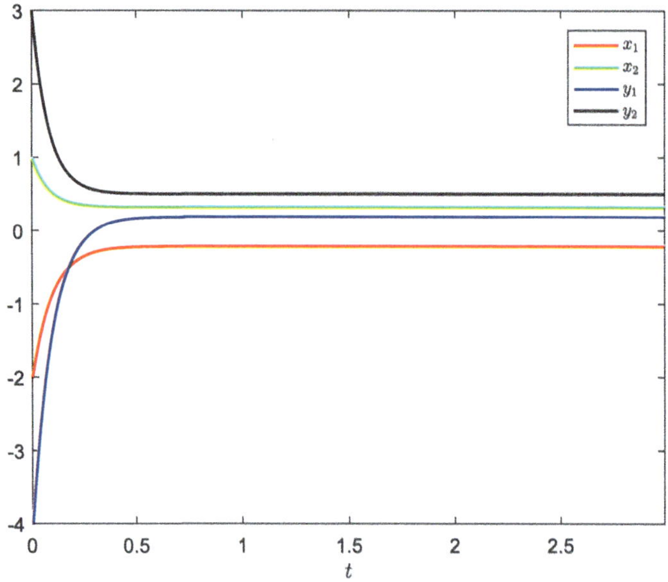

Fig. 3.1 Time responses for the real-imaginary parts of the state variables in Example 3.1

of Corollary 3.1 is feasible and this system is asymptotically stable. And a feasible solution is computed as $\varepsilon_1 = 7.3240$, $\varepsilon_2 = 4.4553$ and

$$M_1 = \begin{bmatrix} 1.1109 & 0.7611 \\ 0.7611 & 1.5116 \end{bmatrix}, \; M_2 = \begin{bmatrix} 0 & 0.4913 \\ -0.4913 & 0 \end{bmatrix}, \; S_1 = \begin{bmatrix} 3.3300 & 1.6871 \\ 1.6871 & 4.1400 \end{bmatrix},$$

$$S_2 = \begin{bmatrix} 0 & 0.0833 \\ -0.0833 & 0 \end{bmatrix}, \; W_1 = \begin{bmatrix} 0.0130 & 0.0071 \\ 0.0071 & 0.0166 \end{bmatrix}, \; W_2 = \begin{bmatrix} 0 & 0.0039 \\ -0.0039 & 0 \end{bmatrix}.$$

The initial conditions are assumed to be $z_1(t) = -1 - 2i$ and $z_2(t) = -2 + 4i$ for $t \in [-3, 0]$ and the corresponding responses of the states z_1 and z_2 are shown in Fig. 3.2.

In order to further indicate the advantages of the obtained results, we also provide the following system with the unbounded activation functions.

Example 3.3 Consider system (2.1) with

$$D = \begin{bmatrix} 10 & 0 \\ 0 & 9 \end{bmatrix}, \; f_j(z_j) = \frac{1}{2} z_j, \; g_j(z_j) = -\bar{z}_j, \; j = 1, 2,$$

and A, B and u are as in Example 3.2. It is easy to obtain

$$L = \begin{bmatrix} \frac{1}{2} & 0 \\ 0 & \frac{1}{2} \end{bmatrix}, \; U = \begin{bmatrix} 1 & 0 \\ 0 & 1 \end{bmatrix}.$$

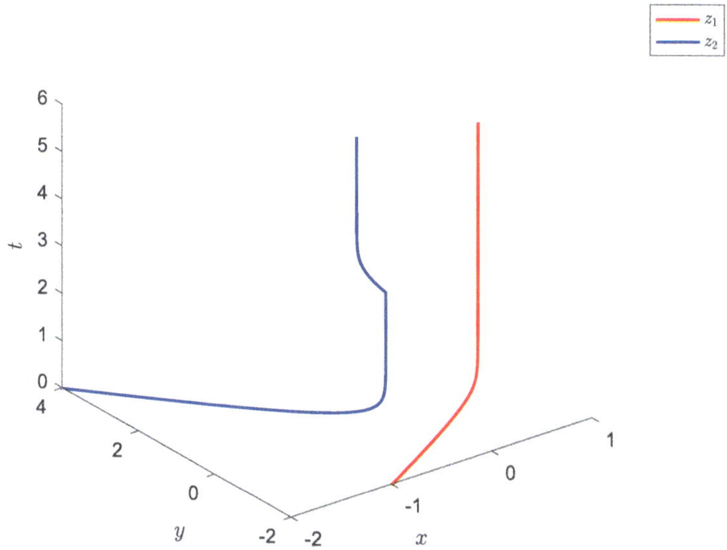

Fig. 3.2 Time responses for the state variables in Example 3.2

By using Corollary 3.1 and let $\tau = 3$, we can verify that the equilibrium point of this system is asymptotically stable. And the accordingly feasible solution to the LMI condition in (3.21) is computed as $\varepsilon_1 = 14.2710$, $\varepsilon_2 = 10.4779$ and

$$M_1 = \begin{bmatrix} 1.9747 & 0.3888 \\ 0.3888 & 2.3083 \end{bmatrix}, \; M_2 = \begin{bmatrix} 0 & 0.6917 \\ -0.6917 & 0 \end{bmatrix}, \; S_1 = \begin{bmatrix} 14.8384 & 3.1892 \\ 3.1892 & 15.9705 \end{bmatrix},$$
$$S_2 = \begin{bmatrix} 0 & 0.8748 \\ -0.8748 & 0 \end{bmatrix}, \; W_1 = \begin{bmatrix} 0.0173 & 0.0039 \\ 0.0039 & 0.0214 \end{bmatrix}, \; W_2 = \begin{bmatrix} 0 & 0.0032 \\ -0.0032 & 0 \end{bmatrix}.$$

Then, the initial conditions are chosen to be $z_1(t) = -3 - 2i$ and $z_2(t) = -1 + 3i$ for $t \in [-3, 0]$ and the time responses of the states z_1 and z_2 are shown in Fig. 3.3.

Remark 3.6 In fact, the activation functions of Example 3.3 can be explicitly expressed by separating real and imaginary parts. So the stability of this system could be checked by Theorem 3.1. However, we find it infeasible by using LMI toolbox. Meanwhile, we have checked many other examples and all verify this fact. On the other hand, for Example 3.1, we also verify that these systems are asymptotically stable by the given conditions of Corollary 3.1. Hence, for the systems with real-imaginary separate-type activation functions, both the results of Corollary 3.1 and Theorem 3.1 can be used to check the stabilities of them, but the former can serve more complex-valued neural networks systems than the latter. Therefore, it would be better that the system is treated as an entirety to study the dynamical behaviors.

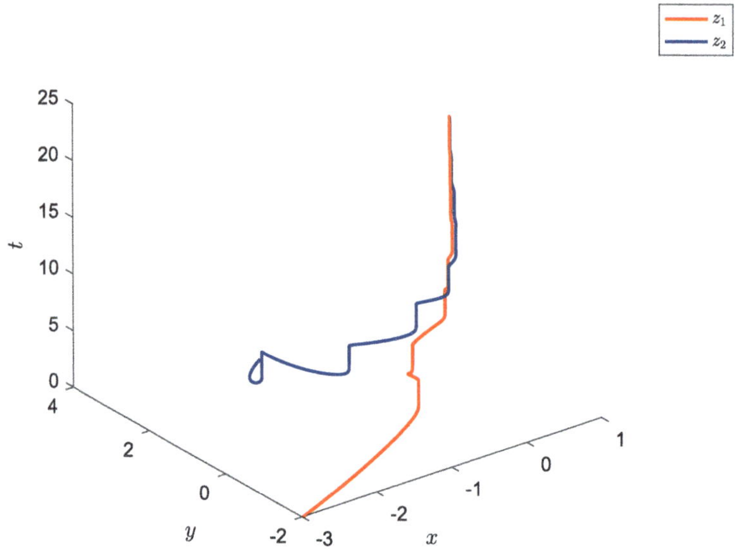

Fig. 3.3 Time responses for the state variables in Example 3.3

Remark 3.7 As we know, the case with real-imaginary separated-type activation functions can be regarded as a special case of the one without explicitly separating real and imaginary parts. But considering the fact that some activation functions similar to those in Example 3.1 often appear in the previous works and can be expressed more clearly by separating real and imaginary parts. Hence, for each case, the stability conditions are given in Theorems 3.1 and 3.2, respectively, which can show the dynamical behaviors of the considered system all sidedly and detailedly. And, in applying the obtained results to numerical simulations, if the activation functions are explicitly real-imaginary separated-type, the condition of Theorem 3.1 can be first used to check the stability. This also paves the way for other dynamical problems of our future work. If the condition of Theorem 3.1 is found infeasible, then, the ones of Theorem 3.2 or Corollary 3.1 can be considered to check the stability of systems.

3.6 Conclusion and Notes

The existence, uniqueness, and the stability problem of the equilibrium point for CVNNs with constant delay have been further studied in this chapter. Based on the fact that whether the complex-valued activation functions are explicitly expressed by separating real and imaginary parts or not, they are always assumed to satisfy the globally Lipschitz condition in the complex domain. Then, via the separable method and nonseparable method, delay-dependent stability criteria have been proposed in terms of real-valued LMI and complex-valued LMI. Several examples have been given to illustrate the effectiveness and advantages of the obtained criteria. Finally, two sufficient results have been simply compared.

Note that other dynamics such as Hopf bifurcation, finite-time stability and Lagrange exponential stability deserve further study. In the following three chapters, we will focus on these items for CVNNs to carry out our work.

References

Chen X, Song Q (2013) Global stability of complex-valued neural networks with both leakage time delay and discrete time delay on time scales. Neurocomputing 121:254–264

Dong T, Liao X, Wang A (2015) Stability and Hopf bifurcation of a complex-valued neural network with two time delays. Nonlinear Dyn 82:173–184

Fang T, Sun J (2014) Further investigate the stability of complex-valued recurrent neural networks with time-delays. IEEE Trans Neural Netw Learn Syst 25(9):1709–1713

Gong W, Liang J, Cao J (2015a) Global μ-stability of complex-valued delayed neural networks with leakage delay. Neurocomputing 168:135–144

Gong W, Liang J, Cao J (2015b) Matrix measure method for global exponential stability of complex-valued recurrent neural networks with time-varying delays. Neural Netw 70:81–89

Gong W, Liang J, Zhang C, Cao J (2016) Nonlinear measure approach for the stability analysis of complex-valued neural networks. Neural Process Lett 44(2):539–554

Gong W, Liang J, Zhang C (2017) Multistability of complex-valued neural networks with distributed delays. Neural Comput Appl 28:1–14

Hu J, Wang J (2012) Global stability of complex-valued recurrent neural networks with time-delays. IEEE Trans Neural Netw Learn Syst 23(6):853–865

Huang Y, Zhang H, Wang Z (2014) Multistability of complex-valued recurrent neural networks with real-imaginary-type activation functions. Appl Math Comput 229:187–200

Kuroe Y, Hashimoto N, Mori T (2002) On energy function for complex-valued neural networks and its applications. In: Proceedings of the 9th international conference on neural information processing, pp 1079–1083

Kuroe Y, Yoshid M, Mori T (2003) On activation functions for complex-valued neural networks-existence of energy functions. In: Artificial neural networks and neural information processing, pp 174–175

Li Y, Liao X, Li H (2016) Global attracting sets of nonautonomous and complex-valued neural networks with time-varying delays. Neurocomputing 173(3):994–1000

Liu X, Chen T (2016) Global exponential stability for complex-valued recurrent neural networks with asynchronous time delays. IEEE Trans Neural Netw Learn Syst 27(3):593–606

Liu X, Fang K, Liu B (2009) A synthesis method based on stability analysis for complex-valued Hopfield neural network. In: Proceeding of 7th Asian control conference, pp 1245–1250

Pan J, Liu X, Xie W (2015) Exponential stability of a class of complex-valued neural networks with time-varying delays. Neurocomputing 164:293–299

Rakkiyappan R, Cao J, Velmurugan G (2015a) Existence and uniform stability analysis of fractional-order complex-valued neural networks with time delays. IEEE Trans Neural Netw Learn Syst 26(1):84–97

Rakkiyappan R, Velmurugan G, Li X (2015b) Complete stability analysis of complex-valued neural networks with time delays and impulses. Neural Process Lett 41:435–468

Rudin W (1987) Real and complex analysis. McGraw-Hill, New York

Song Q, Zhao Z (2016) Stability criterion of complex-valued neural networks with both leakage delay and time-varying delays on time scales. Neurocomputing 171:179–184

Song Q, Zhao Z, Liu Y (2015) Stability analysis of complex-valued neural networks with probabilistic time-varying delays. Neurocomputing 159:96–104
Song Q, Yan H, Zhao Z, Liu Y (2016a) Global exponential stability of complex-valued neural networks with both time-varying delays and impulsive effects. Neural Netw 79:108–116
Song Q, Yan H, Zhao Z, Liu Y (2016b) Global exponential stability of impulsive complex-valued neural networks with both asynchronous time-varying and continuously distributed delays. Neural Netw 81:1–10
Song R, Xiao W, Zhang H, Sun C (2014) Adaptive dynamic programming for a class of complex-valued nonlinear systems. IEEE Trans Neural Netw Learn Syst 25(9):1733–1739
Velmurugan G, Rakkiyappan R, Lakshmanan S (2015a) Passivity analysis of memristor-based complex-valued neural networks with time-varying delays. Neural Process Lett 42:517–540
Velmurugan G, Rakkiyappan R, Cao J (2015b) Further analysis of global μ-stability of complex-valued neural networks with unbounded time-varying delays. Neural Netw 67:14–27
Wang H, Duan S, Huang T, Wang L, Li C (2017) Exponential stability of complex-valued memristive recurrent neural networks. IEEE Trans Neural Netw Learn Syst 28(3):766–771
Wang Z, Huang L, Liu Y (2016) Global stability analysis for delayed complex-valued BAM neural networks. Neurocomputing 173:2083–2089
Xu X, Zhang J, Shi J (2014) Exponential stability of complex-valued neural networks with mixed delays. Neurocomputing 128:483–490
Xu X, Zhang J, Shi J (2017) Dynamical behaviour analysis of delayed complex-valued neural networks with impulsive effect. Int J Syst Sci 48(4):686–694
Zhang W, Li C, Huang T (2014) Global robust stability of complex-valued recurrent neural networks with time-delays and uncertainties. Int J Biomath 7(2):1–24
Zhang Z, Yu S (2016) Global asymptotic stability for a class of complex-valued Cohen-Grossberg neural networks with time delays. Neurocomputing 171:1158–1166
Zhang Z, Lin C, Chen B (2014) Global stability criterion for delayed complex-valued recurrent neural networks. IEEE Trans Neural Netw Learn Syst 25(9):1704–1708
Zhou B, Song Q (2013) Boundedness and complete stability of complex-valued neural networks with time delay. IEEE Trans Neural Netw Learn Syst 24(8):1227–1238

Chapter 4
Hopf Bifurcation Analysis for CVNNs with Discrete and Distributed Delays

Abstract This chapter deals with the stability and Hopf bifurcation problems for a class of CVNNs model with discrete and distributed delays. Regarding the discrete time delay as the bifurcating parameter, based on the separable method, the normal form theory and center manifold theorem, it aims to study the problem of Hopf bifurcation and present some sufficient conditions that determine the direction of the Hopf bifurcation and the stability of the bifurcating periodic solutions.

4.1 Introduction

As we described earlier, sometimes it is more convenient to adopt complex-valued models to describe many real-world problems. The advantage of complex states is that they contain two different kinds of information. When complex signals are involved in some practical applications, complex-valued models are preferable to real-valued models because the amplitude information and phase information of the signals can be simultaneously coped with, which facilitates the information processing and reduces the complexity of information processing (Wang et al. 2017). CVNNs are defined in the complex domain. The information will be processed in the complex one. Generally speaking, it has been suggested that CVNNs have even richer dynamical properties than real-valued ones (Guo et al. 2017). Nowadays, CVNNs have been widely applied to filtering, image processing, speech synthesis and computer vision (Lee 2006; Zhou and Zurada 2009; Dong et al. 2015). Thus, it is more significant and also interesting to investigate the dynamical properties of CVNNs both in theory and practice.

Actually, due to the finite speeds of the switching of amplifiers and transmission of signals in hardware implementation, time delays are inevitable in neural networks. The delayed axonal transmissions in neural networks may destabilize stable equilibria and cause periodic oscillations, bifurcation and even chaos (Yu and Cao 2007; Liu et al. 2016a, b; Olien and Bélair 1997; Liu et al. 2017; Bao et al. 2016a, b). In addition, neural networks usually have a spatial extent due to the presence of a multitude of parallel pathways with a variety of axon sizes and lengths, and hence there will be distribution of propagation delays (Tank and Hopfield 1987; Vries and

Z. Zhang et al., *Complex-Valued Neural Networks Systems with Time Delay*,
Intelligent Control and Learning Systems 4,
https://doi.org/10.1007/978-981-19-5450-4_4

Principe 1992; Principe et al. 1994; Lv and Gan 2016). In such a situation, the signal propagation is not instantaneous, and therefore, cannot be modeled by discrete delays. In other words, the distributed delays should also be incorporated into the neural network models. Hence, in order to describe the dynamical characteristics of neural networks more precisely, both discrete-time delays and distributed time delays should be considered.

Many researchers have devoted their efforts to the investigation of the dynamics of neural networks with both discrete and distributed delays and some excellent results have been presented. For instance, the problems of stability and Hopf bifurcation of a neuron network with discrete and distributed delays have been addressed in Li and Hu (2011). Taking the discrete delay as the bifurcation parameter, the existence of local Hopf bifurcation has been verified by using the Hopf bifurcation theory. The local stability of the equilibrium points and local Hopf bifurcation of a two-neuron network model with multiple discrete and distributed delays have been investigated in Karaoglu et al. (2016). Employing the Routh-Hurwitz criterion, a set of conditions that guarantee the stability of the fixed points and the existence of Hopf bifurcation have been achieved. As for CVNNs, a number of CVNNs models have been proposed and some excellent results have been developed (Zhou and Zurada 2009; Dong et al. 2015). For example, a class of discrete-time recurrent neural networks with complex-valued linear threshold neurons are proposed in Zhou and Zurada (2009), and meanwhile, the boundedness, global attractivity and complete stability have been studied. The stability and Hopf bifurcation of a CVNN model with two time delays have been discussed in Dong et al. (2015). Moreover, it has been found that under certain conditions, the zero solution loses its stability and a Hopf bifurcation occurs when the sum of the delays varies.

From the above discussions, it can be seen that it is meaningful to investigate the Hopf bifurcation of CVNNs with both discrete and distributed delays. To the best of our knowledge, the studies focusing on the Hopf bifurcation in most previously published works are mainly concentrated on neural networks with only discrete delays or distributed delays. There are few results focusing on Hopf bifurcation of CVNNs with both discrete and distributed delays. Inspired by the above-mentioned reasons, this chapter is devoted to studying the Hopf bifurcation of a CVNN model with discrete and distributed delays. Under the assumption that the activation function can be separated into its real and imaginary parts, the existence of Hopf bifurcation is discussed by choosing the discrete-time delay as the bifurcation parameter. Based on the normal form theory and center manifold theorem, some criteria for determining the direction of the Hopf bifurcation and the stability of the bifurcating periodic solutions are derived.

4.2 Problem Formulation

Based on the works Dong et al. (2015), consider a class of CVNN model with discrete and distributed delays, which can be described by

$$\begin{cases} \dot{z}_1(t) = -z_1(t) + b_{11} f_{11}(z_1(t-\tau)) + b_{12} f_{12}\left(\int_{-\infty}^{t} F(t-s) z_2(s) ds\right), \\ \dot{z}_2(t) = -z_2(t) + b_{21} f_{21}(z_1(t-\tau)) + b_{22} f_{22}\left(\int_{-\infty}^{t} F(t-s) z_2(s) ds\right), \end{cases} \tag{4.1}$$

where $z_i(t)$ represents the state of the ith neuron at time t; b_{ij} $(i, j = 1, 2)$ is complex number; $F(\cdot)$ denotes non-negative bounded delay kernel defined on $[0, \infty)$, which reflects the influence of the past states on the current dynamics.

In order to investigate the Hopf bifurcation of system (4.1), the following assumptions on the complex-valued activation function $f_{ij}(\cdot)$ $(i, j = 1, 2)$ are imposed.

$(H1)$ Let $z = x + iy$, the activation function $f_{ij}(z)$ can be separated into its real and imaginary parts as $f_{ij}(z) = f_{ij}^R(x, y) + i f_{ij}^I(x, y)$. The derivatives of $f_{ij}^R(x, y)$ and $f_{ij}^I(x, y)$ with respect to x and y, respectively, exist and continuous. And $f_{ij}^R(0, 0) = 0$, $f_{ij}^I(0, 0) = 0$ for $i, j = 1, 2$.

$(H2)$ The partial derivatives of $f_{ij}^R(x, y)$ and $f_{ij}^I(x, y)$ with respect to x and y, respectively, are bounded.

The distributed delay kernel $F(s)$ defined on $[0, \infty)$ is a non-negative bounded function and satisfies the following condition (Ncube 2013):

$$\int_0^{\infty} F(s) ds = 1, \int_0^{\infty} s F(s) ds < +\infty. \tag{4.2}$$

In general, the kernel $F(s)$ is taken as

$$F(s) = \frac{\alpha^{p+1} s^p}{p!} e^{-\alpha s}, s \geq 0, \alpha > 0, \tag{4.3}$$

where α denotes the mean delay of the kernel and p is a non-negative integer. When $p = 0$, $F(s)$ is referred to as the weak kernel. When $p = 1$, $F(s)$ is referred to as the strong kernel. Here, only the weak kernel is considered, that is

$$F(s) = \alpha e^{-\alpha s}, s \geq 0, \alpha > 0. \tag{4.4}$$

For convenience, a new variable $z_3(t)$ is introduced and defined as

$$z_3(t) = \int_{-\infty}^{t} \alpha e^{-\alpha(t-s)} z_2(s) ds. \tag{4.5}$$

Then, by using the linear chain technique, system (4.1) can be transformed into

$$\begin{cases} \dot{z}_1(t) = -z_1(t) + b_{11} f_{11}(z_1(t-\tau)) + b_{12} f_{12}(z_3(t)), \\ \dot{z}_2(t) = -z_2(t) + b_{21} f_{21}(z_1(t-\tau)) + b_{22} f_{22}(z_3(t)), \\ \dot{z}_3(t) = -\alpha z_3(t) + \alpha z_2(t). \end{cases} \tag{4.6}$$

Let $z = x + iy$ and $b_{ij} = b_{ij}^R + ib_{ij}^I$ $(i, j = 1, 2)$. Then, complex-valued system (4.6) can be expressed by separating it into its real and imaginary parts as

$$\begin{cases} \dot{x}_1(t) = -x_1(t) + b_{11}^R f_{11}^R(x_1(t-\tau), y_1(t-\tau)) - b_{11}^I f_{11}^I(x_1(t-\tau), \\ \qquad y_1(t-\tau)) + b_{12}^R f_{12}^R(x_3(t), y_3(t)) - b_{12}^I f_{12}^I(x_3(t), y_3(t)), \\ \dot{y}_1(t) = -y_1(t) + b_{11}^R f_{11}^I(x_1(t-\tau), y_1(t-\tau)) + b_{11}^I f_{11}^R(x_1(t-\tau), \\ \qquad y_1(t-\tau)) + b_{12}^R f_{12}^I(x_3(t), y_3(t)) + b_{12}^I f_{12}^R(x_3(t), y_3(t)), \\ \dot{x}_2(t) = -x_2(t) + b_{21}^R f_{21}^R(x_1(t-\tau), y_1(t-\tau)) - b_{21}^I f_{21}^I(x_1(t-\tau), \\ \qquad y_1(t-\tau)) + b_{22}^R f_{22}^R(x_3(t), y_3(t)) - b_{22}^I f_{22}^I(x_3(t), y_3(t)), \\ \dot{y}_2(t) = -y_2(t) + b_{21}^R f_{21}^I(x_1(t-\tau), y_1(t-\tau)) + b_{21}^I f_{21}^R(x_1(t-\tau), \\ \qquad y_1(t-\tau)) + b_{22}^R f_{22}^I(x_3(t), y_3(t)) + b_{22}^I f_{22}^R(x_3(t), y_3(t)), \\ \dot{x}_3(t) = -\alpha x_3(t) + \alpha x_2(t), \\ \dot{y}_3(t) = -\alpha y_3(t) + \alpha y_2(t). \end{cases} \tag{4.7}$$

With the substitutions $u_1(t) = x_1(t), u_2(t) = y_1(t), u_3(t) = x_2(t), u_4(t) = y_2(t), u_5(t) = x_3(t), u_6(t) = y_3(t)$, system (4.7) is equivalent to

$$\begin{cases} \dot{u}_1(t) = -u_1(t) + b_{11}^R f_{11}^R(u_1(t-\tau), u_2(t-\tau)) - b_{11}^I f_{11}^I(u_1(t-\tau), \\ \qquad u_2(t-\tau)) + b_{12}^R f_{12}^R(u_5(t), u_6(t)) - b_{12}^I f_{12}^I(u_5(t), u_6(t)), \\ \dot{u}_2(t) = -u_2(t) + b_{11}^R f_{11}^I(u_1(t-\tau), u_2(t-\tau)) + b_{11}^I f_{11}^R(u_1(t-\tau), \\ \qquad u_2(t-\tau)) + b_{12}^R f_{12}^I(u_5(t), u_6(t)) + b_{12}^I f_{12}^R(u_5(t), u_6(t)), \\ \dot{u}_3(t) = -u_3(t) + b_{21}^R f_{21}^R(u_1(t-\tau), u_2(t-\tau)) - b_{21}^I f_{21}^I(u_1(t-\tau), \\ \qquad u_2(t-\tau)) + b_{22}^R f_{22}^R(u_5(t), u_6(t)) - b_{22}^I f_{22}^I(u_5(t), u_6(t)), \\ \dot{u}_4(t) = -u_4(t) + b_{21}^R f_{21}^I(u_1(t-\tau), u_2(t-\tau)) + b_{21}^I f_{21}^R(u_1(t-\tau), \\ \qquad u_2(t-\tau)) + b_{22}^R f_{22}^I(u_5(t), u_6(t)) + b_{22}^I f_{22}^R(u_5(t), u_6(t)), \\ \dot{u}_5(t) = -\alpha u_5(t) + \alpha u_3(t), \\ \dot{u}_6(t) = -\alpha u_6(t) + \alpha u_4(t). \end{cases} \tag{4.8}$$

Lemma 4.1 (Ruan and Wei 2003) *Consider the exponential polynomial*

$$P(\lambda, e^{-\lambda\tau_1}, \cdots, e^{-\lambda\tau_m}) = \lambda^n + p_1^{(0)}\lambda^{n-1} + \cdots + p_{n-1}^{(0)}\lambda + p_n^0 + [p_1^{(1)}\lambda^{n-1} + \cdots + p_{n-1}^{(1)}\lambda + p_n^{(1)}]e^{-\lambda\tau_1} + \cdots + [p_1^{(m)}\lambda^{n-1} + \cdots + p_{n-1}^{(m)}\lambda + p_n^{(m)}]e^{-\lambda\tau_m},$$

where $\tau_i \geq 0$ $(i = 1, 2, \cdots, m)$ and $p_j^{(i)}$ $(i = 0, 1, \cdots, m;\ j = 1, 2, \cdots, n)$ are constants. As $(\tau_1, \tau_2, \cdots, \tau_m)$ varies, the sum of the order of the zeros of $P(\lambda, e^{-\lambda\tau_1}, \cdots, e^{-\lambda\tau_m})$ on the open right halfplane can change only if a zero appears on or crosses the imaginary axis.

4.3 Hopf Bifurcation Result

In this section, we will analyze the Hopf Bifurcation for system (4.1).

Theorem 4.1 *For system (4.8), suppose that the conditions $(H1)$-$(H4)$ and $l_{12} < 0$ hold, then one has*
(1) system (4.8) is local asymptotically stable for $\tau \in [0, \tau_0)$,
(2) system (4.8) is unstable for $\tau > \tau_0$,
(3) system (4.8) undergoes a Hopf bifurcation at the zero equilibrium points when $\tau = \tau_0$.

Proof Evidently, under the hypothesis $(H1)$, the origin $(0, 0, 0, 0, 0, 0)$ is an equilibrium point of system (4.8). The linearization of system (4.8) at the equilibrium point $(0, 0, 0, 0, 0, 0)$ is

$$\begin{cases} \dot{u}_1(t) = -u_1(t) + \alpha_{11}u_1(t-\tau) + \alpha_{12}u_2(t-\tau) + \alpha_{13}u_5(t) + \alpha_{14}u_6(t), \\ \dot{u}_2(t) = -u_2(t) + \alpha_{21}u_1(t-\tau) + \alpha_{22}u_2(t-\tau) + \alpha_{23}u_5(t) + \alpha_{24}u_6(t), \\ \dot{u}_3(t) = -u_3(t) + \alpha_{31}u_1(t-\tau) + \alpha_{32}u_2(t-\tau) + \alpha_{33}u_5(t) + \alpha_{34}u_6(t), \\ \dot{u}_4(t) = -u_4(t) + \alpha_{41}u_1(t-\tau) + \alpha_{42}u_2(t-\tau) + \alpha_{43}u_5(t) + \alpha_{44}u_6(t), \\ \dot{u}_5(t) = -\alpha u_5(t) + \alpha u_3(t), \\ \dot{u}_6(t) = -\alpha u_6(t) + \alpha u_4(t), \end{cases} \tag{4.9}$$

where

$$\begin{aligned}
\alpha_{11} &= b_{11}^R \frac{\partial f_{11}^R(0,0)}{\partial u_1} - b_{11}^I \frac{\partial f_{11}^I(0,0)}{\partial u_1}, \quad \alpha_{12} = b_{11}^R \frac{\partial f_{11}^R(0,0)}{\partial u_2} - b_{11}^I \frac{\partial f_{11}^I(0,0)}{\partial u_2}, \\
\alpha_{13} &= b_{12}^R \frac{\partial f_{12}^R(0,0)}{\partial u_5} - b_{12}^I \frac{\partial f_{12}^I(0,0)}{\partial u_5}, \quad \alpha_{14} = b_{12}^R \frac{\partial f_{12}^R(0,0)}{\partial u_6} - b_{12}^I \frac{\partial f_{12}^I(0,0)}{\partial u_6}, \\
\alpha_{21} &= b_{11}^R \frac{\partial f_{11}^I(0,0)}{\partial u_1} + b_{11}^I \frac{\partial f_{11}^R(0,0)}{\partial u_1}, \quad \alpha_{22} = b_{11}^R \frac{\partial f_{11}^I(0,0)}{\partial u_2} + b_{11}^I \frac{\partial f_{11}^R(0,0)}{\partial u_2}, \\
\alpha_{23} &= b_{12}^R \frac{\partial f_{12}^I(0,0)}{\partial u_5} + b_{12}^I \frac{\partial f_{12}^R(0,0)}{\partial u_5}, \quad \alpha_{24} = b_{12}^R \frac{\partial f_{12}^I(0,0)}{\partial u_6} + b_{12}^I \frac{\partial f_{12}^R(0,0)}{\partial u_6}, \\
\alpha_{31} &= b_{21}^R \frac{\partial f_{21}^R(0,0)}{\partial u_1} - b_{21}^I \frac{\partial f_{21}^I(0,0)}{\partial u_1}, \quad \alpha_{32} = b_{21}^R \frac{\partial f_{21}^R(0,0)}{\partial u_2} - b_{21}^I \frac{\partial f_{21}^I(0,0)}{\partial u_2}, \\
\alpha_{33} &= b_{22}^R \frac{\partial f_{22}^R(0,0)}{\partial u_5} - b_{22}^I \frac{\partial f_{22}^I(0,0)}{\partial u_5}, \quad \alpha_{34} = b_{22}^R \frac{\partial f_{22}^R(0,0)}{\partial u_6} - b_{22}^I \frac{\partial f_{22}^I(0,0)}{\partial u_6}, \\
\alpha_{41} &= b_{21}^R \frac{\partial f_{21}^I(0,0)}{\partial u_1} + b_{21}^I \frac{\partial f_{21}^R(0,0)}{\partial u_1}, \quad \alpha_{42} = b_{21}^R \frac{\partial f_{21}^I(0,0)}{\partial u_2} + b_{21}^I \frac{\partial f_{21}^R(0,0)}{\partial u_2}, \\
\alpha_{43} &= b_{22}^R \frac{\partial f_{22}^I(0,0)}{\partial u_5} + b_{22}^I \frac{\partial f_{22}^R(0,0)}{\partial u_5}, \quad \alpha_{44} = b_{22}^R \frac{\partial f_{22}^I(0,0)}{\partial u_6} + b_{22}^I \frac{\partial f_{22}^R(0,0)}{\partial u_6}.
\end{aligned}$$

Thus, the characteristic equation of system (4.9) is

$$det \begin{pmatrix} \lambda + 1 - \alpha_{11}e^{-\lambda\tau} & -\alpha_{12}e^{-\lambda\tau} & 0 & 0 & -\alpha_{13} & -\alpha_{14} \\ -\alpha_{21}e^{-\lambda\tau} & \lambda + 1 - \alpha_{22}e^{-\lambda\tau} & 0 & 0 & -\alpha_{23} & -\alpha_{24} \\ -\alpha_{31}e^{-\lambda\tau} & -\alpha_{32}e^{-\lambda\tau} & \lambda + 1 & 0 & -\alpha_{33} & -\alpha_{34} \\ -\alpha_{41}e^{-\lambda\tau} & -\alpha_{42}e^{-\lambda\tau} & 0 & \lambda + 1 & -\alpha_{43} & -\alpha_{44} \\ 0 & 0 & -\alpha & 0 & \lambda + \alpha & 0 \\ 0 & 0 & 0 & -\alpha & 0 & \lambda + \alpha \end{pmatrix} = 0. \tag{4.10}$$

By simple calculation, one obtains

$$\begin{aligned}&\lambda^6+d_1\lambda^5+d_2\lambda^4+d_3\lambda^3+d_4\lambda^2+d_5\lambda+d_6\\&+(d_7\lambda^5+d_8\lambda^4+d_9\lambda^3+d_{10}\lambda^2+d_{11}\lambda+d_{12})e^{-\lambda\tau}\\&+(d_{13}\lambda^4+d_{14}\lambda^3+d_{15}\lambda^2+d_{16}\lambda+d_{17})e^{-2\lambda\tau}=0,\end{aligned} \tag{4.11}$$

where

$$\begin{aligned}
d_1&=2\alpha+4,\ d_2=\alpha(\alpha+8-\alpha_{33}-\alpha_{44})+6,\ d_3=(\alpha^2+3\alpha)(4-\alpha_{33}-\alpha_{44})+4,\\
d_4&=\alpha^2(6-3\alpha_{33}-3\alpha_{44}+\alpha_{33}\alpha_{44}-\alpha_{34}\alpha_{43})+\alpha(8-3\alpha_{33}-3\alpha_{44})+1,\\
d_5&=\alpha^2(4-3\alpha_{33}-3\alpha_{44}+2\alpha_{33}\alpha_{44}-2\alpha_{34}\alpha_{43})+\alpha(2-\alpha_{33}-\alpha_{44}),\\
d_6&=\alpha^2(1-\alpha_{33}-\alpha_{44}+\alpha_{33}\alpha_{44}-\alpha_{34}\alpha_{43}),\\
d_7&=-\alpha_{11}-\alpha_{22},\ d_8=(-3-2\alpha)(\alpha_{11}+\alpha_{22}),\\
d_9&=(\alpha_{11}+\alpha_{22})(-3-6\alpha-\alpha^2+\alpha\alpha_{33}+\alpha\alpha_{44})-\alpha(\alpha_{13}\alpha_{31}+\alpha_{14}\alpha_{41}\\
&\quad+\alpha_{23}\alpha_{32}+\alpha_{24}\alpha_{42}),\\
d_{10}&=(\alpha_{11}+\alpha_{22})[-1-6\alpha-3\alpha^2+(2\alpha+\alpha^2)(\alpha_{33}+\alpha_{44})]-(2\alpha+\alpha^2)(\alpha_{13}\alpha_{31}\\
&\quad+\alpha_{14}\alpha_{41}+\alpha_{23}\alpha_{32}+\alpha_{24}\alpha_{42}),\\
d_{11}&=(\alpha_{11}+\alpha_{22})[-2\alpha-3\alpha^2+(2\alpha+\alpha^2)(\alpha_{33}+\alpha_{44})+\alpha^2(\alpha_{34}\alpha_{43}-\alpha_{33}\alpha_{44})]\\
&\quad-(2\alpha^2+\alpha-\alpha^2\alpha_{33})(\alpha_{14}\alpha_{41}+\alpha_{24}\alpha_{42})-(2\alpha^2+\alpha-\alpha^2\alpha_{44})(\alpha_{13}\alpha_{31}\\
&\quad+\alpha_{23}\alpha_{32})-\alpha^2(\alpha_{13}\alpha_{34}\alpha_{41}+\alpha_{14}\alpha_{31}\alpha_{43}+\alpha_{23}\alpha_{34}\alpha_{42}+\alpha_{24}\alpha_{32}\alpha_{43}),\\
d_{12}&=\alpha^2(\alpha_{11}+\alpha_{22})(-1+\alpha_{33}+\alpha_{44}-\alpha_{33}\alpha_{44}+\alpha_{34}\alpha_{43})\\
&\quad+\alpha^2(\alpha_{13}\alpha_{31}+\alpha_{23}\alpha_{32})(\alpha_{44}-1)+\alpha^2(\alpha_{14}\alpha_{41}+\alpha_{24}\alpha_{42})(\alpha_{33}-1)\\
&\quad-\alpha^2(\alpha_{13}\alpha_{34}\alpha_{41}+\alpha_{14}\alpha_{31}\alpha_{43}+\alpha_{23}\alpha_{34}\alpha_{42}+\alpha_{24}\alpha_{32}\alpha_{43}),\\
d_{13}&=\alpha_{11}\alpha_{22}-\alpha_{12}\alpha_{21},\ d_{14}=(2+2\alpha)(\alpha_{11}\alpha_{22}-\alpha_{12}\alpha_{21}),\\
d_{15}&=(\alpha_{11}\alpha_{22}-\alpha_{12}\alpha_{21})(1+4\alpha+\alpha^2-\alpha\alpha_{33}-\alpha\alpha_{44})+\alpha\alpha_{32}(\alpha_{11}\alpha_{23}\\
&\quad-\alpha_{13}\alpha_{21})+\alpha\alpha_{31}(\alpha_{13}\alpha_{22}-\alpha_{12}\alpha_{23})+\alpha\alpha_{42}(\alpha_{11}\alpha_{24}-\alpha_{14}\alpha_{21})\\
&\quad+\alpha\alpha_{41}(\alpha_{14}\alpha_{22}-\alpha_{12}\alpha_{24}),\\
d_{16}&=(\alpha^2+\alpha)(\alpha_{11}\alpha_{22}-\alpha_{12}\alpha_{21})(2-\alpha_{33}-\alpha_{44})+(\alpha^2+\alpha)[\alpha_{31}(\alpha_{13}\alpha_{22}\\
&\quad-\alpha_{12}\alpha_{23})+\alpha_{32}(\alpha_{11}\alpha_{23}-\alpha_{13}\alpha_{21})+\alpha_{42}(\alpha_{11}\alpha_{24}-\alpha_{14}\alpha_{21})\\
&\quad+\alpha_{41}(\alpha_{14}\alpha_{22}-\alpha_{12}\alpha_{24})],\\
d_{17}&=\alpha^2(\alpha_{11}\alpha_{22}-\alpha_{12}\alpha_{21})(1-\alpha_{33}-\alpha_{44}+\alpha_{33}\alpha_{44}-\alpha_{34}\alpha_{43})+\alpha^2(\alpha_{32}\\
&\quad-\alpha_{32}\alpha_{44}+\alpha_{34}\alpha_{42})(\alpha_{11}\alpha_{23}-\alpha_{13}\alpha_{21})+\alpha^2(\alpha_{31}-\alpha_{31}\alpha_{44}+\alpha_{34}\alpha_{41})\\
&\quad\times(\alpha_{13}\alpha_{22}-\alpha_{12}\alpha_{23})+\alpha^2(\alpha_{42}-\alpha_{33}\alpha_{42}+\alpha_{32}\alpha_{43})(\alpha_{11}\alpha_{24}-\alpha_{14}\alpha_{21})\\
&\quad+\alpha^2(\alpha_{41}-\alpha_{33}\alpha_{41}+\alpha_{31}\alpha_{43})(\alpha_{14}\alpha_{22}-\alpha_{12}\alpha_{24})\\
&\quad+\alpha^2(\alpha_{31}\alpha_{42}-\alpha_{32}\alpha_{41})(\alpha_{13}\alpha_{24}-\alpha_{14}\alpha_{23}).
\end{aligned}$$

Multiplying both sides of (4.11) by $e^{\lambda\tau}$, one has

$$\begin{aligned}&(\lambda^6 + d_1\lambda^5 + d_2\lambda^4 + d_3\lambda^3 + d_4\lambda^2 + d_5\lambda + d_6)e^{\lambda\tau}\\&+(d_7\lambda^5 + d_8\lambda^4 + d_9\lambda^3 + d_{10}\lambda^2 + d_{11}\lambda + d_{12})+\\&+(d_{13}\lambda^4 + d_{14}\lambda^3 + d_{15}\lambda^2 + d_{16}\lambda + d_{17})e^{-\lambda\tau} = 0.\end{aligned} \tag{4.12}$$

In what follows, the roots of the transcendental equation (4.12) will be discussed. Suppose iw $(w > 0)$ is a root of (4.12), then one has

$$\begin{aligned}&(-w^6 + id_1w^5 + d_2w^4 - id_3w^3 - d_4w^2 + id_5w + d_6)e^{iw\tau}\\&+(id_7w^5 + d_8w^4 - id_9w^3 - d_{10}w^2 + id_{11}w + d_{12})\\&+(d_{13}w^4 - id_{14}w^3 - d_{15}w^2 + id_{16}w + d_{17})e^{-iw\tau} = 0.\end{aligned} \tag{4.13}$$

Separating the real and imaginary parts of (4.13), one obtains

$$\begin{aligned}&(-w^6 + d_2w^4 - d_4w^2 + d_6 + d_{13}w^4 - d_{15}w^2 + d_{17})cos(w\tau)\\&+(-d_1w^5 + d_3w^3 - d_5w - d_{14}w^3 + d_{16}w)sin(w\tau)\\&= -d_8w^4 + d_{10}w^2 - d_{12}.\\&(d_1w^5 - d_3w^3 + d_5w - d_{14}w^3 + d_{16}w)cos(w\tau)\\&+(-w^6 + d_2w^4 - d_4w^2 + d_6 - d_{13}w^4 + d_{15}w^2 - d_{17})sin(w\tau)\\&= -d_7w^5 + d_9w^3 - d_{11}w.\end{aligned} \tag{4.14}$$

Using Cramer's rule to solve equation (4.14), one gets

$$\begin{aligned}\cos(w\tau) &= \frac{k_{13}w^{10}+k_{14}w^8+k_{15}w^6+k_{16}w^4+k_{17}w^2+k_{18}}{w^{12}+k_1w^{10}+k_2w^8+k_3w^6+k_4w^4+k_5w^2+k_6},\\ \sin(w\tau) &= \frac{k_7w^{11}+k_8w^9+k_9w^7+k_{10}w^5+k_{11}w^3+k_{12}w}{w^{12}+k_1w^{10}+k_2w^8+k_3w^6+k_4w^4+k_5w^2+k_6},\end{aligned} \tag{4.15}$$

where

$k_1 = d_1^2 - 2d_2$, $k_2 = d_2^2 - 2d_1d_3 + 2d_4 - d_{13}^2$,

$k_3 = d_3^2 + 2d_1d_5 - 2d_2d_4 - 2d_6 + 2d_{13}d_{15} - d_{14}^2$,

$k_4 = d_4^2 + 2d_2d_6 - 2d_3d_5 - 2d_{13}d_{17} + 2d_{14}d_{16} - d_{15}^2$,

$k_5 = d_5^2 - 2d_4d_6 + 2d_{15}d_{17} - d_{16}^2$, $k_6 = d_6^2 - d_{17}^2$,

$k_7 = d_7$, $k_8 = -d_9 + d_1d_8 - d_2d_7 - d_7d_{13}$,

$k_9 = d_{11} - d_1d_{10} + d_2d_9 - d_3d_8 + d_4d_7 + d_7d_{15} - d_8d_{14} + d_9d_{13}$,

$k_{10} = d_1d_{12} - d_2d_{11} + d_3d_{10} - d_4d_9 + d_5d_8 - d_6d_7 - d_7d_{17} + d_8d_{16} - d_9d_{15}$
$\quad +d_{10}d_{14} - d_{11}d_{13}$,

$k_{11} = -d_3d_{12} + d_4d_{11} - d_5d_{10} + d_6d_9 + d_9d_{17} - d_{10}d_{16} + d_{11}d_{15} - d_{12}d_{14}$,

$k_{12} = d_5d_{12} - d_6d_{11} - d_{11}d_{17} + d_{12}d_{16},$

$k_{13} = d_8 - d_1d_7, k_{14} = -d_{10} + d_1d_9 - d_2d_8 + d_3d_7 - d_7d_{14} + d_8d_{13},$

$k_{15} = d_{12} - d_1d_{11} + d_2d_{10} - d_3d_9 + d_4d_8 - d_5d_7 + d_7d_{16} - d_8d_{15} + d_9d_{14} - d_{10}d_{13},$

$k_{16} = -d_2d_{12} + d_3d_{11} - d_4d_{10} + d_5d_9 - d_6d_8 + d_8d_{17} - d_9d_{16} + d_{10}d_{15}$

$-d_{11}d_{14} + d_{12}d_{13},$

$k_{17} = d_4d_{12} - d_5d_{11} + d_6d_{10} - d_{10}d_{17} + d_{11}d_{16} - d_{12}d_{15}, k_{18} = d_{12}d_{17} - d_6d_{12}.$
Squaring both sides of (4.15), and adding them together, one has

$$\begin{aligned} &w^{24} + l_1w^{22} + l_2w^{20} + l_3w^{18} + l_4w^{16} + l_5w^{14} + l_6w^{12} \\ &+l_7w^{10} + l_8w^8 + l_9w^6 + l_{10}w^4 + l_{11}w^2 + l_{12} = 0, \end{aligned} \tag{4.16}$$

where
$l_1 = 2k_1 - k_7^2, l_2 = k_1^2 + 2k_2 - 2k_7k_8 - k_{13}^2,$
$l_3 = 2(k_1k_2 + k_3 - k_7k_9 - k_{13}k_{14}) - k_8^2,$
$l_4 = k_2^2 - k_{14}^2 + 2(k_1k_3 + k_4 - k_7k_{10} - k_8k_9 - k_{13}k_{15}),$
$l_5 = 2(k_1k_4 + k_2k_3 + k_5 - k_7k_{11} - k_8k_{10} - k_{13}k_{16} - k_{14}k_{15}) - k_9^2,$
$l_6 = k_3^2 - k_{15}^2 + 2(k_1k_5 + k_2k_4 + k_6 - k_7k_{12} - k_8k_{11} - k_9k_{10} - k_{13}k_{17} - k_{14}k_{16}),$
$l_7 = 2(k_1k_6 + k_2k_5 + k_3k_4 - k_8k_{12} - k_9k_{11} - k_{13}k_{18} - k_{14}k_{17} - k_{15}k_{16}) - k_{10}^2,$
$l_8 = k_4^2 - k_{16}^2 + 2(k_2k_6 + k_3k_5 - k_9k_{12} - k_{10}k_{11} - k_{14}k_{18} - k_{15}k_{17}),$
$l_9 = 2(k_3k_6 + k_4k_5 - k_{10}k_{12} - k_{15}k_{18} - k_{16}k_{17}) - k_{11}^2,$
$l_{10} = k_5^2 - k_{17}^2 + 2(k_4k_6 - k_{11}k_{12} - k_{16}k_{18}),$
$l_{11} = 2(k_5k_6 - k_{17}k_{18}) - k_{12}^2, l_{12} = k_6^2 - k_{18}^2.$

Making the substitution $z = w^2$, (4.16) can be rewritten as

$$\begin{aligned} &z^{12} + l_1z^{11} + l_2z^{10} + l_3z^9 + l_4z^8 + l_5z^7 + l_6z^6 \\ &+l_7z^5 + l_8z^4 + l_9z^3 + l_{10}z^2 + l_{11}z + l_{12} = 0. \end{aligned} \tag{4.17}$$

Define

$$\begin{aligned} f(z) = &z^{12} + l_1z^{11} + l_2z^{10} + l_3z^9 + l_4z^8 + l_5z^7 + l_6z^6 \\ &+l_7z^5 + l_8z^4 + l_9z^3 + l_{10}z^2 + l_{11}z + l_{12}. \end{aligned} \tag{4.18}$$

It is evident that $\lim_{z\to\infty} f(z) = \infty$. Therefore, if $l_{12} < 0$, then (4.18) has at least one positive root.

Without loss of generality, assume that (4.18) has twelve positive roots, denoted by $z_1, z_2, \cdots, z_{12}$. Accordingly, $w_k = \sqrt{z_k}$ $(k = 1, 2, \cdots, 12)$ is the roots of (4.12).

From (4.15), one has

$$\tau_k^j = \frac{1}{w_k}\left\{\arccos\left(\frac{k_{13}w_k^{10} + k_{14}w_k^8 + k_{15}w_k^6 + k_{16}w_k^4 + k_{17}w_k^2 + k_{18}}{w_k^{12} + k_1w_k^{10} + k_2w_k^8 + k_3w_k^6 + k_4w_k^4 + k_5w_k^2 + k_6}\right) + 2j\pi\right\}, \tag{4.19}$$

where $k = 1, 2, \cdots, 12$; $j = 0, 1, 2, \cdots$.

Denote

$$\tau_0 = \tau_{k_0}^0 = \min\{\tau_k^0 | k = 1, 2, \cdots, 12\} \text{ and } w_0 = w_{k_0}. \tag{4.20}$$

Notice that when $\tau = 0$, (4.12) is reduced to

$$\lambda^6 + \rho_1\lambda^5 + \rho_2\lambda^4 + \rho_3\lambda^3 + \rho_4\lambda^2 + \rho_5\lambda + \rho_6 = 0, \tag{4.21}$$

where $\rho_1 = d_1 + d_7$, $\rho_2 = d_2 + d_8 + d_{13}$, $\rho_3 = d_3 + d_9 + d_{14}$, $\rho_4 = d_4 + d_{10} + d_{15}$, $\rho_5 = d_5 + d_{11} + d_{16}$, $\rho_6 = d_6 + d_{12} + d_{17}$.

Using the Routh-Hurwitz criterion, one concludes that equation (4.21) has negative real parts if and only if all of the following inequalities are satisfied:

$$(H3)\quad \begin{aligned}
&\rho_i > 0\ (i = 1, 2, \cdots, 6),\\
&\rho_1\rho_2 - \rho_3 > 0,\\
&\rho_1(\rho_2\rho_3 - \rho_1\rho_4) - \rho_3^2 + \rho_1\rho_5 > 0,\\
&\rho_1\rho_2\rho_3\rho_4 - \rho_1\rho_2^2\rho_5 + \rho_1^2\rho_2\rho_6 - \rho_1^2\rho_4^2 + 2\rho_1\rho_4\rho_5 - \rho_1\rho_3\rho_6 - \rho_3^3\rho_4\\
&+\rho_2\rho_3\rho_5 - \rho_5^2 > 0,\\
&\rho_1\rho_2\rho_3\rho_4\rho_5 - \rho_1\rho_2\rho_3^2\rho_6 - \rho_1\rho_2^2\rho_5^2 + 2\rho_1^2\rho_2\rho_5\rho_6 - \rho_1^2\rho_4^2\rho_5 + \rho_1^2\rho_3\rho_4\rho_6\\
&-\rho_1^3\rho_6^2 + 2\rho_1\rho_4\rho_5^2 - 3\rho_1\rho_3\rho_5\rho_6 - \rho_3^2\rho_4\rho_5 + \rho_3^3\rho_6 + \rho_2\rho_3\rho_5^2 - \rho_5^3 > 0,\\
&(\rho_1\rho_2\rho_3\rho_4\rho_5 - \rho_1\rho_2\rho_3^2\rho_6 - \rho_1\rho_2^2\rho_5^2 + 2\rho_1^2\rho_2\rho_5\rho_6 - \rho_1^2\rho_4^2\rho_5 + \rho_1^2\rho_3\rho_4\rho_6\\
&-\rho_1^3\rho_6^2 + 2\rho_1\rho_4\rho_5^2 - 3\rho_1\rho_3\rho_5\rho_6 - \rho_3^2\rho_4\rho_5 + \rho_3^3\rho_6 + \rho_2\rho_3\rho_5^2 - \rho_5^3)\rho_6 > 0.
\end{aligned}$$

In order to obtain the main results, it is necessary to make the following extra assumption:

$(H4)$ $\left.\frac{d\mathrm{Re}(\lambda(\tau))}{d\tau}\right|_{\tau=\tau_0, w=w_0} \neq 0.$

Differentiating both side of (4.12) with respect to τ yields

$$\left(\frac{d\lambda}{d\tau}\right)^{-1} = \frac{c_3 e^{\lambda\tau} + c_4 + c_1\tau e^{\lambda\tau} + c_5 e^{-\lambda\tau} - c_2\tau e^{-\lambda\tau}}{-\lambda c_1 e^{\lambda\tau} + \lambda c_2 e^{-\lambda\tau}}, \tag{4.22}$$

where

$c_1 = \lambda^6 + d_1\lambda^5 + d_2\lambda^4 + d_3\lambda^3 + d_4\lambda^2 + d_5\lambda + d_6,$
$c_2 = d_{13}\lambda^4 + d_{14}\lambda^3 + d_{15}\lambda^2 + d_{16}\lambda + d_{17},$
$c_3 = 6\lambda^5 + 5d_1\lambda^4 + 4d_2\lambda^3 + 3d_3\lambda^2 + 2d_4\lambda + d_5,$
$c_4 = 5d_7\lambda^4 + 4d_8\lambda^3 + 3d_9\lambda^2 + 2d_{10}\lambda + d_{11},$
$c_5 = 4d_{13}\lambda^3 + 3d_{14}\lambda^2 + 2d_{15}\lambda + d_{16}.$

Hence, one obtains

$$\mathrm{sign}\left\{\mathrm{Re}\left(\frac{d\lambda(\tau)}{d\tau}\right)_{\tau=\tau_0}\right\} = \mathrm{sign}\left\{\mathrm{Re}\left(\frac{d\lambda(\tau)}{d\tau}\right)^{-1}_{\tau=\tau_0}\right\}. \tag{4.23}$$

From Lemma 4.1 and results of the above discussions, one knows that system (4.8) is local asymptotically stable for $\tau \in [0, \tau_0)$ and unstable when $\tau > \tau_0$. Moreover, system (4.8) undergoes a Hopf bifurcation at the zero equilibrium points when $\tau = \tau_0$. □

Remark 4.1 In this chapter, the discrete and distributed delays are both considered. In fact, a neural network model with both discrete and distributed delays is more general than one with only discrete or distributed delay. Compared with the existing model in Dong et al. (2015), our model is more general, which includes discrete and distributed delays.

4.4 Direction of the Hopf Bifurcation

In this section, by employing the normal form theory and center manifold theorem, we will determine the direction and the stability of bifurcating periodic solutions at the critical value τ_0.

Theorem 4.2 *For system (4.8), by employing the normal form theory and center manifold theorem, one can get*

$$\begin{aligned} c_1(0) &= \frac{i}{2w_0}\left(g_{11}g_{20} - 2|g_{11}|^2 - \frac{|g_{02}|^2}{3}\right) + \frac{g_{21}}{2}, \\ \mu_2 &= -\frac{Re\{c_1(0)\}}{Re\{\lambda'(\tau_0)\}}, \\ \kappa_2 &= 2Re\{c_1(0)\}, \\ T_2 &= -\frac{Im\{c_1(0)\} + \mu_2 Im\{\lambda'(\tau_0)\}}{w_0}. \end{aligned} \tag{4.24}$$

where μ_2 determines the direction of the Hopf bifurcation: if $\mu_2 > 0$ ($\mu_2 < 0$), then the Hopf bifurcation is supercritical (subcritical) and the bifurcating periodic solutions exist for $\tau > \tau_0$ ($\tau < \tau_0$); κ_2 determines the stability of the bifurcating periodic solutions: if $\kappa_2 < 0$ ($\kappa_2 > 0$), the bifurcating periodic solutions are stable (unstable); T_2 determines the period of the bifurcating periodic solutions: if $T_2 > 0$ ($T_2 < 0$), the period increases (decreases).

Proof For simplification, transform system (4.8) into a functional differential equation (FDE). Let $t = s\tau$, $x_i(t) = u_i(s\tau)$ and $\tau = \tau_0 + \mu$, where τ_0 is defined by (4.20) and $\mu \in R$. Then, system (4.8) can be transformed into a FDE as

$$x'(t) = L_\mu(x_t) + f(\mu, x_t), \tag{4.25}$$

with

$$L_\mu\phi = (\tau_0 + \mu)[B_1\phi(0) + B_2\phi(-1)], \tag{4.26}$$

where

$$B_1 = \begin{pmatrix} -1 & 0 & 0 & 0 & \alpha_{13} & \alpha_{14} \\ 0 & -1 & 0 & 0 & \alpha_{23} & \alpha_{24} \\ 0 & 0 & -1 & 0 & \alpha_{33} & \alpha_{34} \\ 0 & 0 & 0 & -1 & \alpha_{43} & \alpha_{44} \\ 0 & 0 & \alpha & 0 & -\alpha & 0 \\ 0 & 0 & 0 & \alpha & 0 & -\alpha \end{pmatrix}, \quad B_2 = \begin{pmatrix} \alpha_{11} & \alpha_{12} & 0 & 0 & 0 & 0 \\ \alpha_{21} & \alpha_{22} & 0 & 0 & 0 & 0 \\ \alpha_{31} & \alpha_{32} & 0 & 0 & 0 & 0 \\ \alpha_{41} & \alpha_{42} & 0 & 0 & 0 & 0 \\ 0 & 0 & 0 & 0 & 0 & 0 \\ 0 & 0 & 0 & 0 & 0 & 0 \end{pmatrix},$$

and

$$f(\mu, \phi) = (\tau_0 + \mu) \begin{pmatrix} f_1(\phi) \\ f_2(\phi) \\ f_3(\phi) \\ f_4(\phi) \\ f_5(\phi) \\ f_6(\phi) \end{pmatrix}, \tag{4.27}$$

where

$$\begin{aligned}
f_1(\phi) = {} & m_{11}\phi_1^2(-1) + m_{12}\phi_2^2(-1) + m_{13}\phi_1(-1)\phi_2(-1) + m_{14}\phi_1^3(-1) + \\
& m_{15}\phi_1(-1)\phi_2^2(-1) + m_{16}\phi_1^2(-1)\phi_2(-1) + m_{17}\phi_2^3(-1) + n_{11}\phi_5^2(0) + \\
& n_{12}\phi_6^2(0) + n_{13}\phi_5(0)\phi_6(0) + n_{14}\phi_5^3(0) + n_{15}\phi_5(0)\phi_6^2(0) + \\
& n_{16}\phi_5^2(0)\phi_6(0) + n_{17}\phi_6^3(0) + H.O.T, \\
f_2(\phi) = {} & m_{21}\phi_1^2(-1) + m_{22}\phi_2^2(-1) + m_{23}\phi_1(-1)\phi_2(-1) + m_{24}\phi_1^3(-1) + \\
& m_{25}\phi_1(-1)\phi_2^2(-1) + m_{26}\phi_1^2(-1)\phi_2(-1) + m_{27}\phi_2^3(-1) + n_{21}\phi_5^2(0) + \\
& n_{22}\phi_6^2(0) + n_{23}\phi_5(0)\phi_6(0) + n_{24}\phi_5^3(0) + n_{25}\phi_5(0)\phi_6^2(0) + \\
& n_{26}\phi_5^2(0)\phi_6(0) + n_{27}\phi_6^3(0) + H.O.T, \\
f_3(\phi) = {} & m_{31}\phi_1^2(-1) + m_{32}\phi_2^2(-1) + m_{33}\phi_1(-1)\phi_2(-1) + m_{34}\phi_1^3(-1) + \\
& m_{35}\phi_1(-1)\phi_2^2(-1) + m_{36}\phi_1^2(-1)\phi_2(-1) + m_{37}\phi_2^3(-1) + n_{31}\phi_5^2(0) + \\
& n_{32}\phi_6^2(0) + n_{33}\phi_5(0)\phi_6(0) + n_{34}\phi_5^3(0) + n_{35}\phi_5(0)\phi_6^2(0) + \\
& n_{36}\phi_5^2(0)\phi_6(0) + n_{37}\phi_6^3(0) + H.O.T, \\
f_4(\phi) = {} & m_{41}\phi_1^2(-1) + m_{42}\phi_2^2(-1) + m_{43}\phi_1(-1)\phi_2(-1) + m_{44}\phi_1^3(-1) + \\
& m_{45}\phi_1(-1)\phi_2^2(-1) + m_{46}\phi_1^2(-1)\phi_2(-1) + m_{47}\phi_2^3(-1) + n_{41}\phi_5^2(0) + \\
& n_{42}\phi_6^2(0) + n_{43}\phi_5(0)\phi_6(0) + n_{44}\phi_5^3(0) + n_{45}\phi_5(0)\phi_6^2(0) + \\
& n_{46}\phi_5^2(0)\phi_6(0) + n_{47}\phi_6^3(0) + H.O.T, \\
f_5(\phi) = {} & 0, \\
f_6(\phi) = {} & 0,
\end{aligned}$$

and

$$m_{11} = \left(b_{11}^R \frac{\partial^2 f_{11}^R(0,0)}{\partial u_1^2} - b_{11}^I \frac{\partial^2 f_{11}^I(0,0)}{\partial u_1^2} \right) \Big/ 2, \; m_{12} = \left(b_{11}^R \frac{\partial^2 f_{11}^R(0,0)}{\partial u_2^2} - b_{11}^I \frac{\partial^2 f_{11}^I(0,0)}{\partial u_2^2} \right) \Big/ 2,$$

$$m_{13} = \left(b_{11}^R \frac{\partial^2 f_{11}^R(0,0)}{\partial u_1 \partial u_2} - b_{11}^I \frac{\partial^2 f_{11}^I(0,0)}{\partial u_1 \partial u_2} \right), \; m_{14} = \left(b_{11}^R \frac{\partial^3 f_{11}^R(0,0)}{\partial u_1^3} - b_{11}^I \frac{\partial^3 f_{11}^I(0,0)}{\partial u_1^3} \right) \Big/ 6,$$

$$m_{15} = \left(b_{11}^R \frac{\partial^3 f_{11}^R(0,0)}{\partial u_1 \partial u_2^2} - b_{11}^I \frac{\partial^3 f_{11}^I(0,0)}{\partial u_1 \partial u_2^2} \right) \Big/ 2, \; m_{16} = \left(b_{11}^R \frac{\partial^3 f_{11}^R(0,0)}{\partial u_1^2 \partial u_2} - b_{11}^I \frac{\partial^3 f_{11}^I(0,0)}{\partial u_1^2 \partial u_2} \right) \Big/ 2,$$

$$m_{17} = \left(b_{11}^R \frac{\partial^3 f_{11}^R(0,0)}{\partial u_2^3} - b_{11}^I \frac{\partial^3 f_{11}^I(0,0)}{\partial u_2^3}\right) \Big/ 6,\ n_{11} = \left(b_{12}^R \frac{\partial^2 f_{12}^R(0,0)}{\partial u_5^2} - b_{12}^I \frac{\partial^2 f_{12}^I(0,0)}{\partial u_5^2}\right) \Big/ 2,$$

$$n_{12} = \left(b_{12}^R \frac{\partial^2 f_{12}^R(0,0)}{\partial u_6^2} - b_{12}^I \frac{\partial^2 f_{12}^I(0,0)}{\partial u_6^2}\right) \Big/ 2,\ n_{13} = \left(b_{12}^R \frac{\partial^2 f_{12}^R(0,0)}{\partial u_5 \partial u_6} - b_{12}^I \frac{\partial^2 f_{12}^I(0,0)}{\partial u_5 \partial u_6}\right),$$

$$n_{14} = \left(b_{12}^R \frac{\partial^3 f_{12}^R(0,0)}{\partial u_5^3} - b_{12}^I \frac{\partial^3 f_{12}^I(0,0)}{\partial u_5^3}\right) \Big/ 6,\ n_{15} = \left(b_{12}^R \frac{\partial^3 f_{12}^R(0,0)}{\partial u_5 \partial u_6^2} - b_{12}^I \frac{\partial^3 f_{12}^I(0,0)}{\partial u_5 \partial u_6^2}\right) \Big/ 2,$$

$$n_{16} = \left(b_{12}^R \frac{\partial^3 f_{12}^R(0,0)}{\partial u_5^2 \partial u_6} - b_{12}^I \frac{\partial^3 f_{12}^I(0,0)}{\partial u_5^2 \partial u_6}\right) \Big/ 2,\ n_{17} = \left(b_{12}^R \frac{\partial^3 f_{12}^R(0,0)}{\partial u_6^3} - b_{12}^I \frac{\partial^3 f_{12}^I(0,0)}{\partial u_6^3}\right) \Big/ 6,$$

$$m_{21} = \left(b_{11}^R \frac{\partial^2 f_{11}^I(0,0)}{\partial u_1^2} + b_{11}^I \frac{\partial^2 f_{11}^R(0,0)}{\partial u_1^2}\right) \Big/ 2,\ m_{22} = \left(b_{11}^R \frac{\partial^2 f_{11}^I(0,0)}{\partial u_2^2} + b_{11}^I \frac{\partial^2 f_{11}^R(0,0)}{\partial u_2^2}\right) \Big/ 2,$$

$$m_{23} = \left(b_{11}^R \frac{\partial^2 f_{11}^I(0,0)}{\partial u_1 \partial u_2} + b_{11}^I \frac{\partial^2 f_{11}^R(0,0)}{\partial u_1 \partial u_2}\right),\ m_{24} = \left(b_{11}^R \frac{\partial^3 f_{11}^I(0,0)}{\partial u_1^3} + b_{11}^I \frac{\partial^3 f_{11}^R(0,0)}{\partial u_1^3}\right) \Big/ 6,$$

$$m_{25} = \left(b_{11}^R \frac{\partial^3 f_{11}^I(0,0)}{\partial u_1 \partial u_2^2} + b_{11}^I \frac{\partial^3 f_{11}^R(0,0)}{\partial u_1 \partial u_2^2}\right) \Big/ 2,\ m_{26} = \left(b_{11}^R \frac{\partial^3 f_{11}^I(0,0)}{\partial u_1^2 \partial u_2} + b_{11}^I \frac{\partial^3 f_{11}^R(0,0)}{\partial u_1^2 \partial u_2}\right) \Big/ 2,$$

$$m_{27} = \left(b_{11}^R \frac{\partial^3 f_{11}^I(0,0)}{\partial u_2^3} + b_{11}^I \frac{\partial^3 f_{11}^R(0,0)}{\partial u_2^3}\right) \Big/ 6,\ n_{21} = \left(b_{12}^R \frac{\partial^2 f_{12}^I(0,0)}{\partial u_5^2} + b_{12}^I \frac{\partial^2 f_{12}^R(0,0)}{\partial u_5^2}\right) \Big/ 2,$$

$$n_{22} = \left(b_{12}^R \frac{\partial^2 f_{12}^I(0,0)}{\partial u_6^2} + b_{12}^I \frac{\partial^2 f_{12}^R(0,0)}{\partial u_6^2}\right) \Big/ 2,\ n_{23} = \left(b_{12}^R \frac{\partial^2 f_{12}^I(0,0)}{\partial u_5 \partial u_6} + b_{12}^I \frac{\partial^2 f_{12}^R(0,0)}{\partial u_5 \partial u_6}\right),$$

$$n_{24} = \left(b_{12}^R \frac{\partial^3 f_{12}^I(0,0)}{\partial u_5^3} + b_{12}^I \frac{\partial^3 f_{12}^R(0,0)}{\partial u_5^3}\right) \Big/ 6,\ n_{25} = \left(b_{12}^R \frac{\partial^3 f_{12}^I(0,0)}{\partial u_5 \partial u_6^2} + b_{12}^I \frac{\partial^3 f_{12}^R(0,0)}{\partial u_5 \partial u_6^2}\right) \Big/ 2,$$

$$n_{26} = \left(b_{12}^R \frac{\partial^3 f_{12}^I(0,0)}{\partial u_5^2 \partial u_6} + b_{12}^I \frac{\partial^3 f_{12}^R(0,0)}{\partial u_5^2 \partial u_6}\right) \Big/ 2,\ n_{27} = \left(b_{12}^R \frac{\partial^3 f_{12}^I(0,0)}{\partial u_6^3} + b_{12}^I \frac{\partial^3 f_{12}^R(0,0)}{\partial u_6^3}\right) \Big/ 6,$$

$$m_{31} = \left(b_{21}^R \frac{\partial^2 f_{21}^R(0,0)}{\partial u_1^2} - b_{21}^I \frac{\partial^2 f_{21}^I(0,0)}{\partial u_1^2}\right) \Big/ 2,\ m_{32} = \left(b_{21}^R \frac{\partial^2 f_{21}^R(0,0)}{\partial u_2^2} - b_{21}^I \frac{\partial^2 f_{21}^I(0,0)}{\partial u_2^2}\right) \Big/ 2,$$

$$m_{33} = \left(b_{21}^R \frac{\partial^2 f_{21}^R(0,0)}{\partial u_1 \partial u_2} - b_{21}^I \frac{\partial^2 f_{21}^I(0,0)}{\partial u_1 \partial u_2}\right),\ m_{34} = \left(b_{21}^R \frac{\partial^3 f_{21}^R(0,0)}{\partial u_1^3} - b_{21}^I \frac{\partial^3 f_{21}^I(0,0)}{\partial u_1^3}\right) \Big/ 6,$$

$$m_{35} = \left(b_{21}^R \frac{\partial^3 f_{21}^R(0,0)}{\partial u_1 \partial u_2^2} - b_{21}^I \frac{\partial^3 f_{21}^I(0,0)}{\partial u_1 \partial u_2^2}\right) \Big/ 2,\ m_{36} = \left(b_{21}^R \frac{\partial^3 f_{21}^R(0,0)}{\partial u_1^2 \partial u_2} - b_{21}^I \frac{\partial^3 f_{21}^I(0,0)}{\partial u_1^2 \partial u_2}\right) \Big/ 2,$$

$$m_{37} = \left(b_{21}^R \frac{\partial^3 f_{21}^R(0,0)}{\partial u_2^3} - b_{21}^I \frac{\partial^3 f_{21}^I(0,0)}{\partial u_2^3}\right) \Big/ 6,\ n_{31} = \left(b_{22}^R \frac{\partial^2 f_{22}^R(0,0)}{\partial u_5^2} - b_{22}^I \frac{\partial^2 f_{22}^I(0,0)}{\partial u_5^2}\right) \Big/ 2,$$

$$n_{32} = \left(b_{22}^R \frac{\partial^2 f_{22}^R(0,0)}{\partial u_6^2} - b_{22}^I \frac{\partial^2 f_{22}^I(0,0)}{\partial u_6^2}\right) \Big/ 2,\ n_{33} = \left(b_{22}^R \frac{\partial^2 f_{22}^R(0,0)}{\partial u_5 \partial u_6} - b_{22}^I \frac{\partial^2 f_{22}^I(0,0)}{\partial u_5 \partial u_6}\right),$$

$$n_{34} = \left(b_{22}^R \frac{\partial^3 f_{22}^R(0,0)}{\partial u_5^3} - b_{22}^I \frac{\partial^3 f_{22}^I(0,0)}{\partial u_5^3}\right) \Big/ 6,\ n_{35} = \left(b_{22}^R \frac{\partial^3 f_{22}^R(0,0)}{\partial u_5 \partial u_6^2} - b_{22}^I \frac{\partial^3 f_{22}^I(0,0)}{\partial u_5 \partial u_6^2}\right) \Big/ 2,$$

$$n_{36} = \left(b_{22}^R \frac{\partial^3 f_{22}^R(0,0)}{\partial u_5^2 \partial u_6} - b_{22}^I \frac{\partial^3 f_{22}^I(0,0)}{\partial u_5^2 \partial u_6}\right) \Big/ 2,\ n_{37} = \left(b_{22}^R \frac{\partial^3 f_{22}^R(0,0)}{\partial u_6^3} - b_{22}^I \frac{\partial^3 f_{22}^I(0,0)}{\partial u_6^3}\right) \Big/ 6,$$

$$m_{41} = \left(b_{21}^R \frac{\partial^2 f_{21}^I(0,0)}{\partial u_1^2} + b_{21}^I \frac{\partial^2 f_{21}^R(0,0)}{\partial u_1^2}\right) \Big/ 2,\ m_{42} = \left(b_{21}^R \frac{\partial^2 f_{21}^I(0,0)}{\partial u_2^2} + b_{21}^I \frac{\partial^2 f_{21}^R(0,0)}{\partial u_2^2}\right) \Big/ 2,$$

$$m_{43} = \left(b_{21}^R \frac{\partial^2 f_{21}^I(0,0)}{\partial u_1 \partial u_2} + b_{21}^I \frac{\partial^2 f_{21}^R(0,0)}{\partial u_1 \partial u_2} \right), m_{44} = \left(b_{21}^R \frac{\partial^3 f_{21}^I(0,0)}{\partial u_1^3} + b_{21}^I \frac{\partial^3 f_{21}^R(0,0)}{\partial u_1^3} \right) \Big/ 6,$$

$$m_{45} = \left(b_{21}^R \frac{\partial^3 f_{21}^I(0,0)}{\partial u_1 \partial u_2^2} + b_{21}^I \frac{\partial^3 f_{21}^R(0,0)}{\partial u_1 \partial u_2^2} \right) \Big/ 2, m_{46} = \left(b_{21}^R \frac{\partial^3 f_{21}^I(0,0)}{\partial u_1^2 \partial u_2} + b_{21}^I \frac{\partial^3 f_{21}^R(0,0)}{\partial u_1^2 \partial u_2} \right) \Big/ 2,$$

$$m_{47} = \left(b_{21}^R \frac{\partial^3 f_{21}^I(0,0)}{\partial u_2^3} + b_{21}^I \frac{\partial^3 f_{21}^R(0,0)}{\partial u_2^3} \right) \Big/ 6, n_{41} = \left(b_{22}^R \frac{\partial^2 f_{22}^I(0,0)}{\partial u_5^2} + b_{22}^I \frac{\partial^2 f_{22}^R(0,0)}{\partial u_5^2} \right) \Big/ 2,$$

$$n_{42} = \left(b_{22}^R \frac{\partial^2 f_{22}^I(0,0)}{\partial u_6^2} + b_{22}^I \frac{\partial^2 f_{22}^R(0,0)}{\partial u_6^2} \right) \Big/ 2, n_{43} = \left(b_{22}^R \frac{\partial^2 f_{22}^I(0,0)}{\partial u_5 \partial u_6} + b_{22}^I \frac{\partial^2 f_{22}^R(0,0)}{\partial u_5 \partial u_6} \right),$$

$$n_{44} = \left(b_{22}^R \frac{\partial^3 f_{22}^I(0,0)}{\partial u_5^3} + b_{22}^I \frac{\partial^3 f_{22}^R(0,0)}{\partial u_5^3} \right) \Big/ 6, n_{45} = \left(b_{22}^R \frac{\partial^3 f_{22}^I(0,0)}{\partial u_5 \partial u_6^2} + b_{22}^I \frac{\partial^3 f_{22}^R(0,0)}{\partial u_5 \partial u_6^2} \right) \Big/ 2,$$

$$n_{46} = \left(b_{22}^R \frac{\partial^3 f_{22}^I(0,0)}{\partial u_5^2 \partial u_6} + b_{22}^I \frac{\partial^3 f_{22}^R(0,0)}{\partial u_5^2 \partial u_6} \right) \Big/ 2, n_{47} = \left(b_{22}^R \frac{\partial^3 f_{22}^I(0,0)}{\partial u_6^3} + b_{22}^I \frac{\partial^3 f_{22}^R(0,0)}{\partial u_6^3} \right) \Big/ 6.$$

By the Riesz Representation theorem (Benedetto and Czaja 2009), there exists a function $\eta(\theta, \mu)$ of bounded variation for $\theta \in [-1, 0]$, such that

$$L_\mu \phi = \int_{-1}^{0} d\eta(\theta, \mu)\phi(\theta). \tag{4.28}$$

In fact, one can choose

$$\eta(\theta, \mu) = (\tau_0 + \mu)[B_1 \delta(\theta) + B_2 \delta(\theta + 1)], \tag{4.29}$$

where δ is defined by

$$\delta(\theta) = \begin{cases} 0, & \theta \neq 0, \\ 1, & \theta = 0. \end{cases}$$

For $\phi \in C^1([-1, 0], R^6)$, define

$$A(\mu)\phi = \begin{cases} \frac{d\phi(\theta)}{d\theta}, & \theta \in [-1, 0), \\ \int_{-1}^{0} d\eta(\theta, \mu)\phi(\theta), & \theta = 0, \end{cases} \tag{4.30}$$

and

$$R(\mu)\phi = \begin{cases} 0, & \theta \in [-1, 0), \\ f(\mu, \phi), & \theta = 0. \end{cases} \tag{4.31}$$

Then, the FDE (4.25) is equivalent to

$$x'(t) = A(\mu)x_t + R(\mu)x_t, \tag{4.32}$$

where $x_t = x(t + \theta)$.

The adjoint operator $A^*(\mu)$ of $A(\mu)$ is defined by

$$A^*(\mu)\psi(s) = \begin{cases} -\frac{d\psi(s)}{ds}, & s \in (0, 1], \\ \int_{-1}^{0} d\eta^T(s, \mu)\psi(-s), & s = 0. \end{cases} \tag{4.33}$$

For $\phi \in C^1([-1, 0], R^6)$ and $\psi \in C^1([0, 1], (R^6)^*)$, define

$$\langle \psi(s), \phi(\theta) \rangle = \overline{\psi}(0)^T \phi(0) - \int_{\theta=-1}^{0} \int_{\xi=0}^{\theta} \overline{\psi}^T(\xi - \theta)\, d\eta(\theta)\phi(\xi)\, d\xi, \tag{4.34}$$

where $\eta(\theta) = \eta(\theta, 0)$. Then, $A(0)$ and $A^*(0)$ are adjoint operators as $\langle \psi(s), A(0)\phi(\theta) \rangle = \langle A^*(0)\psi(s), \phi(\theta) \rangle$.
From the previous section, it can be seen that $\pm i\tau_0 w_0$ are the eigenvalues of $A(0)$, so $\pm i\tau_0 w_0$ are also the eigenvalues of $A^*(0)$. Suppose that $q(\theta)$ is the eigenvector of $A(0)$ corresponding to $i\tau_0 w_0$. Then, $A(0)q(\theta) = i\tau_0 w_0 q(\theta)$. By (4.28) and (4.30), one obtains

$$\begin{cases} A(0)q(\theta) = \frac{dq(\theta)}{d\theta} = i\tau_0 w_0 q(\theta), & \theta \in [-1, 0), \\ L_\mu q(0) = i\tau_0 w_0 q(0), & \theta = 0. \end{cases} \tag{4.35}$$

From (4.26) and (4.35), one has

$$q(\theta) = q(0)e^{i\tau_0 w_0 \theta} = [1, q_1, q_2, q_3, q_4, q_5]^T e^{i\tau_0 w_0 \theta}, \tag{4.36}$$

and

$$\begin{aligned} q_1 &= \frac{\beta_7(-\alpha^2\beta_2\beta_3\beta_5+\beta_2^2\beta_6)-\alpha\alpha_{13}\beta_4(-\alpha^2\beta_3\beta_5+\beta_2\beta_6)+\alpha^2\alpha_{13}\beta_5(\beta_1\beta_2-\beta_3\beta_4)-\alpha\alpha_{14}(\beta_1\beta_2^2-\beta_2\beta_3\beta_4)}{\beta_8(-\alpha^2\beta_2\beta_3\beta_5+\beta_2^2\beta_6)}, \\ q_2 &= \frac{(\alpha+iw_0)[\beta_4(-\alpha^2\beta_3\beta_5+\beta_2\beta_6)-\alpha\beta_5(\beta_1\beta_2-\beta_3\beta_4)]}{(-\alpha^2\beta_2\beta_3\beta_5+\beta_2^2\beta_6)}, \\ q_3 &= \frac{(\alpha+iw_0)(\beta_1\beta_2-\beta_3\beta_4)}{(-\alpha^2\beta_3\beta_5+\beta_2\beta_6)}, \\ q_4 &= \frac{\alpha\beta_4(-\alpha^2\beta_3\beta_5+\beta_2\beta_6)-\alpha^2\beta_5(\beta_1\beta_2-\beta_3\beta_4)}{-\alpha^2\beta_2\beta_3\beta_5+\beta_2^2\beta_6}, \\ q_5 &= \frac{\alpha(\beta_1\beta_2-\beta_3\beta_4)}{-\alpha^2\beta_3\beta_5+\beta_2\beta_6}, \end{aligned}$$

where $\beta_1 = -\alpha_{12}\alpha_{41}e^{-i\tau_0 w_0} - \alpha_{42} - iw_0\alpha_{42} + \alpha_{11}\alpha_{42}e^{-i\tau_0 w_0}$, $\beta_2 = \alpha\alpha_{12}\alpha_{33} - \alpha_{12}(1 + iw_0)(\alpha + iw_0) - \alpha\alpha_{13}\alpha_{32}$, $\beta_3 = \alpha_{12}\alpha_{43} - \alpha_{13}\alpha_{42}$, $\beta_4 = -\alpha_{12}\alpha_{31}e^{-i\tau_0 w_0} - \alpha_{32} - iw_0\alpha_{32} + \alpha_{11}\alpha_{32}e^{-i\tau_0 w_0}$, $\beta_5 = \alpha_{12}\alpha_{34} - \alpha_{14}\alpha_{32}$, $\beta_6 = \alpha\alpha_{12}\alpha_{44} - \alpha_{12}(1 + iw_0)(\alpha + iw_0) - \alpha\alpha_{14}\alpha_{42}$, $\beta_7 = 1 + iw_0 - \alpha_{11}e^{-i\tau_0 w_0}$, $\beta_8 = \alpha_{12}e^{-i\tau_0 w_0}$.
On the other hand, suppose that $q^*(s)$ is the eigenvector of $A^*(0)$ corresponding to $-i\tau_0 w_0$. Then $A^*(0)q^*(s) = -i\tau_0 w_0 q^*(s)$. Similarly, by (4.26), (4.28) and (4.33), one obtains

$$q^*(s) = q^*(0)e^{i\tau_0 w_0 s} = B[1, q_1^*, q_2^*, q_3^*, q_4^*, q_5^*]^T e^{i\tau_0 w_0 s}. \tag{4.37}$$

and

$$
\begin{aligned}
q_1^* &= \frac{\iota_7(-\alpha\iota_2\iota_3\iota_5+\iota_2^2\iota_6)-\alpha\alpha_{31}(\alpha\iota_3\iota_4\iota_5+\iota_2\iota_4\iota_6-\iota_1\iota_2\iota_5+\iota_3\iota_4\iota_5)-\alpha_{41}(\iota_1\iota_2^2-\iota_2\iota_3\iota_4)}{\alpha_{21}(-\alpha\iota_2\iota_3\iota_5+\iota_2^2\iota_6)},\\
q_2^* &= \frac{\alpha(\alpha\iota_3\iota_4\iota_5+\iota_2\iota_4\iota_6-\iota_1\iota_2\iota_5+\iota_3\iota_4\iota_5)}{(-\alpha\iota_2\iota_3\iota_5+\iota_2^2\iota_6)},\\
q_3^* &= \frac{(\iota_1\iota_2-\iota_3\iota_4)}{(-\alpha\iota_3\iota_5+\iota_2\iota_6)},\\
q_4^* &= \frac{(1-iw_0)(\alpha\iota_3\iota_4\iota_5+\iota_2\iota_4\iota_6-\iota_1\iota_2\iota_5+\iota_3\iota_4\iota_5)}{(-\alpha\iota_2\iota_3\iota_5+\iota_2^2\iota_6)},\\
q_5^* &= \frac{(1-iw_0)(\iota_1\iota_2-\iota_3\iota_4)}{\alpha(-\alpha\iota_3\iota_5+\iota_2\iota_6)},
\end{aligned}
$$

where $\iota_1 = -\alpha_{14}\alpha_{21} - \alpha_{24}e^{i\tau_0 w_0} + iw_0\alpha_{24}e^{i\tau_0 w_0} + \alpha_{11}\alpha_{24}$, $\iota_2 = \alpha\alpha_{21}\alpha_{33} + \alpha_{21}(1-iw_0)(-\alpha+iw_0) - \alpha\alpha_{23}\alpha_{31}$, $\iota_3 = \alpha_{21}\alpha_{34} - \alpha_{24}\alpha_{31}$, $\iota_4 = -\alpha_{13}\alpha_{21} - \alpha_{23}e^{i\tau_0 w_0} + iw_0\alpha_{23}e^{i\tau_0 w_0} + \alpha_{11}\alpha_{23}$, $\iota_5 = \alpha_{21}\alpha_{43} - \alpha_{23}\alpha_{41}$, $\iota_6 = \alpha_{21}\alpha_{44} + \alpha_{21}(1-iw_0)(-\alpha+iw_0) - \alpha\alpha_{24}\alpha_{41}$, $\iota_7 = e^{i\tau_0 w_0} - iw_0 e^{i\tau_0 w_0} - \alpha_{11}$.

In order to guarantee $\langle q^*(s), q(\theta)\rangle = 1$, the value of B need to be determined. In view of (4.26), (4.28) and (4.34), we have

$$
\begin{aligned}
\langle q^*(s), q(\theta)\rangle &= \overline{B}(1, \overline{q_1^*}, \overline{q_2^*}, \overline{q_3^*}, \overline{q_4^*}, \overline{q_5^*})(1, q_1, q_2, q_3, q_4, q_5)^T\\
&\quad - \int_{-1}^{0}\int_{\xi=0}^{\theta} \overline{q^*}^T(\xi-\theta)d\eta(\theta)q(\xi)d\xi\\
&= \overline{B}\Big\{1 + \sum_{i=1}^{5}\overline{q_i^*}q_i - \int_{-1}^{0}\int_{\xi=0}^{\theta}(1, \overline{q_1^*}, \overline{q_2^*}, \overline{q_3^*}, \overline{q_4^*}, \overline{q_5^*})\cdot\\
&\quad e^{-i\tau_0 w_0(\xi-\theta)}d\eta(\theta)(1, q_1, q_2, q_3, q_4, q_5)^T e^{i\tau_0 w_0 \xi}d\xi\Big\}\\
&= \overline{B}\Big\{1 + \sum_{i=1}^{5}\overline{q_i^*}q_i - (1, \overline{q_1^*}, \overline{q_2^*}, \overline{q_3^*}, \overline{q_4^*}, \overline{q_5^*})\cdot\\
&\quad \int_{-1}^{0} d\eta(\theta)\theta e^{i\tau_0 w_0 \theta}(1, q_1, q_2, q_3, q_4, q_5)^T\Big\}\\
&= \overline{B}\Big\{1 + \sum_{i=1}^{5}\overline{q_i^*}q_i + \tau_0 e^{-i\tau_0 w_0}[\alpha_{11} + \alpha_{21}\overline{q_1^*} + \alpha_{31}\overline{q_2^*}\\
&\quad + \alpha_{41}\overline{q_3^*} + q_1(\alpha_{12} + \alpha_{22}\overline{q_1^*} + \alpha_{32}\overline{q_2^*} + \alpha_{42}\overline{q_3^*})]\Big\}.
\end{aligned}
\tag{4.38}
$$

Thus, in order to ensure $\langle q^*(s), q(\theta)\rangle = 1$, B can be chosen as

$$
\begin{aligned}
B = \Big\{&1 + \sum_{i=1}^{5} q_i^*\overline{q_i} + \tau_0 e^{i\tau_0 w_0}[\alpha_{11} + \alpha_{21}q_1^* + \alpha_{31}q_2^* +\\
&\alpha_{41}q_3^* + \overline{q_1}(\alpha_{12} + \alpha_{22}q_1^* + \alpha_{32}q_2^* + \alpha_{42}q_3^*)]\Big\}^{-1}.
\end{aligned}
\tag{4.39}
$$

In what follows, the method in Hassard et al. (1981) is adopted to compute the coordinates, which describes the center manifold C_0 at $\mu = 0$. Let x_t be the solution of (4.25) when $\mu = 0$.

Define

$$
z(t) = \langle q^*, x_t\rangle, \tag{4.40}
$$

and

$$
W(t, \theta) = x_t - 2Re\{z(t)q(\theta)\}. \tag{4.41}
$$

On the center manifold C_0, one has

$$W(t,\theta)=W(z(t),\overline{z}(t),\theta)=W_{20}(\theta)\frac{z^2}{2}+W_{11}(\theta)z\overline{z}+W_{02}(\theta)\frac{\overline{z}^2}{2}+\cdots, \quad (4.42)$$

where z and $\overline{z}$ are local coordinates for the center manifold C_0 in the direction of q and q^*. For the solution $x_t \in C_0$ and $\mu=0$, one has

$$\begin{aligned}\dot{z}(t)&=\langle q^*,\dot{x}_t\rangle=\langle q^*,Ax_t+Rx_t\rangle=\langle q^*,Ax_t\rangle+\langle q^*,Rx_t\rangle\\&=\langle A^*q^*,x_t\rangle+\langle q^*,Rx_t\rangle=\langle -i\tau_0 w_0 q^*,x_t\rangle+\langle q^*,Rx_t\rangle\\&=i\tau_0 w_0\left\{\overline{q}^{*T}(0)x_t(0)-\int_{-1}^{0}\int_{\xi=0}^{\theta}\overline{q}^{*T}(\xi-\theta)d\eta(\theta)x_t(\xi)d\xi\right\}+\langle q^*,Rx_t\rangle\\&=i\tau_0 w_0\langle q^*,x_t\rangle+\langle q^*,Rx_t\rangle\\&=i\tau_0 w_0 z+\overline{q}^{*T}(0)f\Big(0,W(t,0)+2Re\{z(t)q(0)\}\Big)\\&=i\tau_0 w_0 z+\overline{q}^{*T}(0)f_0(z,\overline{z}).\end{aligned} \quad (4.43)$$

Rewrite (4.43) as

$$\dot{z}(t)=i\tau_0 w_0 z+g(z,\overline{z}), \quad (4.44)$$

in which

$$g(z,\overline{z})=\overline{q}^{*T}(0)f_0(z,\overline{z})=g_{20}\frac{z^2}{2}+g_{11}z\overline{z}+g_{02}\frac{\overline{z}^2}{2}+g_{21}\frac{z^2\overline{z}}{2}+\cdots. \quad (4.45)$$

In view of (4.32) and (4.41), we have

$$\begin{aligned}\dot{W}(t,\theta)&=\dot{x}_t-\dot{z}q-\dot{\overline{z}}\overline{q}\\&=\begin{cases}AW-2Re\{\overline{q^*}(0)f_0(z,\overline{z})q(\theta)\}, & \theta\in[-1,0),\\ AW-2Re\{\overline{q^*}(0)f_0(z,\overline{z})q(0)\}+f_0(z,\overline{z}), & \theta=0,\end{cases}\\&=AW+H(z,\overline{z},\theta),\end{aligned} \quad (4.46)$$

where

$$H(z,\overline{z},\theta)=H_{20}(\theta)\frac{z^2}{2}+H_{11}(\theta)z\overline{z}+H_{02}(\theta)\frac{\overline{z}^2}{2}+\cdots, \quad (4.47)$$

and $f_0(z,\overline{z})$ denotes $f(z,\overline{z})$ at $\mu=0$. In view of (4.46), one gets

$$AW-\dot{W}=-H(z,\overline{z},\theta). \quad (4.48)$$

On the other hand, note that on the center manifold C_0 near the zero equilibrium point,

$$\dot{W}(t,\theta)=\dot{W}(z,\overline{z},\theta)=W_{20}(\theta)z\dot{z}+W_{11}(\theta)(\dot{z}\overline{z}+z\dot{\overline{z}})+W_{02}(\theta)\overline{z}\dot{\overline{z}}+\cdots. \quad (4.49)$$

Substituting (4.42) and (4.49) into (4.48) and comparing the coefficients with (4.47), one obtains

$$\begin{cases} (A - 2i\tau_0 w_0)W_{20}(\theta) = -H_{20}(\theta), \\ AW_{11}(\theta) = -H_{11}(\theta), \\ (A + 2i\tau_0 w_0)W_{02}(\theta) = -H_{02}(\theta). \end{cases} \tag{4.50}$$

Since $x_t(\theta) = x(t+\theta) = W(z, \overline{z}, \theta) + zq + \overline{zq}$ and $q(\theta) = q(0)e^{i\tau_0 w_0 \theta} = (q^{(1)}(\theta), q^{(2)}(\theta), q^{(3)}(\theta), q^{(4)}(\theta), q^{(5)}(\theta), q^{(6)}(\theta))^T$, one has

$$x_t = \begin{pmatrix} W^{(1)}(z, \overline{z}, \theta) \\ W^{(2)}(z, \overline{z}, \theta) \\ W^{(3)}(z, \overline{z}, \theta) \\ W^{(4)}(z, \overline{z}, \theta) \\ W^{(5)}(z, \overline{z}, \theta) \\ W^{(6)}(z, \overline{z}, \theta) \end{pmatrix} + z \begin{pmatrix} q^{(1)} \\ q^{(2)} \\ q^{(3)} \\ q^{(4)} \\ q^{(5)} \\ q^{(6)} \end{pmatrix} + \overline{z} \begin{pmatrix} \overline{q}^{(1)} \\ \overline{q}^{(2)} \\ \overline{q}^{(3)} \\ \overline{q}^{(4)} \\ \overline{q}^{(5)} \\ \overline{q}^{(6)} \end{pmatrix}. \tag{4.51}$$

According to (4.27), (4.43) and (4.44), one obtains

$$g(z, \overline{z}) = \overline{q}^{*T}(0) f_0(z, \overline{z}) = \overline{B}\tau_0(1, \overline{q_1^*}, \overline{q_2^*}, \overline{q_3^*}, \overline{q_4^*}, \overline{q_5^*}) \begin{pmatrix} f_1(x_t) \\ f_2(x_t) \\ f_3(x_t) \\ f_4(x_t) \\ f_5(x_t) \\ f_6(x_t) \end{pmatrix}. \tag{4.52}$$

Substituting (4.51) into (4.52) and comparing the coefficients with those in (4.45), one gets

$$\begin{aligned}
g_{20} &= 2\overline{B}\tau_0\Big[\gamma_1(q^{(1)}(-1))^2 + \gamma_2(q^{(2)}(-1))^2 + \gamma_3 q^{(1)}(-1)q^{(2)}(-1) + \\
&\quad \zeta_1(q^{(5)}(0))^2 + \zeta_2(q^{(6)}(0))^2 + \zeta_3 q^{(5)}(0)q^{(6)}(0)\Big], \\
g_{11} &= 2\overline{B}\tau_0\Big[\gamma_1 q^{(1)}(-1)\overline{q}^{(1)}(-1) + \gamma_2 q^{(2)}(-1)\overline{q}^{(2)}(-1) + \gamma_3\Big(q^{(1)}(-1)\overline{q}^{(1)}(-1) \\
&\quad + q^{(1)}(-1)\overline{q}^{(2)}(-1)\Big)/2 + \zeta_1 q^{(5)}(0)\overline{q}^{(5)}(0) + \zeta_2 q^{(6)}(0)\overline{q}^{(6)}(0) + \\
&\quad \zeta_3\Big(q^{(6)}(0)\overline{q}^{(5)}(0) + q^{(5)}(0)\overline{q}^{(6)}(0)\Big)/2\Big], \\
g_{02} &= 2\overline{B}\tau_0\Big[\gamma_1(\overline{q}^{(1)}(-1))^2 + \gamma_2(\overline{q}^{(2)}(-1))^2 + \gamma_3\overline{q}^{(1)}(-1)\overline{q}^{(2)}(-1) + \\
&\quad \zeta_1(\overline{q}^{(5)}(0))^2 + \zeta_2(\overline{q}^{(6)}(0))^2 + \zeta_3\overline{q}^{(5)}(0)\overline{q}^{(6)}(0)\Big], \\
g_{21} &= 2\overline{B}\tau_0\Big[\gamma_1\Big(W_{20}^{(1)}(-1)\overline{q}^{(1)}(-1) + 2W_{11}^{(1)}(-1)q^{(1)}(-1)\Big) + \\
&\quad \gamma_2\Big(W_{20}^{(2)}(-1)\overline{q}^{(2)}(-1) + 2W_{11}^{(2)}(-1)q^{(2)}(-1)\Big) + \gamma_3\Big(W_{20}^{(2)}(-1)\overline{q}^{(1)}(-1) + \\
&\quad 2W_{11}^{(2)}(-1)q^{(1)}(-1) + W_{20}^{(1)}(-1)\overline{q}^{(2)}(-1) + 2W_{11}^{(1)}(-1)q^{(2)}(-1)\Big) + \\
&\quad 3\gamma_4(q^{(1)}(-1))^2\overline{q}^{(1)}(-1) + \gamma_5\Big((q^{(2)}(-1))^2\overline{q}^{(1)}(-1) + \\
&\quad 2q^{(2)}(-1)\overline{q}^{(2)}(-1)q^{(1)}(-1)\Big) + \gamma_6\Big((q^{(1)}(-1))^2\overline{q}^{(2)}(-1) + \\
&\quad 2q^{(1)}(-1)\overline{q}^{(1)}(-1)q^{(2)}(-1)\Big) + 3\gamma_7(q^{(2)}(-1))^2\overline{q}^{(2)}(-1) + \zeta_1\Big(W_{20}^{(5)}(0)\overline{q}^{(5)}(0) \\
&\quad + 2W_{11}^{5}(0)q^{(5)}(0)\Big) + \zeta_2\Big(W_{20}^{(6)}(0)\overline{q}^{(6)}(0) + 2W_{11}^{(6)}(0)q^{(6)}(0)\Big) + \\
&\quad \zeta_3\Big(W_{20}^{(6)}(0)\overline{q}^{(5)}(0) + 2W_{11}^{(6)}(0)q^{(5)}(0) + W_{20}^{(5)}(0)\overline{q}^{(6)}(0) + 2W_{11}^{(5)}(0)q^{(6)}(0)\Big) \\
&\quad + 3\zeta_4(q^{(5)}(0))^2\overline{q}^{(5)}(0) + \zeta_5\Big((q^{(6)}(0))^2\overline{q}^{(5)}(0) + 2q^{(6)}(0)\overline{q}^{(6)}(0)q^{(5)}(0)\Big) \\
&\quad + \zeta_6\Big((q^{(5)}(0))^2\overline{q}^{(6)}(0) + 2q^{(5)}(0)\overline{q}^{(5)}(0)q^{(6)}(0)\Big) + 3\zeta_7(q^{(6)}(0))^2\overline{q}^{(6)}(0)\Big].
\end{aligned} \tag{4.53}$$

where

$$\gamma_1 = m_{11} + m_{21}\overline{q}_1^* + m_{31}\overline{q}_2^* + m_{41}\overline{q}_3^*, \quad \gamma_2 = m_{12} + m_{22}\overline{q}_1^* + m_{32}\overline{q}_2^* + m_{42}\overline{q}_3^*,$$

$$\gamma_3 = m_{13} + m_{23}\overline{q}_1^* + m_{33}\overline{q}_2^* + m_{43}\overline{q}_3^*, \quad \gamma_4 = m_{14} + m_{24}\overline{q}_1^* + m_{34}\overline{q}_2^* + m_{44}\overline{q}_3^*,$$

$$\gamma_5 = m_{15} + m_{25}\overline{q}_1^* + m_{35}\overline{q}_2^* + m_{45}\overline{q}_3^*, \quad \gamma_6 = m_{16} + m_{26}\overline{q}_1^* + m_{36}\overline{q}_2^* + m_{46}\overline{q}_3^*,$$

$$\gamma_7 = m_{17} + m_{27}\overline{q}_1^* + m_{37}\overline{q}_2^* + m_{47}\overline{q}_3^*, \quad \zeta_1 = n_{11} + n_{21}\overline{q}_1^* + n_{31}\overline{q}_2^* + n_{41}\overline{q}_3^*,$$

$$\zeta_2 = n_{12} + n_{22}\overline{q}_1^* + n_{32}\overline{q}_2^* + n_{42}\overline{q}_3^*, \quad \zeta_3 = n_{13} + n_{23}\overline{q}_1^* + n_{33}\overline{q}_2^* + n_{43}\overline{q}_3^*,$$

$$\zeta_4 = n_{14} + n_{24}\overline{q}_1^* + n_{34}\overline{q}_2^* + n_{44}\overline{q}_3^*, \quad \zeta_5 = n_{15} + n_{25}\overline{q}_1^* + n_{35}\overline{q}_2^* + n_{45}\overline{q}_3^*,$$

$$\zeta_6 = n_{16} + n_{26}\overline{q}_1^* + n_{36}\overline{q}_2^* + n_{46}\overline{q}_3^*, \quad \zeta_7 = n_{17} + n_{27}\overline{q}_1^* + n_{37}\overline{q}_2^* + n_{47}\overline{q}_3^*.$$

In what follows, $W_{20}(\theta)$ and $W_{11}(\theta)$ are calculated. From (4.46), one gets

$$\begin{aligned} H(z, \overline{z}, \theta) &= -\overline{q^*}(0) f_0(z, \overline{z}) q(\theta) - q^*(0) \overline{f_0}(z, \overline{z}) \overline{q}(\theta) \\ &= -g(z, \overline{z}) q(\theta) - \overline{g}(z, \overline{z}) \overline{q}(\theta), \quad \theta \in [-1, 0). \end{aligned} \tag{4.54}$$

Comparing the coefficients of above equations with (4.47), one obtains

$$\begin{aligned} H_{20}(\theta) &= -(g_{20} q(\theta) + \overline{g}_{02} \overline{q}(\theta)), \quad \theta \in [-1, 0), \\ H_{11}(\theta) &= -(g_{11} q(\theta) + \overline{g}_{11} \overline{q}(\theta)), \quad \theta \in [-1, 0). \end{aligned} \tag{4.55}$$

From (4.30), (4.50) and (4.55), one gets

$$W_{20}'(\theta) = 2i\tau_0 w_0 W_{20}(\theta) + g_{20} q(\theta) + \overline{g}_{02} \overline{q}(\theta). \tag{4.56}$$

Then, since $q(\theta) = q(0)e^{i\tau_0 w_0 \theta}$, one has

$$W_{20}(\theta) = \frac{i g_{20}}{w_0 \tau_0} q(0) e^{i\tau_0 w_0 \theta} + \frac{i \overline{g}_{02}}{3 w_0 \tau_0} \overline{q}(0) e^{-i\tau_0 w_0 \theta} + E_1 e^{2i\tau_0 w_0 \theta}. \tag{4.57}$$

Similarly,

$$W_{11}(\theta) = -\frac{i g_{11}}{w_0 \tau_0} q(0) e^{i\tau_0 w_0 \theta} + \frac{i \overline{g}_{11}}{w_0 \tau_0} \overline{q}(0) e^{-i\tau_0 w_0 \theta} + E_2. \tag{4.58}$$

Next, E_1 and E_2 will be determined. From the definition of A and (4.50), we have

$$\int_{-1}^{0} d\eta(\theta) W_{20}(\theta) = 2i\tau_0 w_0 W_{20}(0) - H_{20}(0), \tag{4.59}$$

and

$$\int_{-1}^{0} d\eta(\theta) W_{11}(\theta) = -H_{11}(0). \tag{4.60}$$

From (4.46) and (4.47), one gets

$$\begin{aligned} H(z,\overline{z},0) &= -2Re\{\overline{q^*}(0) f_0(z,\overline{z}) q(0)\} + f_0(z,\overline{z}) \\ &= -g(z,\overline{z})q(0) - \overline{g}(z,\overline{z})\overline{q}(0) + f_0(z,\overline{z}). \end{aligned} \tag{4.61}$$

In view of (4.27), (4.51), one obtains

$$f_0(z,\overline{z}) = \tau_0 \begin{pmatrix} K_{11}z^2 + K_{12}z\overline{z} + K_{13}\overline{z}^2 + H.O.T \\ K_{21}z^2 + K_{22}z\overline{z} + K_{23}\overline{z}^2 + H.O.T \\ K_{31}z^2 + K_{32}z\overline{z} + K_{33}\overline{z}^2 + H.O.T \\ K_{41}z^2 + K_{42}z\overline{z} + K_{43}\overline{z}^2 + H.O.T \\ 0 \\ 0 \end{pmatrix}, \tag{4.62}$$

where

$$\begin{aligned}
K_{11} &= m_{11}(q^{(1)}(-1))^2 + m_{12}(q^{(2)}(-1))^2 + m_{13}q^{(1)}(-1)q^{(2)}(-1) + \\
&\quad n_{11}(q^{(5)}(-1))^2 + n_{12}(q^{(6)}(-1))^2 + n_{13}q^{(5)}(-1)q^{(6)}(-1), \\
K_{12} &= 2m_{11}q^{(1)}(-1)\overline{q}^{(1)}(-1) + 2m_{12}q^{(2)}(-1)\overline{q}^{(2)}(-1) + \\
&\quad m_{13}\Big(q^{(1)}(-1)\overline{q}^{(2)}(-1) + q^{(2)}(-1)\overline{q}^{(1)}(-1)\Big) + 2n_{11}q^{(5)}(0)\overline{q}^{(5)}(0) \\
&\quad + 2n_{12}q^{(6)}(0)\overline{q}^{(6)}(0) + n_{13}\Big(q^{(5)}(0)\overline{q}^{(6)}(0) + q^{(6)}(0)\overline{q}^{5}(0)\Big), \\
K_{13} &= m_{11}(\overline{q}^{(1)}(-1))^2 + m_{12}(\overline{q}^{(2)}(-1))^2 + m_{13}\overline{q}^{(1)}(-1)\overline{q}^{(2)}(-1) + \\
&\quad n_{11}(\overline{q}^{(5)}(-1))^2 + n_{12}(\overline{q}^{(6)}(-1))^2 + n_{13}\overline{q}^{(5)}(-1)\overline{q}^{(6)}(-1), \\
K_{21} &= m_{21}(q^{(1)}(-1))^2 + m_{22}(q^{(2)}(-1))^2 + m_{23}q^{(1)}(-1)q^{(2)}(-1) + \\
&\quad n_{21}(q^{(5)}(-1))^2 + n_{22}(q^{(6)}(-1))^2 + n_{23}q^{(5)}(-1)q^{(6)}(-1), \\
K_{22} &= 2m_{21}q^{(1)}(-1)\overline{q}^{(1)}(-1) + 2m_{22}q^{(2)}(-1)\overline{q}^{(2)}(-1) \\
&\quad + m_{23}\Big(q^{(1)}(-1)\overline{q}^{(2)}(-1) + q^{(2)}(-1)\overline{q}^{(1)}(-1)\Big) + 2n_{21}q^{(5)}(0)\overline{q}^{(5)}(0) \\
&\quad + 2n_{22}q^{(6)}(0)\overline{q}^{(6)}(0) + n_{23}\Big(q^{(5)}(0)\overline{q}^{(6)}(0) + q^{(6)}(0)\overline{q}^{5}(0)\Big), \\
K_{23} &= m_{21}(\overline{q}^{(1)}(-1))^2 + m_{22}(\overline{q}^{(2)}(-1))^2 + m_{23}\overline{q}^{(1)}(-1)\overline{q}^{(2)}(-1) \\
&\quad + n_{21}(\overline{q}^{(5)}(-1))^2 + n_{22}(\overline{q}^{(6)}(-1))^2 + n_{23}\overline{q}^{(5)}(-1)\overline{q}^{(6)}(-1), \\
K_{31} &= m_{31}(q^{(1)}(-1))^2 + m_{32}(q^{(2)}(-1))^2 + m_{33}q^{(1)}(-1)q^{(2)}(-1) + \\
&\quad n_{31}(q^{(5)}(-1))^2 + n_{32}(q^{(6)}(-1))^2 + n_{33}q^{(5)}(-1)q^{(6)}(-1), \\
K_{32} &= 2m_{31}q^{(1)}(-1)\overline{q}^{(1)}(-1) + 2m_{32}q^{(2)}(-1)\overline{q}^{(2)}(-1) + \\
&\quad m_{33}\Big(q^{(1)}(-1)\overline{q}^{(2)}(-1) + q^{(2)}(-1)\overline{q}^{(1)}(-1)\Big) + 2n_{31}q^{(5)}(0)\overline{q}^{(5)}(0) \\
&\quad + 2n_{32}q^{(6)}(0)\overline{q}^{(6)}(0) + n_{33}\Big(q^{(5)}(0)\overline{q}^{(6)}(0) + q^{(6)}(0)\overline{q}^{5}(0)\Big), \\
K_{33} &= m_{31}(\overline{q}^{(1)}(-1))^2 + m_{32}(\overline{q}^{(2)}(-1))^2 + m_{33}\overline{q}^{(1)}(-1)\overline{q}^{(2)}(-1) + \\
&\quad n_{31}(\overline{q}^{(5)}(-1))^2 + n_{32}(\overline{q}^{(6)}(-1))^2 + n_{33}\overline{q}^{(5)}(-1)\overline{q}^{(6)}(-1), \\
K_{41} &= m_{41}(q^{(1)}(-1))^2 + m_{42}(q^{(2)}(-1))^2 + m_{43}q^{(1)}(-1)q^{(2)}(-1) + \\
&\quad n_{41}(q^{(5)}(-1))^2 + n_{42}(q^{(6)}(-1))^2 + n_{43}q^{(5)}(-1)q^{(6)}(-1),
\end{aligned}$$

$$K_{42} = 2m_{41}q^{(1)}(-1)\overline{q}^{(1)}(-1) + 2m_{42}q^{(2)}(-1)\overline{q}^{(2)}(-1)+$$
$$m_{43}\left(q^{(1)}(-1)\overline{q}^{(2)}(-1) + q^{(2)}(-1)\overline{q}^{(1)}(-1)\right) + 2n_{41}q^{(5)}(0)\overline{q}^{(5)}(0)+$$
$$2n_{42}q^{(6)}(0)\overline{q}^{(6)}(0) + n_{43}\left(q^{(5)}(0)\overline{q}^{(6)}(0) + q^{(6)}(0)\overline{q}^{5}(0)\right),$$
$$K_{43} = m_{41}(\overline{q}^{(1)}(-1))^2 + m_{42}(\overline{q}^{(2)}(-1))^2 + m_{43}\overline{q}^{(1)}(-1)\overline{q}^{(2)}(-1)+$$
$$n_{41}(\overline{q}^{(5)}(-1))^2 + n_{42}(\overline{q}^{(6)}(-1))^2 + n_{43}\overline{q}^{(5)}(-1)\overline{q}^{(6)}(-1),$$

and m_{ij} and n_{ij} are defined in (4.27), respectively.

Comparing the coefficients of (4.61) with (4.47), one obtains

$$H_{20}(0) = -g_{20}q(0) - \overline{g}_{02}\overline{q}(0) + \tau_0(K_{11}, K_{21}, K_{31}, K_{41}, 0, 0)^T, \tag{4.63}$$

and

$$H_{11}(0) = -g_{11}q(0) - \overline{g}_{11}\overline{q}(0) + \tau_0(K_{12}, K_{22}, K_{32}, K_{42}, 0, 0)^T. \tag{4.64}$$

Substituting (4.57) and (4.63) into (4.59), one has

$$\left(2i\tau_0 w_0 I - \int_{-1}^{0} d\eta(\theta)e^{2i\tau_0 w_0\theta}\right) E_1 = \tau_0(K_{11}, K_{21}, K_{31}, K_{41}, 0, 0)^T, \tag{4.65}$$

where $2i\tau_0 w_0 I - \int_{-1}^{0} d\eta(\theta)e^{2i\tau_0 w_0\theta} = \tau_0(2iw_0 I - (B_1 + B_2 e^{-2i\tau_0 w_0}))$.

Define $D_1 = 2iw_0 I - (B_1 + B_2 e^{-2i\tau_0 w_0})$, then (4.65) is equivalent to

$$D_1 E_1 = (K_{11}, K_{21}, K_{31}, K_{41}, 0, 0)^T, \tag{4.66}$$

similarly, substituting (4.58) and (4.64) into (4.60), we can obtain

$$D_2 E_2 = (K_{12}, K_{22}, K_{32}, K_{42}, 0, 0)^T, \tag{4.67}$$

where

$$D_1 = \begin{pmatrix} D_{11} & D_{12} & 0 & 0 & -\alpha_{13} & -\alpha_{14} \\ D_{21} & D_{22} & 0 & 0 & -\alpha_{23} & -\alpha_{24} \\ D_{31} & D_{32} & 2iw_0+1 & 0 & -\alpha_{33} & -\alpha_{34} \\ D_{41} & D_{42} & 0 & 2iw_0+1 & -\alpha_{43} & -\alpha_{44} \\ 0 & 0 & -\alpha & 0 & 2iw_0+\alpha & 0 \\ 0 & 0 & 0 & -\alpha & 0 & 2iw_0+\alpha \end{pmatrix},$$

$D_{11} = 2iw_0 + 1 - \alpha_{11}e^{-2iw_0\tau_0}$, $D_{12} = -\alpha_{12}e^{-2iw_0\tau_0}$, $D_{21} = -\alpha_{21}e^{-2iw_0\tau_0}$,
$D_{22} = 2iw_0 + 1 - \alpha_{22}e^{-2iw_0\tau_0}$, $D_{31} = -\alpha_{31}e^{-2iw_0\tau_0}$, $D_{32} = -\alpha_{32}e^{-2iw_0\tau_0}$,
$D_{41} = -\alpha_{41}e^{-2iw_0\tau_0}$, $D_{42} = -\alpha_{42}e^{-2iw_0\tau_0}$,

$$D_2 = \begin{pmatrix} 1-\alpha_{11} & -\alpha_{12} & 0 & 0 & -\alpha_{13} & -\alpha_{14} \\ -\alpha_{21} & 1-\alpha_{22} & 0 & 0 & -\alpha_{23} & -\alpha_{24} \\ -\alpha_{31} & -\alpha_{32} & 1 & 0 & -\alpha_{33} & -\alpha_{34} \\ -\alpha_{41} & -\alpha_{42} & 0 & 1 & -\alpha_{43} & -\alpha_{44} \\ 0 & 0 & -\alpha & 0 & \alpha & 0 \\ 0 & 0 & 0 & -\alpha & 0 & \alpha \end{pmatrix}.$$

It follows that

$$E_1 = D_1^{-1}(K_{11}, K_{21}, K_{31}, K_{41}, 0, 0)^T, \tag{4.68}$$

and

$$E_2 = D_2^{-1}(K_{12}, K_{22}, K_{32}, K_{42}, 0, 0)^T. \tag{4.69}$$

Thus, the formulae for determine

$$\begin{aligned} c_1(0) &= \tfrac{i}{2w_0}\left(g_{11}g_{20} - 2|g_{11}|^2 - \tfrac{|g_{02}|^2}{3}\right) + \tfrac{g_{21}}{2}, \\ \mu_2 &= -\tfrac{Re\{c_1(0)\}}{Re\{\lambda'(\tau_0)\}}, \\ \kappa_2 &= 2Re\{c_1(0)\}, \\ T_2 &= -\tfrac{Im\{c_1(0)\}+\mu_2 Im\{\lambda'(\tau_0)\}}{w_0}. \end{aligned} \tag{4.70}$$

This completes the proof. □

4.5 Illustrative Example

In this section, we will present some numerical results to illustrate the effectiveness of our results.

Example 4.1 Choose the activation functions $f_{ij}(z) = \tanh(x) + \tanh(y) + i\Big(\tanh(x) + \tanh(y)\Big)$ $(i, j = 1, 2)$. It is easy to see that system (4.1) has a zero equilibrium. Let $b_{11} = -1 - 2i$, $b_{12} = -1 + 2i$, $b_{21} = 2 - i$, $b_{22} = -1 - 2i$, $\alpha = 0.1$.

Using the results in Sect. 4.3, we obtain $w_0 = 1.5012$ and $\tau_0 = 1.2234$. It follows from Theorem 4.2 that

$$\mu_2 = 7.4539, \kappa_2 = -7.2228, T_2 = 4.9098.$$

Obviously, by Theorem 4.1, the equilibrium of system (4.1) is asymptotically stable when $\tau = 1.12 < \tau_0$. This is supported by the numerical simulation shown in Fig. 4.1. Since $\mu_2 > 0$, the bifurcation is supercritical. When τ is increased to the critical value τ_0, the origin loses its stability and a Hopf bifurcation occurs., i.e., a periodic solution bifurcate from the origin. The individual periodic solution is stable since $\kappa_2 < 0$. There are stable limit cycles with $\tau = 1.45$ as shown in Fig. 4.2. Since $T_2 > 0$, the period of the periodic solutions increases as τ increases. Take $\tau = 1.55$, the corresponding numerical simulation is shown in Fig. 4.3. Compared Fig. 4.2 with Fig. 4.3, we can conclude that the period for $\tau = 1.55$ is longer than that for $\tau = 1.45$.

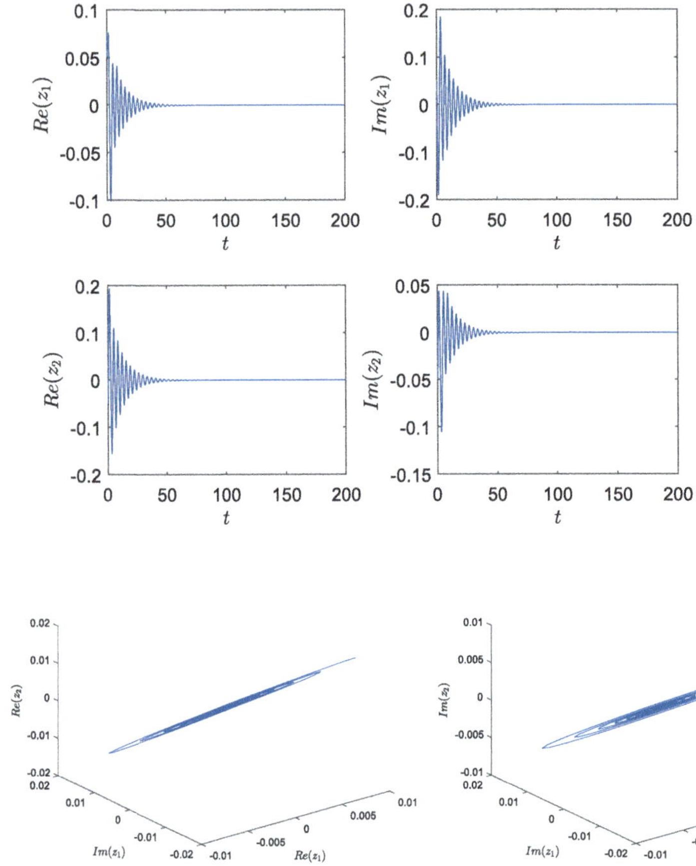

Fig. 4.1 Phase diagram of system (4.8) when $\tau = 1.12 < \tau_0$. The origin is asymptotically stable

4.6 Conclusion

In this chapter, we have studied the Hopf bifurcation for a class of CVNNs with discrete and distributed delays. It has been shown that under certain conditions, the zero solution loses its stability and a Hopf bifurcation occurs. Taking the discrete-time delay as the bifurcation parameter, the critical value of the time delay, which guarantees the occurrence of the Hopf bifurcation, is determined. Moreover, by employing

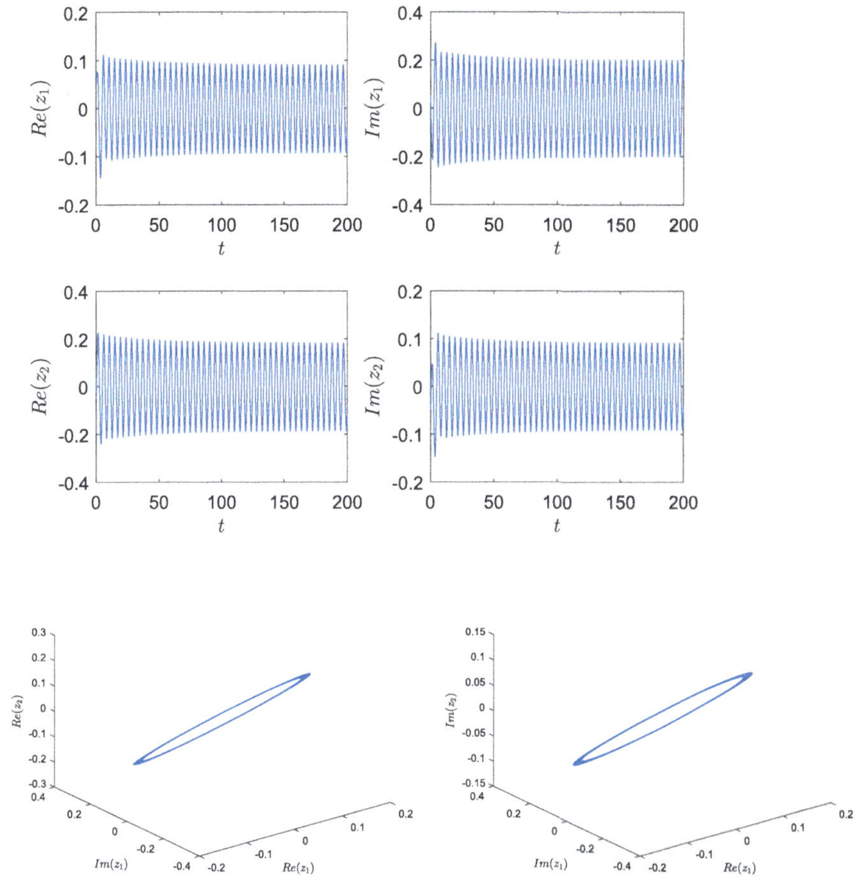

Fig. 4.2 The bifurcation periodic solution when $\tau = 1.45 > \tau_0$

the normal form theory and the center manifold theorem, we have determined the direction of Hopf bifurcation and the stability and period of the bifurcating periodic solutions. Some numerical simulation results have demonstrated the correctness of our theoretical analysis.

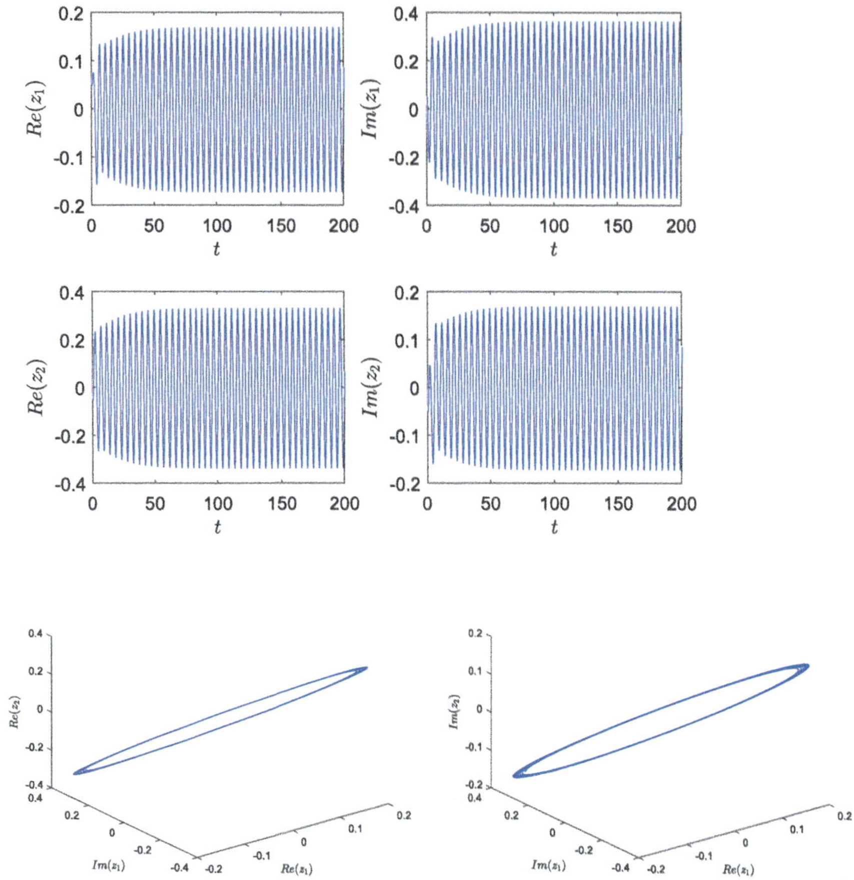

Fig. 4.3 The bifurcation periodic solution when $\tau = 1.55 > \tau_0$

References

Wang Z, Wang X, Li Y, Huang X (2017) Stability and Hopf bifurcation of fractional-order complex-valued single neuron model with time delay. Int J Bifurc Chaos 27(13):1750209

Guo R, Zhang Z, Liu X, Lin C (2017) Existence, uniqueness, and exponential stability analysis for complex-value d memristor-based BAM neural networks with time delays. Appl Math Comput 311:100–117

Lee DL (2006) Improvement of complex-valued Hopfield associative memory by using generalized projection rules. IEEE Trans Neural Netw 17(5):1341–1347

Zhou W, Zurada JM (2009) Discrete-time recurrent neural networks with complex-valued linear threshold neurons. IEEE Trans Circuits Syst II Express Briefs 56(8):669–673

Dong T, Liao X, Wang A (2015) Stability and Hopf bifurcation of a complex-valued neural network with two time delays. Nonlinear Dyn 82:173–184

Yu W, Cao J (2007) Stability and Hopf bifurcation on a two-neuron system with time delay in the frequency domain. Int J Bifurc Chaos 17(4):1355–1366

Liu Y, Zhang D, Lu J, Cao J (2016) Global μ-stability criteria for quaternion-valued neural networks with unbounded time-varying delays. Inf Sci 360:273–288

Liu Y, Xu P, Lu J, Liang J (2016) Global stability of Clifford-valued recurrent neural networks with time delays. Nonlinear Dyn 84(2):767–777

Olien L, Bélair J (1997) Bifurcations, stability, and monotonicity properties of a delayed neural network model. Phys D Nonlinear Phenom 102(3–4):349–363

Liu Y, Zhang D, Lu J (2017) Global exponential stability for quaternion-valued recurrent neural networks with time-varying delays. Nonlinear Dyn 87(1):553–565

Bao H, Park JH, Cao J (2016) Synchronization of fractional-order complex-valued neural networks with time delay. Neural Netw 81:16–28

Bao H, Park JH, Cao J (2016) Exponential synchronization of coupled stochastic memristor-based neural networks with time-varying probabilistic delay coupling and impulsive delay. IEEE Trans Neural Netw Learn Syst 27(1):190–201

Tank DW, Hopfield JJ (1987) Neural computation by concentrating information in time. Proc Natl Acad Sci 84(7):1896–1900

Vries BD, Principe JC (1992) The Gamma modelCa new neural model for temporal processing. Neural Netw 5(4):565–576

Principe JC, Kuo JM, Celebi S (1994) An analysis of the Gamma memory in dynamic neural networks. IEEE Trans Neural Netw 5(2):331–337

Lv T, Gan Q (2016) Zhu Q (2016) Stability and bifurcation analysis for a class of generalized reaction-diffusion neural networks with time delay. Discr Dyn Nature Soc 4:1–9

Li X, Hu G (2011) Stability and Hopf bifurcation on a neuron network with discrete and distributed delays. Appl Math Sci 5(42):2077–2084

Karaoglu E, Yilmaz E, Merdan H (2016) Stability and bifurcation analysis of two-neuron network with discrete and distributed delays. Neurocomputing 182:102–110

Ncube I (2013) Stability switching and Hopf bifurcation in a multiple-delayed neural network with distributed delay. J Math Anal Appl 407(1):141–146

Ruan S, Wei J (2003) On the zeros of transcendental functions with applications to stability of delay differential equations with two delays. Dyn Contin Discr Impul Syst 10(6):863–874

Benedetto JJ, Czaja W (2009) Riesz representation theorem. Birkhäuser, Boston, pp 321–357

Hassard BD, Kazarinoff ND, Wan YH (1981) Theory and applications of Hopf bifurcation. Cambridge University Press

Chapter 5
Finite-Time Stability Analysis for CVBAMNNs with Constant Delay

Abstract This chapter focuses on the finite-time stability problem of complex-valued bidirectional associative memory neural networks (CVBAMNNs) with time delay. First, a tractable entire real-valued system is formed based on the original system. Then, it presents a condition to ensure the existence and uniqueness of the equilibrium point for this system by using the nonlinear measure approach. It also establishes a finite-time stability criterion of the equilibrium point by Lyapunov function approach.

5.1 Introduction

It is well known that several kinds of recurrent neural network models have been extensively investigated quite recently because of their wide applications in many fields. Among which, the bidirectional associative memory neural networks (BAMNNs) model is regarded as an important network model, which was first proposed by Kosko (1987, 1988) with two-layer (the U-layer and the V-layer) hetero associative neurons. Because of iterations of forward and backward information flows between the two layers, this model owns information memory and information association ability. As a result, BAMNNs model has been applied to various crucial areas such as pattern recognition, associative memories and and automatic control engineering problems. Moreover, the analysis of BAMNNs has attracted considerable attention, and especially, various dynamical behaviors of this kind of model have been studied extensively (Li et al. 2016; Zhu and Cao 2012; Zhu et al. 2014; Tian et al. 2015; Sakthivel et al. 2015; Li and Li 2016), which involve many aspects including exponential stability, stochastic stability, Hopf bifurcation, state estimator, stabilization and so forth.

Quite recently, due to the fruitful practical applications, the research on complex dynamical systems draws many scholars' attention. Among which, complex-valued neural networks (CVNNs) have become a very popular research object, and many efforts have been made to develop the dynamical behaviors of complex-valued neural

Z. Zhang et al., *Complex-Valued Neural Networks Systems with Time Delay*,
Intelligent Control and Learning Systems 4,
https://doi.org/10.1007/978-981-19-5450-4_5

networks and numerous achievements have sprung up (Hu and Wang 2012; Liu and Chen 2016; Rakkiyappan et al. 2015; Zhang et al. 2014; Chen et al. 2017; Gong et al. 2016; Wang et al. 2017; Wei et al. 2017). With introducing complex signal into BAMNNs model, the studies on complex-valued bidirectional associative memory neural networks (CVBAMNNs) have begun. For example, global stability analysis for delayed CVBAMNNs and Cohen-Grossberg CVBAMNNs in Wang et al. (2016), Guo et al. (2017), Subramanian and Muthukumar (2018).

It should be noticed that all the dynamical behaviors above are running in an infinite time interval. While, in practical engineering applications, the dynamical behavior of a system over a finite-time interval is often required. This leads to the occurrence of finite-time stability of the system. It was originally introduced in Kamenkov (1953), which means that as soon as a finite-time interval is fixed, the state of system will start within a prescribed bound and will not exceed some bounds during the given interval. Then, some valuable achievements about the dynamical behaviors for neural networks models in a finite-time interval have been reported in succession (Yan et al. 2015; Li and Cao 2017; Zhao et al. 2016; Chen et al. 2013; Wu et al. 2013; Rakkiyappan et al. 2014; Wang and Shen 2015; Wang et al. 2017, 2016; Liu et al. 2013; Velmurugan et al. 2016; Abdurahman et al. 2016; Xiao et al. 2017). Among these, the problems of finite-time stability analysis for various neural networks systems are considered in Li and Cao (2017), Zhao et al. (2016), Chen et al. (2013), Wu et al. (2013) and Rakkiyappan et al. (2014). Finite-time stabilization and finite-time synchronization for neural network models are studied in Wang and Shen (2015), Wang et al. (2017), Wang et al. (2016), Liu et al. (2013) and in Velmurugan et al. (2016), Abdurahman et al. (2016), Xiao et al. (2017), respectively. Nevertheless, hitherto, only Liu et al. (2013) and Xiao et al. (2017) engage in the finite-time stabilization and synchronization research for BAMNNs, and only Rakkiyappan et al. (2014) deals with the finite-time stability problem of CVNNs. Furthermore, for the topic of the dynamics within a finite-time interval about CVBAMNNs, there is no result published.

From the discussions above on CVBAMNNs, we know that it is essential to explore the finite-time stability for this kind of system. In addition, in practical systems, time delay is unavoidable and it's existence often causes the instability or performance descent of systems. Hence, in this chapter, the finite-time stability problem of delayed CVBAMNNs model is studied. The main contributions are summarized as follows: (1) For delayed CVBAMNNs model, it is the first time that the finite-time stability problem is addressed; (2) For the considered system, a tractable entire system is formed by separating the real and imaginary parts; (3) Via nonlinear measure approach, a condition to ensure the existence and uniqueness of the equilibrium point of the considered system is derived. Based on this result, a finite-time stability criterion of the equilibrium point is presented in terms of LMIs by Lyapunov function approach.

5.2 Problem Formulation and Preliminaries

Consider a nonlinear system represented by the following CVBAMNNs:

$$\begin{cases} \dot{u}(t) = -D_1 u(t) + A_1 f(v(t)) + B_1 f(v(t-\tau)) + J_1, \\ \dot{v}(t) = -D_2 v(t) + A_2 g(u(t)) + B_2 g(u(t-\tau)) + J_2, \\ u(s) = \varphi(s), \ s \in [-\tau, 0], \\ v(s) = \psi(s), \ s \in [-\tau, 0], \end{cases} \tag{5.1}$$

where $u(t) = (u_1(t), u_2(t), ..., u_n(t))^T \in \mathbb{C}^n$ and $v(t) = (v_1(t), v_2(t), ..., v_m(t))^T \in \mathbb{C}^m$ are the neuron state vectors; $D_1 = \text{diag}\{d_1^1, d_2^1, ..., d_n^1\} \in \mathbb{R}^{n\times n}$ with $d_k^1 > 0$ $(k = 1, 2, ..., n)$ and $D_2 = \text{diag}\{d_1^2, d_2^2, ..., d_m^2\} \in \mathbb{R}^{m\times m}$ with $d_j^2 > 0$ $(j = 1, 2, ..., m)$ are the self-feedback connection weight matrices, respectively; $J_1 \in \mathbb{C}^n$ and $J_2 \in \mathbb{C}^m$ denote the external input vectors; $A_1 \in \mathbb{C}^{n\times m}$, $A_2 \in \mathbb{C}^{m\times n}$ and $B_1 \in \mathbb{C}^{n\times m}$, $B_2 \in \mathbb{C}^{m\times n}$ are the connection weight matrices without and with time delay, respectively; τ is constant time delay; $f(v(t)) = [f_1(v_1(t)), f_2(v_2(t)), ..., f_m(v_m(t))]^T \in \mathbb{C}^m$, $f(v(t-\tau)) = [f_1(v_1(t-\tau)), f_2(v_2(t-\tau)), ..., f_m(v_m(t-\tau))]^T \in \mathbb{C}^m$, $g(u(t)) = [g_1(u_1(t)), g_2(u_2(t)), ..., g_n(u_n(t))]^T \in \mathbb{C}^n$, and $g(u(t-\tau)) = [g_1(u_1(t-\tau)), g_2(u_2(t-\tau)), ..., g_n(u_n(t-\tau))]^T \in \mathbb{C}^n$ are the complex-valued activation functions; the continuous functions $\varphi(s) \in \mathbb{C}^n$ and $\psi(s) \in \mathbb{C}^m$ are the initial conditions.

Here, we assume that the component activation functions $f_j(v_j(t))$, $f_j(v_j(t-\tau))$ $(j = 1, 2, ..., m)$ and $g_k(u_k(t))$, $g_k(u_k(t-\tau))$ $(k = 1, 2, ..., n)$ are real-imaginary separate-type and they satisfy the following assumption.

Assumption 5.1 The functions $f_j(.)$ and $g_k(.)$ are Lipschitz continuous in the complex domain, i.e., there exist constants $l_j > 0$ and $m_k > 0$ such that the following inequalities hold:

$$\begin{aligned} |f_j(v) - f_j(v')| &\le l_j |v - v'|, \quad \forall v, v' \in \mathbb{C}, \\ |g_k(u) - g_k(u')| &\le m_k |u - u'|, \quad \forall u, u' \in \mathbb{C}. \end{aligned} \tag{5.2}$$

Let $u(t) = x_1(t) + iy_1(t)$, $v(t) = x_2(t) + iy_2(t)$, $f(v(t)) = f^R(x_2(t), y_2(t)) + if^I(x_2(t), y_2(t))$, $f(v(t-\tau)) = f^R(x_2(t-\tau), y_2(t-\tau)) + if^I(x_2(t-\tau), y_2(t-\tau))$, $g(u(t)) = g^R(x_1(t), y_1(t)) + ig^I(x_1(t), y_1(t))$, $g(u(t-\tau)) = g^R(x_1(t-\tau), y_1(t-\tau)) + ig^I(x_1(t-\tau), y_1(t-\tau))$, $A_1 = A_1^R + iA_1^I$, $A_2 = A_2^R + iA_2^I$, $B_1 = B_1^R + iB_1^I$, $B_2 = B_2^R + iB_2^I$, $J_1 = J_1^R + iJ_1^I$, and $J_2 = J_2^R + iJ_2^I$. For the sake of convenience, set $x_1 = x_1(t)$, $y_1 = y_1(t)$, $x_2 = x_2(t)$, $y_2 = y_2(t)$, $x_1^\tau = x_1(t-\tau)$, $y_1^\tau = y_1(t-\tau)$, $x_2^\tau = x_2(t-\tau)$, and $y_2^\tau = y_2(t-\tau)$. Then, system (5.1) can be equally changed as

$$\begin{cases} \dot{x}_1 = -D_1x_1 + A_1^R f^R(x_2, y_2) - A_1^I f^I(x_2, y_2) + B_1^R f^R(x_2^\tau, y_2^\tau) \\ \qquad - B_1^I f^I(x_2^\tau, y_2^\tau) + J_1^R, \\ \dot{y}_1 = -D_1y_1 + A_1^I f^R(x_2, y_2) + A_1^R f^I(x_2, y_2) + B_1^I f^R(x_2^\tau, y_2^\tau) \\ \qquad + B_1^R f^I(x_2^\tau, y_2^\tau) + J_1^I, \\ \dot{x}_2 = -D_2x_2 + A_2^R g^R(x_1, y_1) - A_2^I g^I(x_1, y_1) + B_2^R g^R(x_1^\tau, y_1^\tau) \\ \qquad - B_2^I g^I(x_1^\tau, y_1^\tau) + J_2^R, \\ \dot{y}_2 = -D_2y_2 + A_2^I g^R(x_1, y_1) + A_2^R g^I(x_1, y_1) + B_2^I g^R(x_1^\tau, y_1^\tau) \\ \qquad + B_2^R g^I(x_1^\tau, y_1^\tau) + J_2^I. \end{cases} \tag{5.3}$$

The initial conditions associated with (5.3) are the following form:

$$\begin{cases} x_1(s) = \varphi^R(s),\ y_1(s) = \varphi^I(s), \\ x_2(s) = \psi^R(s),\ y_2(s) = \psi^I(s),\ s \in [-\tau, 0], \end{cases} \tag{5.4}$$

where $\varphi^R(s), \varphi^I(s) \in \mathbb{R}^n$ and $\psi^R(s), \psi^I(s) \in \mathbb{R}^m$.

Let

$$\omega_1 = \begin{bmatrix} x_1 \\ y_1 \end{bmatrix},\ \bar{J}_1 = \begin{bmatrix} J_1^R \\ J_1^I \end{bmatrix},\ \bar{D}_1 = \begin{bmatrix} D_1 & 0 \\ 0 & D_1 \end{bmatrix},\ \bar{f}(\omega_2) = \begin{bmatrix} f^R(x_2, y_2) \\ f^I(x_2, y_2) \end{bmatrix},$$
$$\bar{f}(\omega_2^\tau) = \begin{bmatrix} f^R(x_2^\tau, y_2^\tau) \\ f^I(x_2^\tau, y_2^\tau) \end{bmatrix},\ \bar{A}_1 = \begin{bmatrix} A_1^R & -A_1^I \\ A_1^I & A_1^R \end{bmatrix},\ \bar{B}_1 = \begin{bmatrix} B_1^R & -B_1^I \\ B_1^I & B_1^R \end{bmatrix},$$
$$\omega_2 = \begin{bmatrix} x_2 \\ y_2 \end{bmatrix},\ \bar{J}_2 = \begin{bmatrix} J_2^R \\ J_2^I \end{bmatrix},\ \bar{D}_2 = \begin{bmatrix} D_2 & 0 \\ 0 & D_2 \end{bmatrix},\ \bar{g}(\omega_1) = \begin{bmatrix} g^R(x_1, y_1) \\ g^I(x_1, y_1) \end{bmatrix},$$
$$\bar{g}(\omega_1^\tau) = \begin{bmatrix} g^R(x_1^\tau, y_1^\tau) \\ g^I(x_1^\tau, y_1^\tau) \end{bmatrix},\ \bar{A}_2 = \begin{bmatrix} A_2^R & -A_2^I \\ A_2^I & A_2^R \end{bmatrix},\ \bar{B}_2 = \begin{bmatrix} B_2^R & -B_2^I \\ B_2^I & B_2^R \end{bmatrix}.$$

Then, Eq. (5.3) can be rewritten as

$$\begin{cases} \dot{\omega}_1 = -\bar{D}_1\omega_1 + \bar{A}_1\bar{f}(\omega_2) + \bar{B}_1\bar{f}(\omega_2^\tau) + \bar{J}_1, \\ \dot{\omega}_2 = -\bar{D}_2\omega_2 + \bar{A}_2\bar{g}(\omega_1) + \bar{B}_2\bar{g}(\omega_1^\tau) + \bar{J}_2 \end{cases} \tag{5.5}$$

with

$$\omega_1(s) = \phi_1(s),\ \omega_2(s) = \phi_2(s),\ s \in [-\tau, 0], \tag{5.6}$$

where

$$\phi_1(s) = \begin{bmatrix} \varphi^R(s) \\ \varphi^I(s) \end{bmatrix} \in \mathbb{R}^{2n},\ \phi_2(s) = \begin{bmatrix} \psi^R(s) \\ \psi^I(s) \end{bmatrix} \in \mathbb{R}^{2m}.$$

Define

$$\omega = \begin{bmatrix} \omega_1 \\ \omega_2 \end{bmatrix}, \ \bar{J} = \begin{bmatrix} \bar{J}_1 \\ \bar{J}_2 \end{bmatrix}, \ \bar{D} = \begin{bmatrix} \bar{D}_1 & 0 \\ 0 & \bar{D}_2 \end{bmatrix}, \ \bar{F}(\omega) = \begin{bmatrix} \bar{f}(\omega_2) \\ \bar{g}(\omega_1) \end{bmatrix},$$

$$\bar{F}(\omega^\tau) = \begin{bmatrix} \bar{f}(\omega_2^\tau) \\ \bar{g}(\omega_1^\tau) \end{bmatrix}, \ \bar{A} = \begin{bmatrix} \bar{A}_1 & 0 \\ 0 & \bar{A}_2 \end{bmatrix}, \ \bar{B} = \begin{bmatrix} \bar{B}_1 & 0 \\ 0 & \bar{B}_2 \end{bmatrix}.$$

Equation (5.5) can be described as

$$\dot{\omega} = -\bar{D}\omega + \bar{A}\bar{F}(\omega) + \bar{B}\bar{F}(\omega^\tau) + \bar{J} \tag{5.7}$$

with the initial condition

$$\omega(s) = \phi(s) = \begin{bmatrix} \phi_1(s) \\ \phi_2(s) \end{bmatrix} \in \mathbb{R}^{2n+2m}. \tag{5.8}$$

On the other hand, it can be shown from (5.2) that for any positive matrices X and Y, the following inequalities hold:

$$\begin{aligned} &(f(v) - f(v'))^* X (f(v) - f(v')) \le (v - v')^* L X L (v - v'), \\ &(g(u) - g(u'))^* Y (g(u) - g(u')) \le (u - u')^* M Y M (u - u'), \end{aligned} \tag{5.9}$$

where $L = \text{diag}\{l_1, l_2, ..., l_m\}$ and $M = \text{diag}\{m_1, m_2, ..., m_n\}$. By separating real and imaginary parts, (5.9) can be changed as

$$\begin{aligned} &(\bar{f}(\omega_2) - \bar{f}(\omega_2'))^T \bar{X} (\bar{f}(\omega_2) - \bar{f}(\omega_2')) \le (\omega_2 - \omega_2')^T \bar{L} \bar{X} \bar{L} (\omega_2 - \omega_2'), \\ &(\bar{g}(\omega_1) - \bar{g}(\omega_1'))^T \bar{Y} (\bar{g}(\omega_1) - \bar{g}(\omega_1')) \le (\omega_1 - \omega_1')^T \bar{M} \bar{Y} \bar{M} (\omega_1 - \omega_1'), \end{aligned} \tag{5.10}$$

where

$$\bar{X} = \begin{bmatrix} X & 0 \\ 0 & X \end{bmatrix}, \ \bar{Y} = \begin{bmatrix} Y & 0 \\ 0 & Y \end{bmatrix}, \ \bar{L} = \begin{bmatrix} L & 0 \\ 0 & L \end{bmatrix}, \ \bar{M} = \begin{bmatrix} M & 0 \\ 0 & M \end{bmatrix}.$$

Then, (5.10) can be combined into one inequality as follows:

$$(\bar{F}(\omega) - \bar{F}(\omega'))^T \begin{bmatrix} \bar{X} & 0 \\ 0 & \bar{Y} \end{bmatrix} (\bar{F}(\omega) - \bar{F}(\omega')) \le (\omega - \omega')^T \mathcal{M} \begin{bmatrix} \bar{Y} & 0 \\ 0 & \bar{X} \end{bmatrix} \mathcal{M}(\omega - \omega') \tag{5.11}$$

with $\mathcal{M} = \begin{bmatrix} \bar{M} & 0 \\ 0 & \bar{L} \end{bmatrix}$.

Remark 5.1 In the most existing references for CVBAMNNs (Wang et al. 2016; Guo et al. 2017; Subramanian and Muthukumar 2018), the considered systems can be transformed into equivalent real-valued ones similar to (5.5) above by separating real and imaginary parts. Then, equivalent systems are used to study the dynamical behaviors. Here, we continue to combine equivalent real-valued system (5.5) into a tractable entirety (5.7). Meanwhile, Assumption 5.1 can be optimized into (5.11). It is obvious that the equilibrium point of (5.1) is also the equilibrium point of (5.7), and the equilibrium point of (5.1) is finite-time stable if and only if the equilibrium point of (5.7) is finite-time stable. Hence, in the sequel, we will make efforts on system (5.7) with the initial condition (5.8).

The following definitions and lemmas are necessary in the derivation of the main results.

Definition 5.1 (Rakkiyappan et al. 2014) The equilibrium point $\hat{\omega}$ of system (5.7) is said to be finite-time stable with respect to $\{\varepsilon_1, \varepsilon_2, T\}$, if

$$\|\phi - \hat{\omega}\| < \varepsilon_1$$

implies

$$\|\omega(t) - \hat{\omega}\| < \varepsilon_2, \ \forall t \in J,$$

where $J = [t_0, t_0 + T)$, t_0 is the initial time of observation, ε_1, ε_2, T are given positive numbers with $\|\omega(t) - \hat{\omega}\| = \sqrt{\sum_{k=1}^{2n}(\omega_{1k} - \hat{\omega}_{1k})^2 + \sum_{k=1}^{2m}(\omega_{2k} - \hat{\omega}_{2k})^2}$, $\|\phi - \hat{\omega}\| = \sup_{s\in[-\tau,0]}\sqrt{\sum_{k=1}^{2n}(\phi_{1k}(s) - \hat{\omega}_{1k})^2 + \sum_{k=1}^{2m}(\phi_{2k}(s) - \hat{\omega}_{2k})^2}$, and $\varepsilon_1 < \varepsilon_2$.

Definition 5.2 (Li and Cao 2006) Suppose that Ξ is an open set of $\mathbb{R}^n$, and $H : \Xi \to \mathbb{R}^n$ is an operator. The nonlinear measure of H on Ξ with the Euclidean norm $\|\cdot\|_2$ is given as follows:

$$\begin{aligned} m_\Xi(H) &\triangleq \sup_{x,y\in\Xi, x\neq y} \frac{\langle H(x) - H(y), x - y\rangle}{\|x - y\|_2^2} \\ &= \sup_{x,y\in\Xi, x\neq y} \frac{(x - y)^T(H(x) - H(y))}{\|x - y\|_2^2}. \end{aligned}$$

Lemma 5.1 (Li and Cao 2006) *If $m_\Xi(H) < 0$, then H is an injective mapping on Ξ. In addition, if $\Xi = \mathbb{R}^n$, then H is homeomorphic in $\mathbb{R}^n$.*

Lemma 5.2 (Zhang et al. 2014) *For any vectors $x, y \in \mathbb{R}^n$, positive definite matrix $P \in \mathbb{R}^{n\times n}$ and a constant $\varepsilon > 0$, the following inequality holds:*

$$2x^T y \le \varepsilon x^T P x + \frac{1}{\varepsilon} y^T P^{-1} y.$$

Remark 5.2 In this chapter, the main difficulty is how to establish the sufficient condition to ensure the existence and uniqueness of the equilibrium point for delayed CVBAMNNs and derive finite-time stability criterion of the equilibrium point of the considered system. Hence, in order to achieve the desired aims, a tractable entire system (5.7) is firstly formed by separating the real and imaginary parts. Then, by making full use of the nonlinear measure approach and Lyapunov function approach, the expected conditions will be obtained in the following sections.

5.3 Sufficient Criterion for the Existence and Uniqueness

In this section, by using the nonlinear measure approach, we give a sufficient condition to assure system (5.7) which has a unique equilibrium point.

Theorem 5.1 *Under the Assumption 5.1, if there exist positive definite diagonal matrices X_1, X_2, Y_1, Y_2, and a positive matrix P such that the following LMI hold:*

$$\Omega = \begin{bmatrix} -P\bar{D} - \bar{D}P & P\bar{A} & P\bar{B} & \mathcal{M}\mathcal{Y}_1 & \mathcal{M}\mathcal{Y}_2 \\ * & -\mathcal{X}_1 & 0 & 0 & 0 \\ * & * & -\mathcal{X}_2 & 0 & 0 \\ * & * & * & -\mathcal{Y}_1 & 0 \\ * & * & * & * & -\mathcal{Y}_2 \end{bmatrix} < 0, \tag{5.12}$$

$$\mathcal{Y}_k = \begin{bmatrix} \bar{Y}_k & 0 \\ 0 & \bar{X}_k \end{bmatrix}, \mathcal{X}_k = \begin{bmatrix} \bar{X}_k & 0 \\ 0 & \bar{Y}_k \end{bmatrix},$$
$$\bar{Y}_k = \begin{bmatrix} Y_k & 0 \\ 0 & Y_k \end{bmatrix}, \bar{X}_k = \begin{bmatrix} X_k & 0 \\ 0 & X_k \end{bmatrix}, \ (k = 1, 2),$$

then system (5.7) has a unique equilibrium point.

Proof Define the following operator $H : \mathbb{R}^{2n+2m} \to \mathbb{R}^{2n+2m}$:

$$H(\omega) = -P\bar{D}\omega + P\bar{A}\bar{F}(\omega) + P\bar{B}\bar{F}(\omega) + P\bar{J}. \tag{5.13}$$

Since the matrix $P > 0$, it can be concluded that the zero points of $H(\omega)$ are the same as the equilibrium points of system (5.7). Next, we will show that $m_{\mathbb{R}^{2n+2m}}(H) < 0$.

According to Definition 5.2, one has

$$m_{\mathbb{R}^{2n+2m}}(H) = \sup_{\omega,\omega' \in \mathbb{R}^{2n+2m}, \omega \neq \omega'} \frac{(\omega - \omega')^T (H(\omega) - H(\omega'))}{\|\omega - \omega'\|_2^2}. \tag{5.14}$$

By (5.11), (5.13) and Lemma 5.2, for diagonal matrices $X_1 > 0$, $X_2 > 0$, $Y_1 > 0$, and $Y_2 > 0$, we have

$$
\begin{aligned}
&2(\omega-\omega')^T(H(\omega)-H(\omega'))\\
&=2(\omega-\omega')^T(-P\bar{D}(\omega-\omega')+P\bar{A}(\bar{F}(\omega)-\bar{F}(\omega'))+P\bar{B}(\bar{F}(\omega)-\bar{F}(\omega')))\\
&\leq(\omega-\omega')^T(-P\bar{D}-\bar{D}P)(\omega-\omega')+(\omega-\omega')^TP\bar{A}\begin{bmatrix}\bar{X}_1 & 0\\ 0 & \bar{Y}_1\end{bmatrix}^{-1}\bar{A}^TP(\omega-\omega')\\
&+(\bar{F}(\omega)-\bar{F}(\omega'))^T\begin{bmatrix}\bar{X}_1 & 0\\ 0 & \bar{Y}_1\end{bmatrix}(\bar{F}(\omega)-\bar{F}(\omega'))\\
&+(\omega-\omega')^TP\bar{B}\begin{bmatrix}\bar{X}_2 & 0\\ 0 & \bar{Y}_2\end{bmatrix}^{-1}\bar{B}^TP(\omega-\omega')\\
&+(\bar{F}(\omega)-\bar{F}(\omega'))^T\begin{bmatrix}\bar{X}_2 & 0\\ 0 & \bar{Y}_2\end{bmatrix}(\bar{F}(\omega)-\bar{F}(\omega'))\\
&\leq(\omega-\omega')^T\Big(-P\bar{D}-\bar{D}P+P\bar{A}\begin{bmatrix}\bar{X}_1 & 0\\ 0 & \bar{Y}_1\end{bmatrix}^{-1}\bar{A}^TP+P\bar{B}\begin{bmatrix}\bar{X}_2 & 0\\ 0 & \bar{Y}_2\end{bmatrix}^{-1}\bar{B}^TP+\\
&\begin{bmatrix}\bar{M} & 0\\ 0 & \bar{L}\end{bmatrix}\begin{bmatrix}\bar{Y}_1 & 0\\ 0 & \bar{X}_1\end{bmatrix}\begin{bmatrix}\bar{M} & 0\\ 0 & \bar{L}\end{bmatrix}+\begin{bmatrix}\bar{M} & 0\\ 0 & \bar{L}\end{bmatrix}\begin{bmatrix}\bar{Y}_2 & 0\\ 0 & \bar{X}_2\end{bmatrix}\begin{bmatrix}\bar{M} & 0\\ 0 & \bar{L}\end{bmatrix}\Big)(\omega-\omega')\\
&=(\omega-\omega')^T\Psi(\omega-\omega').
\end{aligned}\tag{5.15}
$$

Using Schur Complement, it follows from (5.12) that $\Psi<0$. Thus, for $\omega\neq\omega'$, $(\omega-\omega')^T(H(\omega)-H(\omega'))<0$. By (5.14), one can get that $m_{\mathbb{R}^{2n+2m}}(H)<0$. Then, based on Lemma 5.1, $H(\omega)$ is a homeomorphism of $\mathbb{R}^{2n+2m}$. Hence, $H(\omega)$ has a unique zero point, i.e., system (5.7) has a unique equilibrium point. □

Suppose that system (5.7) has a unique equilibrium point $\hat{\omega}$, let $\tilde{\omega}=\omega-\hat{\omega}$, $\hat{\omega}$ can be shifted to the origin for the following system:

$$\dot{\tilde{\omega}}=-\bar{D}\tilde{\omega}+\bar{A}\tilde{F}(\tilde{\omega})+\bar{B}\tilde{F}(\tilde{\omega}^{\tau})\tag{5.16}$$

with the initial condition

$$\tilde{\omega}(s)=\tilde{\phi}(s)=\phi(s)-\hat{\omega}\in C([-\tau,0],R^{2n+2m}),\tag{5.17}$$

and $\tilde{F}(\tilde{\omega})=\bar{F}(\tilde{\omega}+\hat{\omega})-\bar{F}(\hat{\omega})$ and $\tilde{F}(\tilde{\omega}^{\tau})=\bar{F}(\tilde{\omega}^{\tau}+\hat{\omega})-\bar{F}(\hat{\omega})$.

In the following section, we discuss the finite-time stability problem for the equilibrium point $\hat{\omega}$ of system (5.7), i.e., the finite-time stability problem for the origin of system (5.16).

5.4 Finite-Time Stability Criterion

Theorem 5.2 *Under the Assumption 5.1, the equilibrium point $\hat{\omega}$ of system (5.7) is finite-time stable with respect to $\{\varepsilon_1,\varepsilon_2,T\}$ if there exist positive definite matrices Q, R, W, Z, and positive definite diagonal matrices X_3, X_4, Y_3, Y_4, for any given scalar $\gamma>0$, such that*

$$\Theta = \begin{bmatrix} \Theta_{11} & \frac{1}{\tau}Z & \Theta_{13} & \Theta_{14} & \mathcal{M}\mathcal{Y}_3 & 0 \\ * & -R-\frac{1}{\tau}Z & 0 & 0 & 0 & \mathcal{M}\mathcal{Y}_4 \\ * & * & \Theta_{33} & \tau\bar{A}^T Z\bar{B} & 0 & 0 \\ * & * & * & \Theta_{44} & 0 & 0 \\ * & * & * & * & -\mathcal{Y}_3 & 0 \\ * & * & * & * & * & -\mathcal{Y}_4 \end{bmatrix} < 0, \tag{5.18}$$

$$\Theta_{11} = -Q\bar{D} - \bar{D}Q + R - \frac{1}{\tau}Z + \tau\bar{D}^T Z\bar{D} - \gamma Q,$$

$$\Theta_{13} = Q\bar{A} - \tau\bar{D}Z\bar{A},\ \Theta_{14} = Q\bar{B} - \tau\bar{D}Z\bar{B},$$

$$\Theta_{33} = W - \mathcal{X}_3 + \tau\bar{A}^T Z\bar{A},\ \Theta_{44} = -W - \mathcal{X}_4 + \tau\bar{B}^T Z\bar{B},$$

$$\mathcal{Y}_k = \begin{bmatrix} \bar{Y}_k & 0 \\ 0 & \bar{X}_k \end{bmatrix},\ \mathcal{X}_k = \begin{bmatrix} \bar{X}_k & 0 \\ 0 & \bar{Y}_k \end{bmatrix},$$

$$\bar{Y}_k = \begin{bmatrix} Y_k & 0 \\ 0 & Y_k \end{bmatrix},\ \bar{X}_k = \begin{bmatrix} X_k & 0 \\ 0 & X_k \end{bmatrix},\ (k = 3, 4),$$

and

$$\sqrt{\frac{e^{\gamma T}\left(\lambda_{\max}(Q) + \tau(\lambda_{\max}(R) + \lambda_{\max}(W)\lambda_{\max}(\mathcal{M}^2)) + \tau^2\lambda_{\max}(Z)\right)}{\lambda_{\min}(Q)}} < \frac{\varepsilon_2}{\varepsilon_1}. \tag{5.19}$$

Proof Choose the following Lyapunov functional:

$$V(t) = \tilde{\omega}^T(t)Q\tilde{\omega}(t) + \int_{t-\tau}^{t} \tilde{\omega}^T(s)R\tilde{\omega}(s)ds + \int_{t-\tau}^{t} \tilde{F}^T(\tilde{\omega}(s))W\tilde{F}(\tilde{\omega}(s))ds$$
$$+ \int_{-\tau}^{0}\int_{t+\theta}^{t} \dot{\tilde{\omega}}^T(s)Z\dot{\tilde{\omega}}(s)dsd\theta. \tag{5.20}$$

The time derivative of $V(t)$ along the trajectory of (5.16) can be bounded as

$$\begin{aligned} \dot{V}(t) &= 2\tilde{\omega}^T Q\dot{\tilde{\omega}} + \tilde{\omega}^T R\tilde{\omega} - \tilde{\omega}^{\tau T} R\tilde{\omega}^{\tau} + \tilde{F}^T(\tilde{\omega})W\tilde{F}(\tilde{\omega}) - \tilde{F}^T(\tilde{\omega}^{\tau})W\tilde{F}(\tilde{\omega}^{\tau}) \\ &\quad + \tau\dot{\tilde{\omega}}^T Z\dot{\tilde{\omega}} - \int_{t-\tau}^{t} \dot{\tilde{\omega}}^T(s)Z\dot{\tilde{\omega}}(s)ds \\ &\le 2\tilde{\omega}^T Q(-\bar{D}\tilde{\omega} + \bar{A}\tilde{F}(\tilde{\omega}) + \bar{B}\tilde{F}(\tilde{\omega}^{\tau})) + \tilde{\omega}^T R\tilde{\omega} - \tilde{\omega}^{\tau T} R\tilde{\omega}^{\tau} \\ &\quad + \tilde{F}^T(\tilde{\omega})W\tilde{F}(\tilde{\omega}) - \tilde{F}^T(\tilde{\omega}^{\tau})W\tilde{F}(\tilde{\omega}^{\tau}) - \frac{1}{\tau}(\tilde{\omega} - \tilde{\omega}^{\tau})^T Z(\tilde{\omega} - \tilde{\omega}^{\tau}) \\ &\quad + \tau(-\bar{D}\tilde{\omega} + \bar{A}\tilde{F}(\tilde{\omega}) + \bar{B}\tilde{F}(\tilde{\omega}^{\tau}))^T Z(-\bar{D}\tilde{\omega} + \bar{A}\tilde{F}(\tilde{\omega}) + \bar{B}\tilde{F}(\tilde{\omega}^{\tau})). \end{aligned} \tag{5.21}$$

Notice that from (5.11), the following inequalities are correct,

$$\begin{aligned} &\tilde{\omega}^T \mathcal{M}\mathcal{Y}_3\mathcal{M}\tilde{\omega} - \tilde{F}^T(\tilde{\omega})\mathcal{X}_3\tilde{F}(\tilde{\omega}) \ge 0, \\ &\tilde{\omega}^{\tau T} \mathcal{M}\mathcal{Y}_4\mathcal{M}\tilde{\omega}^{\tau} - \tilde{F}^T(\tilde{\omega}^{\tau})\mathcal{X}_4\tilde{F}(\tilde{\omega}^{\tau}) \ge 0. \end{aligned} \tag{5.22}$$

Then, by combining (5.21) with (5.22), we have

$$\dot{V}(t) \leq \xi^T(t) \begin{bmatrix} \Theta_{11} + \mathcal{M}\mathcal{Y}_3\mathcal{M} + \gamma Q & \frac{1}{\tau}Z & \Theta_{13} & \Theta_{14} \\ * & -R - \frac{1}{\tau}Z + \mathcal{M}\mathcal{Y}_4\mathcal{M} & 0 & 0 \\ * & * & \Theta_{33} & \tau\bar{A}^T Z\bar{B} \\ * & * & * & \Theta_{44} \end{bmatrix} \xi(t)$$

$$= \xi^T(t)\left(\Phi + \begin{bmatrix} \gamma Q & 0 & 0 & 0 \\ * & 0 & 0 & 0 \\ * & * & 0 & 0 \\ * & * & * & 0 \end{bmatrix}\right)\xi(t)$$

$$= \xi^T(t)\Phi\xi(t) + \gamma\tilde{\omega}^T(t)Q\tilde{\omega}(t), \tag{5.23}$$

where $\xi(t) = \left[\tilde{\omega}^T \ \tilde{\omega}^{\tau T} \ \tilde{F}^T(\tilde{\omega}) \ \tilde{F}^T(\tilde{\omega}^\tau)\right]^T$.

On the other hand, by Schur Complement, $\Theta < 0$ if and only if $\Phi < 0$. Thus, one gets that

$$\dot{V}(t) < \gamma\tilde{\omega}^T(t)Q\tilde{\omega}(t) \leq \gamma V(t), \tag{5.24}$$

then, we have

$$\lambda_{min}(Q)\tilde{\omega}^T(t)\tilde{\omega}(t) \leq \tilde{\omega}^T(t)Q\tilde{\omega}(t) \leq V(t) \leq e^{\gamma t}V(0) \leq e^{\gamma T}V(0), \tag{5.25}$$

and

$$V(0) = \tilde{\omega}^T(0)Q\tilde{\omega}(0) + \int_{-\tau}^{0} \tilde{\omega}^T(s)R\tilde{\omega}(s)ds + \int_{-\tau}^{0} \tilde{F}^T(\tilde{\omega}(s))W\tilde{F}(\tilde{\omega}(s))ds$$
$$+ \int_{-\tau}^{0}\int_{\theta}^{0} \dot{\tilde{\omega}}^T(s)Z\dot{\tilde{\omega}}(s)dsd\theta$$
$$\leq \Big(\lambda_{max}(Q) + \tau(\lambda_{max}(R) + \lambda_{max}(W)\lambda_{max}(\mathcal{M}^2)) + \tau^2\lambda_{max}(Z)\Big)\|\tilde{\phi}\|^2. \tag{5.26}$$

Along with (5.25) and (5.26), one can obtain that

$$\tilde{\omega}^T(t)\tilde{\omega}(t)$$
$$\leq \frac{e^{\gamma T}\Big(\lambda_{max}(Q) + \tau(\lambda_{max}(R) + \lambda_{max}(W)\lambda_{max}(\mathcal{M}^2)) + \tau^2\lambda_{max}(Z)\Big)}{\lambda_{min}(Q)}\|\tilde{\phi}\|^2. \tag{5.27}$$

Obviously, when $\|\tilde{\phi}\| < \varepsilon_1$, if the condition (5.19) holds, then $\|\tilde{\omega}(t)\| < \varepsilon_2$, from Definition 5.1, the equilibrium point of system (5.7) is finite-time stable with respect to $\{\varepsilon_1, \varepsilon_2, T\}$. □

Remark 5.3 Recently, many results on the finite-time stability analysis for real-valued neural networks have been developed (Li and Cao 2017; Zhao et al. 2016; Chen et al. 2013; Wu et al. 2013; Rakkiyappan et al. 2014). However, for CVNNs, only finite-time stability problem of fractional-order complex-valued memristor-based neural networks with time delays was discussed in Rakkiyappan et al. (2014). Until now, for CVBAMNNs, this topic has not been addressed. In this chapter, we analyze it and propose a sufficient condition to ascertain the finite-time stability of CVBAMNNs with time delay.

Remark 5.4 It should be pointed out that some existing works often omit the procedure of proving the existence of the equilibrium point in dealing with the finite-time stability problem. For completeness, we preserve this part and involve the nonlinear measure approach which has been seldom used in complex-valued systems besides (Gong et al. 2016; Wei et al. 2017). Based on this approach, a tractable condition is derived, which builds a strong foundation on studying the finite-time stability problem of the considered system.

5.5 Illustrative Examples

In this section, we provide two examples to illustrate the effectiveness and advantages of our proposed results.

Example 5.1 Consider system (5.1) with

$$D_1 = \begin{bmatrix} 5 & 0 \\ 0 & 9 \end{bmatrix}, \; A_1 = \begin{bmatrix} 1+0.5i & 0.2+i \\ 2-2i & 1+0.3i \end{bmatrix}, \; B_1 = \begin{bmatrix} 3+2i & 0.2+2i \\ 4+i & 3+0.2i \end{bmatrix},$$

$$D_2 = \begin{bmatrix} 11 & 0 \\ 0 & 8 \end{bmatrix}, \; A_2 = \begin{bmatrix} 0.5+0.1i & 0.6+0.3i \\ 2+0.2i & 3-2i \end{bmatrix}, \; B_2 = \begin{bmatrix} 2+3i & -3+i \\ 0.6+0.4i & 2-2i \end{bmatrix},$$

$$f_j(v_j) = \frac{1-e^{-2x_{2j}-y_{2j}}}{1+e^{-2x_{2j}-y_{2j}}} + i\frac{1}{1+e^{-x_{2j}+2y_{2j}}}, \; v_j = x_{2j} + iy_{2j} (j = 1, 2),$$

$$g_k(u_k) = \frac{1}{1+e^{-x_{1k}+2y_{1k}}} + i\frac{1-e^{-2x_{1k}-y_{1k}}}{1+e^{-2x_{1k}-y_{1k}}}, \; u_k = x_{1k} + iy_{1k} (k = 1, 2),$$

$$J_1 = \begin{bmatrix} -1.5+i \\ 0.7-0.5i \end{bmatrix}, \; J_2 = \begin{bmatrix} -2.1+2.6i \\ 1.4+3.4i \end{bmatrix}.$$

Moreover, it is easy to obtain that

$$L = M = \begin{bmatrix} \sqrt{41}/4 & 0 \\ 0 & \sqrt{41}/4 \end{bmatrix}.$$

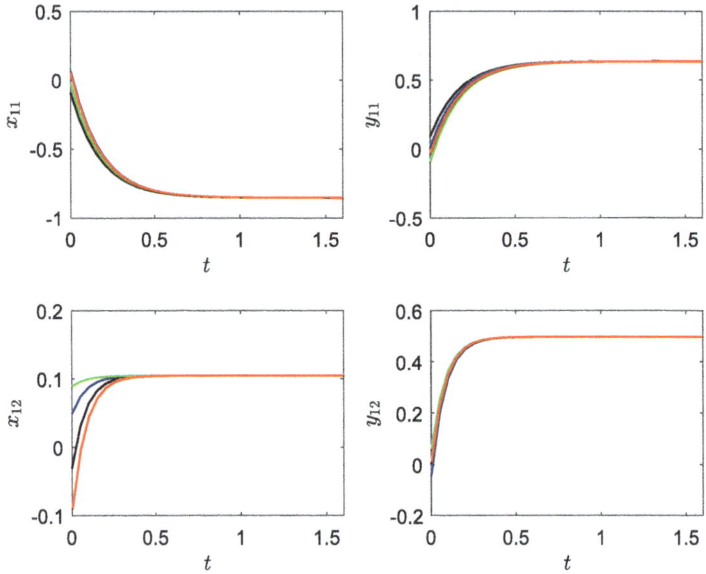

Fig. 5.1 Time responses of the real and imaginary parts of the state u in Example 5.1

By means of Theorem 5.1, we know that this system has a unique equilibrium point. If $\varepsilon_1 = 0.1$, $\varepsilon_2 = 1.5$, $t_0 = 0$ and $\gamma = 0.8$, and let $\tau = 4$, it can be verified that the LMI condition in Theorem 5.2 holds and the estimated time of finite-time stability is $T_e = 1.5396$. These show that the equilibrium point is finite-time stable for a given T less than 1.5396. Figures 5.1 and 5.2 show the time responses of the real and imaginary parts of the states u and v for this system under multiple initial conditions. Obviously, the validity of Theorems 5.1 and 5.2 are verified by these simulations.

Example 5.2 Consider system (5.1) with

$$D_1 = \begin{bmatrix} 9 & 0 \\ 0 & 8 \end{bmatrix}, \; A_1 = \begin{bmatrix} 1-i & -1+2i \\ 1+i & 2+3i \end{bmatrix}, \; B_1 = \begin{bmatrix} -1+2i & 2+2i \\ 1-2i & 4+2i \end{bmatrix},$$

$$D_2 = \begin{bmatrix} 7 & 0 \\ 0 & 6 \end{bmatrix}, \; A_2 = \begin{bmatrix} 3-i & 1+i \\ 2+2i & -2-i \end{bmatrix}, \; B_2 = \begin{bmatrix} -3-i & 2-3i \\ i & -2-i \end{bmatrix},$$

$$f_j(v_j) = \frac{|x_{2j}+1| - |x_{2j}-1|}{2} + i\frac{|y_{2j}+1| - |y_{2j}-1|}{2}, \; v_j = x_{2j} + iy_{2j},$$

$$g_k(u_k) = \frac{|x_{1k}|}{2} + i\frac{|y_{1k}|}{2}, \; u_k = x_{1k} + iy_{1k} (j,k = 1,2),$$

$$J_1 = \begin{bmatrix} 2-i \\ -4+i \end{bmatrix}, \; J_2 = \begin{bmatrix} 2-3i \\ -2+i \end{bmatrix}.$$

After a simple calculation, we have

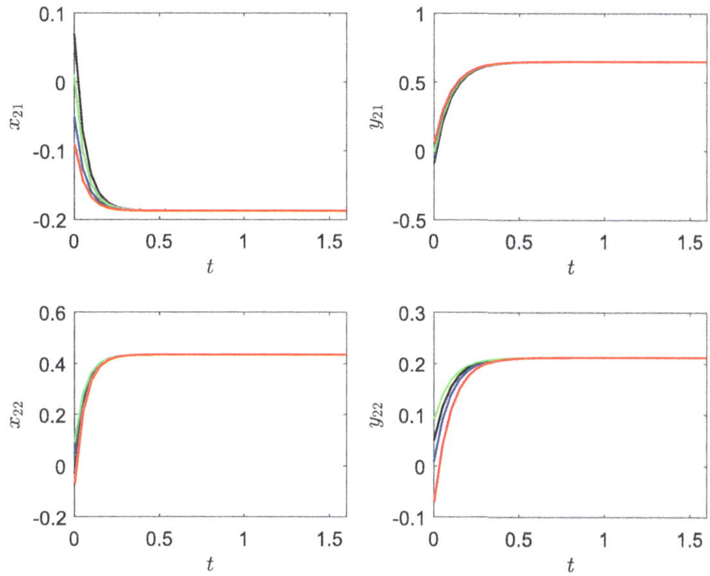

Fig. 5.2 Time responses of the real and imaginary parts of the state υ in Example 5.1

$$L = \begin{bmatrix} \sqrt{2} & 0 \\ 0 & \sqrt{2} \end{bmatrix}, \ M = \begin{bmatrix} \sqrt{\frac{1}{2}} & 0 \\ 0 & \sqrt{\frac{1}{2}} \end{bmatrix}.$$

From Theorem 5.1, it can be concluded that this system has a unique equilibrium point. If $\varepsilon_1 = 0.1, \varepsilon_2 = 2, t_0 = 0$ and $\gamma = 0.6$, and let $\tau = 4$, it can be verified that the LMI condition in Theorem 5.2 holds and the estimated time of finite-time stability is $T_e = 3.0093$. Thus, the equilibrium point is finite-time stable for given T less than 3.0093. Figures 5.3 and 5.4 show the time responses of the real and imaginary parts of the states u and υ for this system under multiple initial conditions, which further confirm the validity of Theorems 5.1 and 5.2.

Remark 5.5 In some existing results, such as the references (Hu and Wang 2012; Liu and Chen 2016; Rakkiyappan et al. 2015; Wang et al. 2017; Subramanian and Muthukumar 2018; Rakkiyappan et al. 2014), the partial derivatives of the real and imaginary parts of complex-valued activation functions are always assumed to exist. Obviously, in Example 5.2, $\frac{\partial}{\partial x_{2j}}(\frac{|x_{2j}+1|-|x_{2j}-1|}{2})|_{x_{2j}=\pm 1}$, $\frac{\partial}{\partial y_{2j}}(\frac{|y_{2j}+1|-|y_{2j}-1|}{2})|_{y_{2j}=\pm 1}$, $\frac{\partial}{\partial x_{1k}}(\frac{|x_{1k}|}{2})|_{x_{1k}=0}$ and $\frac{\partial}{\partial y_{1k}}(\frac{|y_{1k}|}{2})|_{y_{1k}=0}$ do not exist. Hence, the obtained results in the references (Hu and Wang 2012; Liu and Chen 2016; Rakkiyappan et al. 2015; Wang et al. 2017; Subramanian and Muthukumar 2018; Rakkiyappan et al. 2014) are not applicable to this kind of numerical example. But, here, we remove these constraints and the proposed results in this chapter can be verified by the system with such complex-valued activation functions. This shows that our work can serve a wider

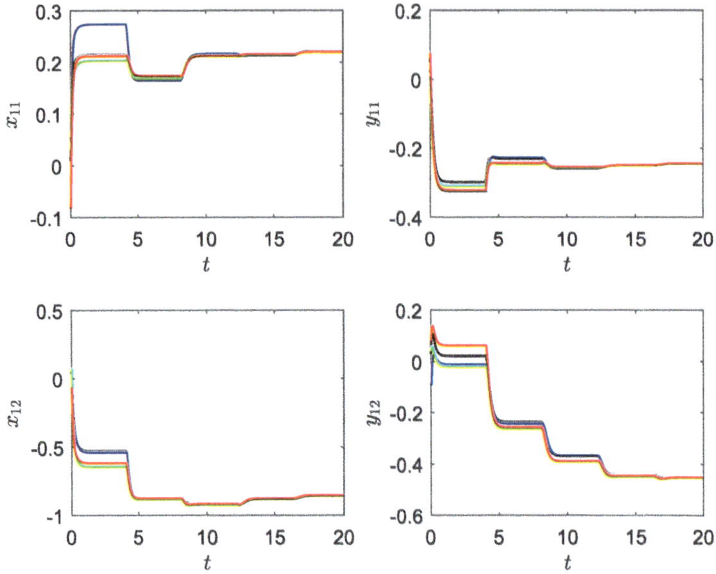

Fig. 5.3 Time responses of the real and imaginary parts of the state u in Example 5.2

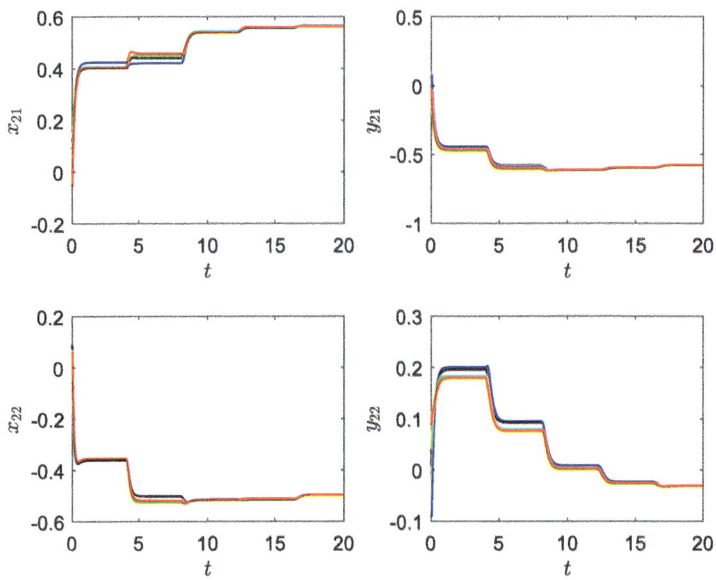

Fig. 5.4 Time responses of the real and imaginary parts of the state v in Example 5.2

class of systems than those in Hu and Wang (2012), Liu and Chen (2016), Rakkiyappan et al. (2015), Wang et al. (2017), Subramanian and Muthukumar (2018) and Rakkiyappan et al. (2014).

5.6 Conclusion

In this chapter, we have investigated the finite-time stability problem of CVBAMNNs with time delay. In order to handle this issue easily, we have formed an entire real-valued system by separating the real and imaginary parts. We have derived a condition to ensure the existence and uniqueness of the equilibrium point for this system and a finite-time stability criterion of the equilibrium point based on the nonlinear measure approach and Lyapunov function approach, respectively. The validity and advantages of the proposed results have been illustrated by two simulation examples.

References

Kosko B (1987) Adaptive bidirectional associative memories. Appl Opt 26(23):4947–4960

Kosko B (1988) Bidirectional associative memories. IEEE Trans Syst Man Cybern 18(1):49–60

Li H, Jiang H, Hu C (2016) Existence and global exponential stability of periodic solution of memristor-based BAM neural networks with time-varying delays. Neural Netw 75:97–109

Zhu Q, Cao J (2012) Stability analysis of Markovian jump stochastic BAM neural networks with impulse control and mixed time delays. IEEE Trans Neural Netw Learn Syst 23(3):467–479

Zhu Q, Rakkiyappan R, Chandrasekar A (2014) Stochastic stability of Markovian jump BAM neural networks with leakage delays and impulse control. Neurocomputing 136:136–151

Tian X, Xu R, Gan Q (2015) Hopf bifurcation analysis of a BAM neural network with multiple time delays and diffusion. Appl Math Comput 266:909–926

Sakthivel R, Vadivel P, Mathiyalagan K, Arunkumar A, Sivachitra M (2015) Design of state estimator for bidirectional associative memory neural networks with leakage delays. Inf Sci 296:263–274

Li Y, Li C (2016) Matrix measure strategies for stabilization and synchronization of delayed BAM neural networks. Nonlinear Dyn 84(3):1759–1770

Hu J, Wang J (2012) Global stability of complex-valued recurrent neural networks with time-delays. IEEE Trans Neural Netw Learn Syst 23(6):853–865

Liu X, Chen T (2016) Global exponential stability for complex-valued recurrent neural networks with asynchronous time delays. IEEE Trans Neural Netw Learn Syst 27(3):593–606

Rakkiyappan R, Cao J, Velmurugan G (2015) Existence and uniform stability analysis of fractional-order complex-valued neural networks with time delays. IEEE Trans Neural Netw Learn Syst 26(1):84–97

Zhang Z, Lin C, Chen B (2014) Global stability criterion for delayed complex-valued recurrent neural networks. IEEE Trans Neural Netw Learn Syst 25(9):1704–1708

Chen X, Zhao Z, Song Q, Hu J (2017) Multistability of complex-valued neural networks with time-varying delays. Appl Math Comput 294:18–35

Gong W, Liang J, Zhang C, Cao J (2016) Nonlinear measure approach for the stability analysis of complex-valued neural networks. Neural Process Lett 44(2):539–554

Wang H, Duan S, Huang T, Wang L, Li C (2017) Exponential stability of complex-valued memristive recurrent neural networks. IEEE Trans Neural Netw Learn Syst 28:766–771

Wei H, Li R, Chen C, Tu Z (2017) Stability analysis of fractional order complex-valued memristive neural networks with time delays. Neural Process Lett 45:379–399
Wang Z, Huang L, Liu Y (2016) Global stability analysis for delayed complex-valued BAM neural networks. Neurocomputing 173:2083–2089
Guo R, Zhang Z, Liu X, Lin C (2017) Existence, uniqueness, and exponential stability analysis for complex-valued memristor-based BAM neural networks with time delays. Appl Math Comput 311:100–117
Subramanian K, Muthukumar P (2018) Existence, uniqueness, and global asymptotic stability analysis for delayed complex-valued Cohen-Grossberg BAM neural networks. Neural Comput Applic 29:565–584
Kamenkov G (1953) On stability of motion over a finite interval of time. Akad Nauk SSSR Prikl Mat Meh 17:529–540
Yan Z, Zhang G, Wang J, Zhang W (2015) State and output feedback finite-time guaranteed cost control of linear Ito stochastic systems. J Syst Sci Complex 28:813–829
Li R, Cao J (2017) Finite-time stability analysis for Markovian jump memristive neural networks with partly unknown transition probabilities. IEEE Trans Neural Netw Learn Syst 28(12):2924–2935
Zhao H, Li L, Peng H, Xiao J, Yang Y (2016) Finite-time boundedness analysis of memristive neural network with time-varying delay. Neural Process Lett 44(3):665–679
Chen X, Huang L, Guo Z (2013) Finite time stability of periodic solution for Hopfield neural networks with discontinuous activations. Neurocomputing 103:43–49
Wu R, Hei X, Chen L (2013) Finite-time stability of fractional-order neural networks with delay. Commun Theor Phys 60:189–193
Rakkiyappan R, Velmurugan G, Cao J (2014) Finite-time stability analysis of fractional-order complex-valued memristor-based neural networks with time delays. Nonlinear Dyn 78(4):2823–2836
Wang L, Shen Y (2015) Finite-time stabilizability and instabilizability of delayed memristive neural networks with nonlinear discontinuous controller. IEEE Trans Neural Netw Learn Syst 26(11):2914–2924
Wang L, Shen Y, Zhang G (2017) Finite-time stabilization and adaptive control of memristor-based delayed neural networks. IEEE Trans Neural Netw Learn Syst 28(11):2648–2659
Wang L, Shen Y, Sheng Y (2016) Finite-time robust stabilization of uncertain delayed neural networks with discontinuous activations via delayed feedback control. Neural Netw 76:46–54
Liu X, Jiang N, Cao J, Wang S, Wang Z (2013) Finite-time stochastic stabilization for BAM neural networks with uncertainties. J Frankl Inst 350(8):2109–2123
Velmurugan G, Rakkiyappan R, Cao J (2016) Finite-time synchronization of fractional-order memristor-based neural networks with time delays. Neural Netw 73:36–46
Abdurahman A, Jiang H, Teng Z (2016) Finite-time synchronization for fuzzy cellular neural networks with time-varying delays. Fuzzy Sets Syst 297:96–111
Xiao J, Zhong S, Li Y, Xu F (2017) Finite-time Mittag-Leffler synchronization of fractional-order memristive BAM neural networks with time delays. Neurocomputing 219:431–439
Li P, Cao J (2006) Stability in static delayed neural networks: a nonlinear measure approach. Neurocomputing 69(13–15):1776–1781

Chapter 6
Lagrange Exponential Stability for CVBAMNNs with Time-Varying Delays

Abstract This chapter investigates the Lagrange exponential stability problem of complex-valued bidirectional associative memory neural networks (CVBAMNNs) with time-varying delays. Based on different assumption conditions for activation functions, by combining the Lyapunov function approach with some inequalities techniques, different sufficient criteria including algebraic conditions and the condition in terms of LMI are given to guarantee Lagrange exponential stability of the addressed system, respectively. Moreover, the estimations of different globally attractive sets named the convergence balls are also provided.

6.1 Introduction

Complex-valued recurrent neural networks (CVNNs) have become strongly desired because they have wider practical applications in practical systems dealing with image processing, antenna design, computer vision, remote sensing, sonic waves and radar imaging contrast to real-valued neural networks. As everyone knows, different from activation functions of real-valued neural networks, the choice of activation functions in studying complex-valued ones is the main challenge and great difficulty. When activation functions can be explicitly expressed by separating real and imaginary parts, some achievements have been provided for various dynamical behaviors of CVNNs (Hu and Wang 2012; Zhou and Song 2013; Zhang et al. 2014; Liu and Chen 2016; Wang et al. 2017; Zhang et al. 2018; Wang et al. 2017; Guo et al. 2018). When activation functions can't be explicitly expressed by separating real and imaginary parts, some results have been also developed for this class of system under the globally Lipschitz continuity condition in the complex domain required for activation functions (Fang and Sun 2014; Zhang et al. 2017; Song et al. 2016a, b; Zhang and Yu 2016).

Moreover, in the study of CVNNs, complex-valued bidirectional associative memory neural networks (CVBAMNNs) model also arouses the researchers' interests. For example, the stability and dissipativity problems for common delayed CVBAMNNs are discussed in Wang et al. (2016), Rajivganthi et al. (2019). The asymptotic stability for delayed Cohen-Grossberg CVBAMNNs and delayed neutral-type CVBAMNNs

Z. Zhang et al., *Complex-Valued Neural Networks Systems with Time Delay*,
Intelligent Control and Learning Systems 4,
https://doi.org/10.1007/978-981-19-5450-4_6

is analyzed in Subramanian and Muthukumar (2018) and Xu and Tan (2017), respectively. The exponential stability and exponential input-to-state stability for memristor-based CVBAMNNs with time delays are considered in Guo et al. (2017) and Guo et al. (2018), respectively.

It should be noted that in most existing results for neural networks the Lyapunov global stability is the main dynamical behavior feature to be desired, which shows that a global stable system in Lyapunov sense is a monostable system. But, monostable systems can lead to computational limitations in some practical applications, such as multiple target decision processing. To deal with such problems, it is essential to involve multistable dynamics. For example, when a neural network is used for pattern recognition or an associative memory, it is necessary to ensure the existence of multiple equilibrium points (Foss et al. 1996; Huang and Cao 2008). Therefore, an interesting dynamical behavior occurs, named as the Lagrange stability. Different from Lyapunov stability, Lagrange stability focuses on the stability of a total system and it involves the boundedness of solutions and the estimation of globally attractive set. As stated as in Liao et al. (2008), Wang et al. (2009), no equilibrium point, periodic state or chaos attractor exists outside the global attractive set. That is to say, once this set is determined, the bound of periodic states and chaotic attractors can be estimated, which greatly simplifies the dynamic behavior analysis of system. In the past few years, some researchers have always been devoted to this issue for neural networks and many results have been reported (Liao et al. 2008; Wang et al. 2009; Wu and Zeng 2014a, b; Wang et al. 2010; Jian and Wang 2015; Tu et al. 2016, 2011; Xiong et al. 2008), concerning Cohen-Grossberg neural networks, memristive neural networks, BAM neural networks, etc. Moreover, for CVNNs, Jian and Wan (2017) considers Lagrange α-exponential stability of fractional-order CVNNs and Song et al. (2017), Tu et al. (2016) study Lagrange stability of delayed CVNNs. However, to the best of our knowledge, for CVBAMNNs model, there is no information on Lagrange stability.

Meanwhile, considering the inevitability of time delays in practical applications and motivated by the above work, we will investigate the exponential stability in Lagrange sense for CVBAMNNs with time-varying delays in this chapter. Our main contributions lie in the following points: (1) It is the first time that the Lagrange exponential stability problem for CVBAMNNs with time-varying delays is studied; (2) Based on activation functions satisfying different assumption conditions, different sufficient criteria are established for the considered system, which containing algebraic conditions and the condition in terms of LMI, respectively; (3) As for different sufficient conditions, their effectiveness and superiority-inferiority are verified by illustrative examples.

6.2 Problem Formulation and Preliminaries

Consider the following CVBAMNNs with time-varying delays:

$$\begin{cases} \dot{u}_j(t) = -d_{1j}u_j(t) + \sum_{k=1}^{m} a_{jk} f_k(v_k(t)) + \sum_{k=1}^{m} b_{jk} f_k(v_k(t-\tau_{2k}(t))) + U_j(t), \\ \dot{v}_k(t) = -d_{2k}v_k(t) + \sum_{j=1}^{n} \tilde{a}_{kj} g_j(u_j(t)) + \sum_{j=1}^{n} \tilde{b}_{kj} g_j(u_j(t-\tau_{1j}(t))) + V_k(t), \end{cases} \tag{6.1}$$

or equivalently

$$\begin{cases} \dot{u}(t) = -D_1 u(t) + Af(v(t)) + Bf(v(t-\tau_2(t))) + U(t), \\ \dot{v}(t) = -D_2 v(t) + \tilde{A}g(u(t)) + \tilde{B}g(u(t-\tau_1(t))) + \mathcal{V}(t), \end{cases} \tag{6.2}$$

where $u(t) = (u_1(t), u_2(t), ..., u_n(t))^T \in \mathbb{C}^n$ and $v(t) = (v_1(t), v_2(t), ..., v_m(t))^T \in \mathbb{C}^m$ are the neuron state vectors; $D_1 = \text{diag}\{d_{11}, d_{12}, ..., d_{1n}\} \in \mathbb{R}^{n\times n}$ and $D_2 = \text{diag}\{d_{21}, d_{22}, ..., d_{2m}\} \in \mathbb{R}^{m\times m}$ with $D_1 > 0$, $D_2 > 0$ are the self-feedback connection weight matrix; $A = (a_{jk})_{n\times m} \in \mathbb{C}^{n\times m}$, $\tilde{A} = (\tilde{a}_{kj})_{m\times n} \in \mathbb{C}^{m\times n}$, and $B = (b_{jk})_{n\times m} \in \mathbb{C}^{n\times m}$, $\tilde{B} = (\tilde{b}_{kj})_{m\times n} \in \mathbb{C}^{m\times n}$ are the connection weight matrices without and with time delays, respectively; $f(v(t)) = (f_1(v_1(t)), f_2(v_2(t)), \ldots, f_m(v_m(t)))^T \in \mathbb{C}^m$, $f(v(t-\tau_2(t))) = (f_1(v_1(t-\tau_{21}(t))), f_2(v_2(t-\tau_{22}(t))), \ldots, f_m(v_m(t-\tau_{2m}(t)))) \in \mathbb{C}^m$, $g(u(t)) = (g_1(u_1(t)), g_2(u_2(t)), \ldots, g_n(u_n(t)))^T \in \mathbb{C}^n$, and $g(u(t-\tau_1(t))) = (g_1(u_1(t-\tau_{11}(t))), g_2(u_1(t-\tau_{12}(t))), \ldots, g_n(u_n(t-\tau_{1n}(t)))) \in \mathbb{C}^n$ are the complex-valued neuron activation functions; $U(t) = (U_1(t), U_2(t), \ldots, U_n(t)) \in \mathbb{C}^n$ and $\mathcal{V}(t) = (V_1(t), V_2(t), \ldots, V_m(t)) \in \mathbb{C}^m$ are the external input vectors. Moreover, time-varying delays $\tau_{1j}(t)$ and $\tau_{2k}(t)$ satisfy the inequalities below

$$\begin{aligned} &0 \le \tau_{1j}(t) \le \tau_1,\ \dot{\tau}_{1j}(t) \le \rho_1 < 1, \\ &0 \le \tau_{2k}(t) \le \tau_2,\ \dot{\tau}_{2k}(t) \le \rho_2 < 1. \end{aligned} \tag{6.3}$$

Moreover, to facilitate our research, we give the following assumptions.

Assumption 6.1 For $u_j = x_j + iy_j$ and $v_k = \tilde{x}_k + i\tilde{y}_k$, $f_k(.)$ and $g_j(.)$ can be separated into the real and imaginary parts as

$$\begin{aligned} f_k(v_k) &= f_k^R(\tilde{x}_k, \tilde{y}_k) + if_k^I(\tilde{x}_k, \tilde{y}_k), \\ g_j(u_j) &= g_j^R(x_j, y_j) + ig_j^I(x_j, y_j), \end{aligned}$$

and satisfy

$$\begin{cases} |f_k^\sigma(\tilde{x}_k, \tilde{x}_k) - f_k^\sigma(\tilde{x}_k', \tilde{y}_k')| \le \lambda_k^{\sigma R}|\tilde{x}_k - \tilde{x}_k'| + \lambda_k^{\sigma I}|\tilde{y}_k - \tilde{y}_k'|, \\ |g_j^\sigma(x_j, y_j) - g_j^\sigma(x_j', y_j')| \le \chi_j^{\sigma R}|x_j - x_j'| + \chi_j^{\sigma I}|y_j - y_j'|, \end{cases}$$

for $\sigma = R, I$.

Assumption 6.2 There exist $l_k > 0$ and $\mu_j > 0$ or positive diagonal matrices $L = \text{diag}\{l_1, l_2, \ldots, l_k\}$ and $K = \text{diag}\{\mu_1, \mu_2, \ldots, \mu_j\}$ such that

$$\begin{cases} |f_k(v_k) - f_k(v'_k)| \le l_k |v_k - v'_k|, \\ |g_j(u_j) - g_k(u'_j)| \le \mu_j |u_j - u'_j|. \end{cases}$$

Assumption 6.3 The inputs $U_j(t)$ and $V_k(t)$ can be separated into the real and imaginary parts as

$$U_j(t) = U_j^R(t) + iU_j^I(t), \ V_k(t) = V_k^R(t) + iV_k^I(t),$$

and satisfy $|U_j^R(t)| \le \tilde{U}_j^R$, $|U_j^I(t)| \le \tilde{U}_j^I$, $|V_k^R(t)| \le \tilde{V}_k^R$, $|V_k^I(t)| \le \tilde{V}_k^I$.

Assumption 6.4 For $U(t) \in \mathbb{C}^n$, $\mathcal{V}(t) \in \mathbb{C}^m$, there exist positive diagonal matrices $\tilde{U} = \text{diag}\{\tilde{U}_1, \tilde{U}_2, \ldots, \tilde{U}_n\}$ and $\tilde{V} = \text{diag}\{\tilde{V}_1, \tilde{V}_2, \ldots, \tilde{V}_m\}$ such that $|U_j(t)| \le \tilde{U}_j$, $|V_k(t)| \le \tilde{V}_k$.

Remark 6.1 In the most existing results, there are two types of common activation functions, which have good properties. One type is that they can be expressed by separating real and imaginary parts and satisfy Assumption 6.1. The other type is that they can't be expressed explicitly by real-imaginary parts but satisfy the Lipschitz continuity condition of Assumption 6.2. So as to show the obtained results more accurately and detailedly, in this chapter, based on different classes of activation functions satisfying corresponding assumptions, different criteria will be derived later.

Remark 6.2 In fact, it is easy to verify that Assumption 6.1 implies Assumption 6.2. Thus, for the real-imaginary separate-type activation functions, the approach similar to the other type with activation functions which can't be expressed explicitly by real-imaginary parts can also be used to study the dynamics of the considered system. That is, for this case, suitable theoretical result can be selected to apply to relative issue in different areas.

Remark 6.3 To deal with different cases, the input vectors $U(t)$ and $\mathcal{V}(t)$ satisfy different forms of conditions, i.e., Assumptions 6.3 and 6.4. In fact, they are equivalent. Here, we provide different forms which are to be used conveniently in the derivation process of main results.

Let $\omega(t) = \text{col}\{u(t), v(t)\}$, the continuous initial condition associated with system (6.1) or (6.2) is $\omega(s) = \vartheta(s)$, $\forall s \in [-\tau, 0]$, where $\tau = \max\{\tau_1, \tau_2\}$. Moreover, for $\omega = \text{col}\{\omega_1, \omega_2, ..., \omega_n\} \in \mathbb{C}^n$ with $\omega_k = \omega_k^R + i\omega_k^I$, $\|\omega\|$ denotes the norm of ω, defined by $\|\omega\| = \sum_{k=1}^n (|\omega_k^R| + |\omega_k^I|)$ or $\|\omega\| = \sqrt{\sum_{k=1}^n ((\omega_k^R)^2 + (\omega_k^I)^2)}$.

Definition 6.1 (Jian and Wan 2017) If for any $H > 0$, there exists a constant $K = K(H) > 0$ such that $\|\omega(t)\| < K$ for any $\vartheta \in \mathcal{C}_H = \{\vartheta \in \mathcal{C}([-\tau, 0], \mathbb{C}^{n+m}) | \|\vartheta\| \leq H\}$, $t \geq 0$, here, $\mathcal{C}([-\tau, 0], \mathbb{C}^{n+m})$ denotes the family of continuous functions from $[-\tau, 0]$ to $\mathbb{C}^{n+m}$, then system (6.1) or (6.2) is said to be uniformly stable in Lagrange sense.

Definition 6.2 (Jian and Wan 2017) If there exist positive constants β and ε, for any given positive number H there exists $K = K(H) > 0$ such that $\|\omega(t)\| < \beta + Ke^{-\varepsilon t}$ for any $t \geq 0$, $\vartheta \in \mathcal{C}_H$, then system (6.1) or (6.2) is said to be globally exponentially stable in Lagrange sense. Meanwhile, system (6.1) or (6.2) is also said to be globally uniformly exponentially convergent to the ball $B(\beta) = \{\omega(t) \in \mathbb{C}^{n+m} | \|\omega(t)\| \leq \beta\}$ with a rate ε.

Lemma 6.1 (Zhang et al. 2017) *For any vectors $X, Y \in \mathbb{C}^n$, a positive definite Hermitian matrix $P \in \mathbb{C}^{n\times n}$ and a positive constant ε, the following inequality holds:*

$$X^*Y + Y^*X \leq \varepsilon X^*PX + \frac{1}{\varepsilon}Y^*P^{-1}Y.$$

6.3 Stability Criteria Based on Algebraic Structure

In this section, we will derive sufficient criteria based on algebraic structure for the Lagrange exponential stability of system (6.1) or (6.2) with (6.3).

6.3.1 Stability Criterion Dependent on Separable Method

In this subsection, we will consider real-imaginary parts separable case and propose a sufficient criterion based on algebraic structure.

Theorem 6.1 *Assume that Assumptions 6.1 and 6.3 hold, system (6.1) or (6.2) with (6.3) is globally exponentially stable in Lagrange sense, if there exist positive constants α_j, β_j, $\tilde{\alpha}_k$, $\tilde{\beta}_k$ such that the following inequalities hold*

$$\kappa_{1j} = d_{1j} - \sum_{k=1}^{m} \frac{\tilde{\alpha}_k}{\alpha_j}(|\tilde{a}_{kj}^R|\chi_j^{RR} + |\tilde{a}_{kj}^I|\chi_j^{IR}) - \sum_{k=1}^{m} \frac{\tilde{\beta}_k}{\alpha_j}(|\tilde{a}_{kj}^R|\chi_j^{IR} + |\tilde{a}_{kj}^I|\chi_j^{RR}) -$$
$$\sum_{k=1}^{m} \frac{1}{1-\rho_1}\frac{\tilde{\alpha}_k}{\alpha_j}(|\tilde{b}_{kj}^R|\chi_j^{RR} + |\tilde{b}_{kj}^I|\chi_j^{IR}) - \sum_{k=1}^{m} \frac{1}{1-\rho_1}\frac{\tilde{\beta}_k}{\alpha_j}(|\tilde{b}_{kj}^R|\chi_j^{IR} + |\tilde{b}_{kj}^I|\chi_j^{RR}) > 0,$$
$$\kappa_{2j} = d_{1j} - \sum_{k=1}^{m} \frac{\tilde{\alpha}_k}{\beta_j}(|\tilde{a}_{kj}^R|\chi_j^{RI} + |\tilde{a}_{kj}^I|\chi_j^{II}) - \sum_{k=1}^{m} \frac{\tilde{\beta}_k}{\beta_j}(|\tilde{a}_{kj}^R|\chi_j^{II} + |\tilde{a}_{kj}^I|\chi_j^{RI}) -$$
$$\sum_{k=1}^{m} \frac{1}{1-\rho_1}\frac{\tilde{\alpha}_k}{\beta_j}(|\tilde{b}_{kj}^R|\chi_j^{RI} + |\tilde{b}_{kj}^I|\chi_j^{II}) - \sum_{k=1}^{m} \frac{1}{1-\rho_1}\frac{\tilde{\beta}_k}{\beta_j}(|\tilde{b}_{kj}^I|\chi_j^{RI} + |\tilde{b}_{kj}^R|\chi_j^{II}) > 0,$$

$$\kappa_{3k} = d_{2k} - \sum_{j=1}^{n} \frac{\alpha_j}{\tilde{\alpha}_k} (|a_{jk}^R|\lambda_k^{RR} + |a_{jk}^I|\lambda_k^{IR}) - \sum_{j=1}^{n} \frac{\alpha_j}{\tilde{\alpha}_k} (|a_{jk}^R|\lambda_k^{IR} + |a_{jk}^I|\lambda_k^{RR}) -$$

$$\sum_{j=1}^{n} \frac{1}{1-\rho_2} \frac{\alpha_j}{\tilde{\alpha}_k} (|b_{jk}^R|\lambda_k^{RR} + |b_{jk}^I|\lambda_k^{IR}) - \sum_{j=1}^{n} \frac{1}{1-\rho_2} \frac{\beta_j}{\tilde{\alpha}_k} (|b_{jk}^R|\lambda_k^{IR} + |b_{jk}^I|\lambda_k^{RR}) > 0,$$

$$\kappa_{4k} = d_{2k} - \sum_{j=1}^{n} \frac{\alpha_j}{\tilde{\beta}_k} (|a_{jk}^R|\lambda_k^{RI} + |a_{jk}^I|\lambda_k^{II}) - \sum_{j=1}^{n} \frac{\alpha_j}{\tilde{\beta}_k} (|a_{jk}^R|\lambda_k^{II} + |a_{jk}^I|\lambda_k^{RI}) -$$

$$\sum_{j=1}^{n} \frac{1}{1-\rho_2} \frac{\alpha_j}{\tilde{\beta}_k} (|b_{jk}^R|\lambda_k^{RI} + |b_{jk}^I|\lambda_k^{II}) - \sum_{j=1}^{n} \frac{1}{1-\rho_2} \frac{\beta_j}{\tilde{\beta}_k} (|b_{jk}^R|\lambda_k^{II} + |b_{jk}^I|\lambda_k^{RI}) > 0, \quad (6.4)$$

where $a_{jk}^R = \mathrm{Re}(a_{jk})$, $a_{jk}^I = \mathrm{Im}(a_{jk})$, $b_{jk}^R = \mathrm{Re}(b_{jk})$, $b_{jk}^I = \mathrm{Im}(b_{jk})$, $\tilde{a}_{kj}^R = \mathrm{Re}(\tilde{a}_{kj})$, $\tilde{a}_{kj}^I = \mathrm{Im}(\tilde{a}_{kj})$, *and* $\tilde{b}_{kj}^R = \mathrm{Re}(\tilde{b}_{kj})$, $\tilde{b}_{kj}^I = \mathrm{Im}(\tilde{b}_{kj})$. *Moreover, system (6.1) or (6.2) globally exponentially converges to the ball*

$$\Omega_1 = \Big\{u(t) = x(t) + y(t) \in \mathbb{C}^n, v(t) = \tilde{x}(t) + \tilde{y}(t) \in \mathbb{C}^m \;\Big|\; \sum_{j=1}^{n}(|x_j| + |y_j|) + \sum_{k=1}^{m}(|\tilde{x}_k| + |\tilde{y}_k|) \le \frac{\varphi}{\delta \varrho}\Big\}, \quad (6.5)$$

where

$$\varrho = \min_{1\le j\le n, 1\le k\le m}\{\alpha_j, \beta_j, \tilde{\alpha}_k, \tilde{\beta}_k\},$$

$$\varphi = \sum_{j=1}^{n}\sum_{k=1}^{m}\Big((\alpha_j(|a_{jk}^R| + |b_{jk}^R|) + \beta_j(|a_{jk}^I| + |b_{jk}^I|))f_k^R(0,0) + (\alpha_j(|a_{jk}^I| +$$

$$|b_{jk}^I|) + \beta_j(|a_{jk}^R| + |b_{jk}^R|))f_k^I(0,0)\Big) + \sum_{k=1}^{m}\sum_{j=1}^{n}\Big((\tilde{\alpha}_k(|\tilde{a}_{kj}^R| + |\tilde{b}_{kj}^R|) + \tilde{\beta}_k(|\tilde{a}_{kj}^I|$$

$$+|\tilde{b}_{kj}^I|))g_j^R(0,0) + (\tilde{\alpha}_k(|\tilde{a}_{kj}^I| + |\tilde{b}_{kj}^I|) + \tilde{\beta}_k(|\tilde{a}_{kj}^R| + |\tilde{b}_{kj}^R|))g_j^I(0,0)\Big)$$

$$+\sum_{j=1}^{n}\alpha_j\tilde{U}_j^R + \sum_{j=1}^{n}\beta_j\tilde{U}_j^I + \sum_{k=1}^{m}\tilde{\alpha}_k\tilde{V}_k^R + \sum_{k=1}^{m}\tilde{\beta}_k\tilde{V}_k^I,$$

and δ *will be determined later.*

Proof Let $u_j(t) = x_j(t) + iy_j(t)$, $v_k(t) = \tilde{x}_k(t) + i\tilde{y}_k(t)$. To simplify, we denote $x_j(t)$, $y_j(t)$, $\tilde{x}_k(t)$, $\tilde{y}_k(t)$, $x_j(t-\tau_{1j}(t))$, $y_j(t-\tau_{1j}(t))$, $\tilde{x}_k(t-\tau_{2k}(t))$, []$\tilde{y}_k(t-\tau_{2k}(t))$, $U_j^R(t)$, $U_j^I(t)$, $V_k^R(t)$, $V_k^I(t)$ as x_j, y_j, $\tilde{x}_k$, $\tilde{y}_k$, $x_j^{\tau_{1j}}$, $y_j^{\tau_{1j}}$, $\tilde{x}_k^{\tau_{2k}}$, $\tilde{y}_k^{\tau_{2k}}$, U_j^R, U_j^I, V_k^R and V_k^I, respectively. Then, we separate system (6.1) into the real and imaginary parts as

$$
\begin{aligned}
\dot{x}_j &= -d_{1j}x_j + \sum_{k=1}^{m} a_{jk}^R f_k^R(\tilde{x}_k, \tilde{y}_k) - \sum_{k=1}^{m} a_{jk}^I f_k^I(\tilde{x}_k, \tilde{y}_k) \\
&\quad + \sum_{k=1}^{m} b_{jk}^R f_k^R(\tilde{x}_k^{\tau_{2k}}, \tilde{y}_k^{\tau_{2k}}) - \sum_{k=1}^{m} b_{jk}^I f_k^I(\tilde{x}_k^{\tau_{2k}}, \tilde{y}_k^{\tau_{2k}}) + U_j^R, \\
\dot{y}_j &= -d_{1j}y_j + \sum_{k=1}^{m} a_{jk}^I f_k^R(\tilde{x}_k, \tilde{y}_k) + \sum_{k=1}^{m} a_{jk}^R f_k^I(\tilde{x}_k, \tilde{y}_k) \\
&\quad + \sum_{k=1}^{m} b_{jk}^I f_k^R(\tilde{x}_k^{\tau_{2k}}, \tilde{y}_k^{\tau_{2k}}) + \sum_{k=1}^{m} b_{jk}^R f_k^I(\tilde{x}_k^{\tau_{2k}}, \tilde{y}_k^{\tau_{2k}}) + U_j^I, \\
\dot{\tilde{x}}_k &= -d_{2k}\tilde{x}_k + \sum_{j=1}^{n} \tilde{a}_{kj}^R g_j^R(x_j, y_j) - \sum_{j=1}^{n} \tilde{a}_{kj}^I g_j^I(x_j, y_j) \\
&\quad + \sum_{j=1}^{n} \tilde{b}_{kj}^R g_j^R(x_j^{\tau_{1j}}, y_j^{\tau_{1j}}) - \sum_{j=1}^{n} \tilde{b}_{kj}^I g_j^I(x_j^{\tau_{1j}}, y_j^{\tau_{1j}}) + V_k^R, \\
\dot{\tilde{y}}_k &= -d_{2k}\tilde{y}_k + \sum_{j=1}^{n} \tilde{a}_{kj}^I (g_j^R(x_j, y_j) + \sum_{j=1}^{n} \tilde{a}_{kj}^R (g_j^I(x_j, y_j) \\
&\quad + \sum_{j=1}^{n} \tilde{b}_{kj}^I g_j^R(x_j^{\tau_{1j}}, y_j^{\tau_{1j}}) + \sum_{j=1}^{n} \tilde{b}_{kj}^I g_j^I(x_j^{\tau_{1j}}, y_j^{\tau_{1j}}) + V_k^I.
\end{aligned}
\tag{6.6}
$$

Construct a candidate Lyapunov functional as follows

$$
\begin{aligned}
V(t) &= \sum_{j=1}^{n} \alpha_j |x_j(t)| + \sum_{j=1}^{n} \beta_j |y_j(t)| + \sum_{k=1}^{m} \tilde{\alpha}_k |\tilde{x}_k(t)| + \sum_{k=1}^{m} \tilde{\beta}_k |\tilde{y}_k(t)| \\
&+ \sum_{j=1}^{n} \frac{1}{1-\rho_1}\theta_j \int_{t-\tau_{1j}(t)}^{t} |x_j(s)|ds + \sum_{j=1}^{n} \frac{1}{1-\rho_1}\phi_j \int_{t-\tau_{1j}(t)}^{t} |y_j(s)|ds \\
&+ \sum_{k=1}^{m} \frac{1}{1-\rho_2}\tilde{\theta}_k \int_{t-\tau_{2k}(t)}^{t} |\tilde{x}_k(s)|ds + \sum_{k=1}^{m} \frac{1}{1-\rho_2}\tilde{\phi}_k \int_{t-\tau_{2k}(t)}^{t} |\tilde{y}_k(s)|ds,
\end{aligned}
\tag{6.7}
$$

where

$$\theta_j = \sum_{k=1}^{m} \tilde{\alpha}_k(|\tilde{b}_{kj}^R|\chi_j^{RR} + |\tilde{b}_{kj}^I|\chi_j^{IR}) + \sum_{k=1}^{n} \tilde{\beta}_k(|\tilde{b}_{kj}^I|\chi_j^{RR} + |\tilde{b}_{kj}^R|\chi_j^{IR}),$$

$$\phi_j = \sum_{k=1}^{m} \tilde{\alpha}_k(|\tilde{b}_{kj}^R|\chi_j^{RI} + |\tilde{b}_{kj}^I|\chi_j^{II}) + \sum_{k=1}^{n} \tilde{\beta}_k(|\tilde{b}_{kj}^I|\chi_j^{RI} + |\tilde{b}_{kj}^R|\chi_j^{II}),$$

$$\tilde{\theta}_k = \sum_{j=1}^{n} \alpha_j(|b_{jk}^R|\lambda_k^{RR} + |b_{jk}^I|\lambda_k^{IR}) + \sum_{j=1}^{n} \beta_j(|b_{jk}^I|\lambda_k^{RR} + |b_{jk}^R|\lambda_k^{IR}),$$

$$\tilde{\phi}_k = \sum_{j=1}^{n} \alpha_j(|b_{jk}^R|\lambda_k^{RI} + |b_{jk}^I|\lambda_k^{II}) + \sum_{j=1}^{n} \beta_j(|b_{jk}^I|\lambda_k^{RI} + |b_{jk}^R|\lambda_k^{II}).$$

Computing the upper Dini-derivative of $V(t)$ along the solution trajectories of system (6.6) with (6.3), we have

$$\begin{aligned}
D^+V(t) &= \sum_{j=1}^{n} \alpha_j(\mathrm{sgn} x_j)\dot{x}_j + \sum_{j=1}^{n} \frac{1}{1-\rho_1}\theta_j(|x_j| - (1-\dot{\tau}_{1j}(t))|x_j^{\tau_{1j}}|) \\
&+ \sum_{j=1}^{n} \beta_j(\mathrm{sgn} y_j)\dot{y}_j + \sum_{j=1}^{n} \frac{1}{1-\rho_1}\phi_j(|y_j| - (1-\dot{\tau}_{1j}(t))|y_j^{\tau_{1j}}|) \\
&+ \sum_{k=1}^{m} \tilde{\alpha}_k(\mathrm{sgn}\tilde{x}_k)\dot{\tilde{x}}_k + \sum_{k=1}^{m} \frac{1}{1-\rho_2}\tilde{\theta}_k(|\tilde{x}_k| - (1-\dot{\tau}_{2k}(t))|\tilde{x}_k^{\tau_{2k}}|) \\
&+ \sum_{k=1}^{m} \tilde{\beta}_k(\mathrm{sgn}\tilde{y}_k)\dot{\tilde{y}}_k + \sum_{k=1}^{m} \frac{1}{1-\rho_2}\tilde{\phi}_k(|\tilde{y}_k| - (1-\dot{\tau}_{2k}(t))|\tilde{y}_k^{\tau_{2k}}|) \\
&\le \sum_{j=1}^{n} \alpha_j\Big(- d_{1j}|x_j| + \sum_{k=1}^{m} |a_{jk}^R|(\lambda_k^{RR}|\tilde{x}_k| + \lambda_k^{RI}|\tilde{y}_k|) + \sum_{k=1}^{m} |a_{jk}^I|(\lambda_k^{IR}|\tilde{x}_k| \\
&+ \lambda_k^{II}|\tilde{y}_k|) + \sum_{k=1}^{m} |b_{jk}^R|(\lambda_k^{RR}|\tilde{x}_k^{\tau_{2k}}| + \lambda_k^{RI}|\tilde{y}_k^{\tau_{2k}}|) + \sum_{k=1}^{m} |b_{jk}^I|(\lambda_k^{IR}|\tilde{x}_k^{\tau_{2k}}| \\
&+ \lambda_k^{II}|\tilde{y}_k^{\tau_{2k}}|)\Big) + \sum_{j=1}^{n} \beta_j\Big(- d_{1j}|y_j| + \sum_{k=1}^{m} |a_{jk}^R|(\lambda_k^{IR}|\tilde{x}_k| + \lambda_k^{II}|\tilde{y}_k|) \\
&+ \sum_{k=1}^{m} |a_{jk}^I|(\lambda_k^{RR}|\tilde{x}_k| + \lambda_k^{RI}|\tilde{y}_k|) + \sum_{k=1}^{m} |b_{jk}^R|(\lambda_k^{IR}|\tilde{x}_k^{\tau_{2k}}| + \lambda_k^{II}|\tilde{y}_k^{\tau_{2k}}|) \\
&+ \sum_{k=1}^{m} |b_{jk}^I|(\lambda_k^{RR}|\tilde{x}_k^{\tau_{2k}}| + \lambda_k^{RI}|\tilde{y}_k|^{\tau_{2k}})\Big) + \sum_{k=1}^{m} \tilde{\alpha}_k\Big(- d_{2k}|\tilde{x}_k|
\end{aligned}$$

$$+\sum_{j=1}^{n}|\tilde{a}_{kj}^{R}|(\chi_j^{RR}|x_j|+\chi_j^{RI}|y_j|)+\sum_{j=1}^{n}|\tilde{a}_{kj}^{I}|(\chi_j^{IR}|x_j|+\chi_j^{II}|y_j|)$$

$$+\sum_{j=1}^{n}|\tilde{b}_{kj}^{R}|(\chi_j^{RR}|x_j^{\tau_{1j}}|+\chi_j^{RI}|y_j^{\tau_{1j}}|)+\sum_{j=1}^{n}|\tilde{b}_{kj}^{I}|(\chi_j^{IR}|x_j^{\tau_{1j}}|+\chi_j^{II}|y_j^{\tau_{1j}}|)\Big)$$

$$+\sum_{k=1}^{m}\tilde{\beta}_k\Big(-d_{2k}|\tilde{x}_k|+\sum_{j=1}^{n}|\tilde{a}_{kj}^{R}|(\chi_j^{IR}|x_j|+\chi_j^{II}|y_j|)+\sum_{j=1}^{n}|\tilde{a}_{kj}^{I}|(\chi_j^{RR}|x_j|$$

$$+\chi_j^{RI}|y_j|)+\sum_{j=1}^{n}|\tilde{b}_{kj}^{R}|(\chi_j^{IR}|x_j^{\tau_{1j}}|+\chi_j^{II}|y_j^{\tau_{1j}}|)+\sum_{j=1}^{n}|\tilde{b}_{kj}^{I}|(\chi_j^{RR}|x_j^{\tau_{1j}}|$$

$$+\chi_j^{RI}|y_j^{\tau_{1j}}|)\Big)+\sum_{j=1}^{n}\sum_{k=1}^{m}\frac{1}{1-\rho_1}\Big(\tilde{\alpha}_k(|\tilde{b}_{kj}^{R}|\chi_j^{RR}+|\tilde{b}_{kj}^{I}|\chi_j^{IR})+\tilde{\beta}_k(|\tilde{b}_{kj}^{I}|\chi_j^{RR}$$

$$+|\tilde{b}_{kj}^{R}|\chi_j^{IR})\Big)(|x_j|-(1-\rho_1)|x_j^{\tau_{1j}}|)+\sum_{j=1}^{n}\sum_{k=1}^{m}\frac{1}{1-\rho_1}\Big(\tilde{\alpha}_k(|\tilde{b}_{kj}^{R}|\chi_j^{RI}$$

$$+|\tilde{b}_{kj}^{I}|\chi_j^{II})+\tilde{\beta}_k(|\tilde{b}_{kj}^{I}|\chi_j^{RI}+|\tilde{b}_{kj}^{R}|\chi_j^{II})\Big)(|y_j|-(1-\rho_1)|y_j^{\tau_{1j}}|)$$

$$+\sum_{k=1}^{m}\sum_{j=1}^{n}\frac{1}{1-\rho_2}\Big(\alpha_j(|b_{jk}^{R}|\lambda_k^{RR}+|b_{jk}^{I}|\lambda_k^{IR})+\beta_j(|b_{jk}^{I}|\lambda_k^{RR}+|b_{jk}^{R}|\lambda_k^{IR})\Big)$$

$$\times(|\tilde{x}_k|-(1-\rho_2)|\tilde{x}_k^{\tau_{2k}}|)+\sum_{k=1}^{m}\sum_{j=1}^{n}\frac{1}{1-\rho_2}\Big(\alpha_j(|b_{jk}^{R}|\lambda_k^{RI}+|b_{jk}^{I}|\lambda_k^{II})$$

$$+\beta_j(|b_{jk}^{I}|\lambda_k^{RI}+|b_{jk}^{R}|\lambda_k^{II})\Big)(|\tilde{y}_k|-(1-\rho_2)|\tilde{y}_k^{\tau_{2k}}|)+\varphi. \tag{6.8}$$

Via exchanging the order of the sum, we can get

$$D^+V(t)\le-\sum_{j=1}^{n}\alpha_j\Big[d_{1j}-\sum_{k=1}^{m}\frac{\tilde{\alpha}_k}{\alpha_j}(|\tilde{a}_{kj}^{R}|\chi_j^{RR}+|\tilde{a}_{kj}^{I}|\chi_j^{IR})-\sum_{k=1}^{m}\frac{\tilde{\beta}_k}{\alpha_j}(|\tilde{a}_{kj}^{R}|\chi_j^{IR}$$

$$+|\tilde{a}_{kj}^{I}|\chi_j^{RR})-\sum_{k=1}^{m}\frac{1}{1-\rho_1}\frac{\tilde{\alpha}_k}{\alpha_j}(|\tilde{b}_{kj}^{R}|\chi_j^{RR}+|\tilde{b}_{kj}^{I}|\chi_j^{IR})$$

$$-\sum_{k=1}^{m}\frac{1}{1-\rho_1}\frac{\tilde{\beta}_k}{\alpha_j}(|\tilde{b}_{kj}^{R}|\chi_j^{IR}+|\tilde{b}_{kj}^{I}|\chi_j^{RR})\Big]|x_j|-\sum_{j=1}^{n}\beta_j\Big[d_{1j}$$

$$-\sum_{k=1}^{m}\frac{\tilde{\alpha}_k}{\beta_j}(|\tilde{a}_{kj}^{R}|\chi_j^{RI}+|\tilde{a}_{kj}^{I}|\chi_j^{II})-\sum_{k=1}^{m}\frac{\tilde{\beta}_k}{\beta_j}(|\tilde{a}_{kj}^{R}|\chi_j^{II}+\tilde{a}_{kj}^{I}\chi_j^{RI})$$

$$-\sum_{k=1}^{m}\frac{1}{1-\rho_1}\frac{\tilde{\alpha}_k}{\beta_j}(|\tilde{b}_{kj}^{R}|\chi_j^{RI}+|\tilde{b}_{kj}^{I}|\chi_j^{II})-\sum_{k=1}^{m}\frac{1}{1-\rho_1}\frac{\tilde{\beta}_k}{\beta_j}(|\tilde{b}_{kj}^{R}|\chi_j^{II}$$

$$
\begin{aligned}
&+|\tilde{b}_{kj}^{I}||\chi_j^{RI})\Big]|y_j| - \sum_{k=1}^{m}\tilde{\alpha}_k\Big[d_{2k} - \sum_{j=1}^{n}\frac{\alpha_j}{\tilde{\alpha}_k}(|a_{jk}^R|\lambda_k^{RR} + |a_{jk}^I|\lambda_k^{IR}) \\
&-\sum_{j=1}^{n}\frac{\beta_j}{\tilde{\alpha}_k}(|a_{jk}^R|\lambda_k^{IR} + |a_{jk}^I|\lambda_k^{RR}) - \sum_{j=1}^{n}\frac{1}{1-\rho_2}\frac{\alpha_j}{\tilde{\alpha}_k}(|b_{jk}^R|\lambda_k^{RR} + |b_{jk}^I|\lambda_k^{IR}) \\
&-\sum_{j=1}^{n}\frac{1}{1-\rho_2}\frac{\beta_j}{\tilde{\alpha}_k}(|b_{jk}^R|\lambda_k^{IR} + |b_{jk}^I|\lambda_k^{RR})\Big]|\tilde{x}_k| - \sum_{k=1}^{m}\tilde{\beta}_k\Big[d_{2k} \\
&-\sum_{j=1}^{n}\frac{\alpha_j}{\tilde{\beta}_k}(|a_{jk}^R|\lambda_k^{RI} + |a_{jk}^I|\lambda_k^{II}) - \sum_{j=1}^{n}\frac{\beta_j}{\tilde{\beta}_k}(|a_{jk}^R|\lambda_k^{II} + |a_{jk}^I|\lambda_k^{RI}) \\
&-\sum_{j=1}^{n}\frac{1}{1-\rho_2}\frac{\alpha_j}{\tilde{\beta}_k}(|b_{jk}^R|\lambda_k^{RI} + |b_{jk}^I|\lambda_k^{II}) \\
&-\sum_{j=1}^{n}\frac{1}{1-\rho_2}\frac{\beta_j}{\tilde{\beta}_k}(|b_{jk}^R|\lambda_k^{II} + |b_{jk}^I|\lambda_k^{RI})\Big]|\tilde{y}_k| + \varphi \\
&= -\sum_{j=1}^{n}\alpha_j\kappa_{1j}|x_j| - \sum_{j=1}^{n}\beta_j\kappa_{2j}|y_j| - \sum_{k=1}^{m}\tilde{\alpha}_k\kappa_{3k}|\tilde{x}_k| - \sum_{k=1}^{m}\tilde{\beta}_k\kappa_{4k}|\tilde{y}_k| + \varphi \\
&\le -\kappa(\sum_{j=1}^{n}\alpha_j|x_j| + \sum_{j=1}^{n}\beta_j|y_j| + \sum_{k=1}^{m}\tilde{\alpha}_k|\tilde{x}_k| + \sum_{k=1}^{m}\tilde{\beta}_k|\tilde{y}_k|) + \varphi, \qquad (6.9)
\end{aligned}
$$

where $\kappa = \min_{1\le j\le n, 1\le k\le m}\{\kappa_{1j}, \kappa_{2j}, \kappa_{3k}, \kappa_{4k}\}$.

On the other hand, for $j = 1, 2, \dots, n, k = 1, 2, \dots, m$, there must exist positive constants δ_{1j}, δ_{2j}, $\tilde{\delta}_{1k}$ and $\tilde{\delta}_{2k}$ such that

$$
(-\kappa + \delta_{1j})\alpha_j + \frac{\delta_{1j}\theta_j}{1-\rho_1}\tau_1 e^{\delta_{1j}\tau_1} = 0, \ (-\kappa + \delta_{2j})\beta_j + \frac{\delta_{2j}\phi_j}{1-\rho_1}\tau_1 e^{\delta_{2j}\tau_1} = 0,
$$

$$
(-\kappa + \tilde{\delta}_{1k})\tilde{\alpha}_k + \frac{\tilde{\delta}_{1k}\tilde{\theta}_k}{1-\rho_2}\tau_2 e^{\tilde{\delta}_{1k}\tau_2} = 0, \ (-\kappa + \tilde{\delta}_{2k})\tilde{\beta}_k + \frac{\tilde{\delta}_{1k}\tilde{\phi}_k}{1-\rho_2}\tau_2 e^{\tilde{\delta}_{2k}\tau_2} = 0.
$$

Let $\delta = \min_{\{1\le j\le n, 1\le k\le m\}}\{\delta_{1j}, \delta_{2j}, \tilde{\delta}_{1k}, \tilde{\delta}_{2k}\}$, it is easy to get

$$
\begin{aligned}
&(-\kappa + \delta)\alpha_j + \frac{\delta\theta_j}{1-\rho_1}\tau_1 e^{\delta\tau_1} \le 0, \ (-\kappa + \delta)\beta_j + \frac{\delta\phi_j}{1-\rho_1}\tau_1 e^{\delta\tau_1} \le 0, \\
&(-\kappa + \delta)\tilde{\alpha}_k + \frac{\delta\tilde{\theta}_k}{1-\rho_2}\tau_2 e^{\delta\tau_2} \le 0, \ (-\kappa + \delta)\tilde{\beta}_k + \frac{\delta\tilde{\phi}_k}{1-\rho_2}\tau_2 e^{\delta\tau_2} \le 0. \qquad (6.10)
\end{aligned}
$$

Next, let $V'(t) = (V(t) - \frac{\varphi}{\delta})e^{\delta t}$, one has

$$
\begin{aligned}
D^+V'(t) &= \delta e^{\delta t}(V(t) - \frac{\varphi}{\delta}) + e^{\delta t}D^+V(t) \\
&\le (-\kappa + \delta)e^{\delta t}\Big(\sum_{j=1}^{n}\alpha_j|x_j(t)| + \sum_{j=1}^{n}\beta_j|y_j(t)| + \sum_{k=1}^{m}\tilde{\alpha}_k|\tilde{x}_k(t)| + \sum_{k=1}^{m}\tilde{\beta}_k|\tilde{y}_k(t)|\Big) \\
&+\delta e^{\delta t}\Big(\sum_{j=1}^{n}\frac{1}{1-\rho_1}\theta_j\int_{t-\tau_{1j}(t)}^{t}|x_j(s)|ds + \sum_{j=1}^{n}\frac{1}{1-\rho_1}\phi_j\int_{t-\tau_{1j}(t)}^{t}|y_j(s)|ds \\
&+\sum_{k=1}^{m}\frac{1}{1-\rho_2}\tilde{\theta}_k\int_{t-\tau_{2k}(t)}^{t}|\tilde{x}_k(s)|ds + \sum_{k=1}^{m}\frac{1}{1-\rho_2}\tilde{\phi}_k\int_{t-\tau_{2k}(t)}^{t}|\tilde{y}_k(s)|ds\Big).
\end{aligned}
\tag{6.11}
$$

Integrating both sides of (6.11) on $[0, t]$, we can get

$$
\begin{aligned}
V'(t) - V'(0) &\le (-\kappa+\delta)\int_0^t e^{\delta s}\Big(\sum_{j=1}^{n}\alpha_j|x_j(s)| + \sum_{j=1}^{n}\beta_j|y_j(s)| \\
&+\sum_{k=1}^{m}\tilde{\alpha}_k|\tilde{x}_k(s)| + \sum_{k=1}^{m}\tilde{\beta}_k|\tilde{y}_k(s)|\Big)ds + \sum_{j=1}^{n}\frac{\delta\theta_j}{1-\rho_1}\int_0^t e^{\delta s}\int_{s-\tau_1}^{s}|x_j(\psi)|d\psi ds \\
&+\sum_{j=1}^{n}\frac{\delta\phi_j}{1-\rho_1}\int_0^t e^{\delta s}\int_{s-\tau_1}^{s}|y_j(\psi)|d\psi ds + \sum_{k=1}^{m}\frac{\delta\tilde{\theta}_k}{1-\rho_2}\int_0^t e^{\delta s}\int_{s-\tau_2}^{s}|\tilde{x}_k(\psi)|d\psi ds \\
&+\sum_{k=1}^{m}\frac{\delta\tilde{\phi}_k}{1-\rho_2}\int_0^t e^{\delta s}\int_{s-\tau_2}^{s}|\tilde{y}_k(\psi)|d\psi ds.
\end{aligned}
\tag{6.12}
$$

Among them, the first double integral term can be estimated as

$$
\begin{aligned}
&\sum_{j=1}^{n}\frac{\delta\theta_j}{1-\rho_1}\int_0^t e^{\delta s}\int_{s-\tau_1}^{s}|x_j(\psi)|d\psi ds \\
&= \sum_{j=1}^{n}\frac{\delta\theta_j}{1-\rho_1}\int_{-\tau_1}^{t}|x_j(\psi)|\int_{\max\{0,\psi\}}^{\min\{t,\psi+\tau_1\}} e^{\delta s}ds d\psi \\
&\le \sum_{j=1}^{n}\frac{\delta\theta_j}{1-\rho_1}\int_{-\tau_1}^{t}\tau_1 e^{\delta(\psi+\tau_1)}|x_j(\psi)|d\psi \\
&\le \sum_{j=1}^{n}\frac{\delta\theta_j}{1-\rho_1}\tau_1 e^{\delta\tau_1}\int_{-\tau_1}^{0}|x_j(s)|ds + \sum_{j=1}^{n}\frac{\delta\theta_j}{1-\rho_1}\tau_1 e^{\delta\tau_1}\int_0^t e^{\delta s}|x_j(s)|ds.
\end{aligned}
\tag{6.13}
$$

Similarly,

$$\sum_{j=1}^{n}\frac{\delta\phi_j}{1-\rho_1}\int_0^t e^{\delta s}\int_{s-\tau_1}^{s}|y_j(\psi)|d\psi ds$$
$$\leq\sum_{j=1}^{n}\frac{\delta\phi_j}{1-\rho_1}\tau_1 e^{\delta\tau_1}\int_{-\tau_1}^{0}|y_j(s)|ds+\sum_{j=1}^{n}\frac{\delta\phi_j}{1-\rho_1}\tau_1 e^{\delta\tau_1}\int_0^t e^{\delta s}|y_j(s)|ds,$$
$$\sum_{k=1}^{m}\frac{\delta\tilde{\theta}_k}{1-\rho_2}\int_0^t e^{\delta s}\int_{s-\tau_2}^{s}|\tilde{x}_k(\psi)|d\psi ds$$
$$\leq\sum_{k=1}^{m}\frac{\delta\tilde{\theta}_k}{1-\rho_2}\tau_2 e^{\delta\tau_2}\int_{-\tau_2}^{0}|\tilde{x}_k(s)|ds+\sum_{k=1}^{m}\frac{\delta\tilde{\theta}_k}{1-\rho_2}\tau_2 e^{\delta\tau_2}\int_0^t e^{\delta s}|\tilde{x}_k(s)|ds,$$
$$\sum_{k=1}^{m}\frac{\delta\tilde{\phi}_k}{1-\rho_2}\int_0^t e^{\delta s}\int_{s-\tau_2}^{s}|\tilde{y}_k(\psi)|d\psi ds$$
$$\leq\sum_{k=1}^{m}\frac{\delta\tilde{\phi}_k}{1-\rho_2}\tau_2 e^{\delta\tau_2}\int_{-\tau_2}^{0}|\tilde{y}_k(s)|ds+\sum_{k=1}^{m}\frac{\delta\tilde{\phi}_k}{1-\rho_2}\tau_2 e^{\delta\tau_2}\int_0^t e^{\delta s}|\tilde{y}_k(s)|ds. \quad (6.14)$$

Combining (6.12)–(6.14), we have

$$\begin{aligned}
&e^{\delta t}(V(t)-\frac{\varphi}{\delta})-(V(0)-\frac{\varphi}{\delta})\\
&\leq\sum_{j=1}^{n}\Big((-\kappa+\delta)\alpha_j+\frac{\delta\theta_j}{1-\rho_1}\tau_1 e^{\delta\tau_1}\Big)\int_0^t e^{\delta s}|x_j(s)|ds\\
&+\sum_{j=1}^{n}\Big((-\kappa+\delta)\beta_j+\frac{\delta\phi_j}{1-\rho_1}\tau_1 e^{\delta\tau_1}\Big)\int_0^t e^{\delta s}|y_j(s)|ds\\
&+\sum_{k=1}^{m}\Big((-\kappa+\delta)\tilde{\alpha}_k+\frac{\delta\tilde{\theta}_k}{1-\rho_2}\tau_2 e^{\delta\tau_2}\Big)\int_0^t e^{\delta s}|\tilde{x}_k(s)|ds\\
&+\sum_{k=1}^{m}\Big((-\kappa+\delta)\tilde{\beta}_k+\frac{\delta\tilde{\phi}_k}{1-\rho_2}\tau_2 e^{\delta\tau_2}\Big)\int_0^t e^{\delta s}|\tilde{y}_k(s)|ds\\
&+\sum_{j=1}^{n}\frac{\delta\theta_j}{1-\rho_1}\tau_1 e^{\delta\tau_1}\int_{-\tau_1}^{0}|x_j(s)|ds+\sum_{j=1}^{n}\frac{\delta\phi_j}{1-\rho_1}\tau_1 e^{\delta\tau_1}\int_{-\tau_1}^{0}|y_j(s)|ds\\
&+\sum_{k=1}^{m}\frac{\delta\tilde{\theta}_k}{1-\rho_2}\tau_2 e^{\delta\tau_2}\int_{-\tau_2}^{0}|\tilde{x}_k(s)|ds+\sum_{k=1}^{m}\frac{\delta\tilde{\phi}_k}{1-\rho_2}\tau_2 e^{\delta\tau_2}\int_{-\tau_2}^{0}|\tilde{y}_k(s)|ds. \quad (6.15)
\end{aligned}$$

It follows from (6.10) immediately that

$$V(t)-\frac{\varphi}{\delta}\leq(V(0)-\frac{\varphi}{\delta}+\Pi)e^{-\delta t}\leq(V(0)+\Pi)e^{-\delta t}, \quad (6.16)$$

where

$$\Pi = \sum_{j=1}^{n} \frac{\delta\theta_j}{1-\rho_1}\tau_1 e^{\delta\tau_1} \int_{-\tau_1}^{0} |x_j(s)|ds + \sum_{j=1}^{n} \frac{\delta\phi_j}{1-\rho_1}\tau_1 e^{\delta\tau_1} \int_{-\tau_1}^{0} |y_j(s)|ds$$
$$+ \sum_{k=1}^{m} \frac{\delta\tilde{\theta}_k}{1-\rho_2}\tau_2 e^{\delta\tau_2} \int_{-\tau_2}^{0} |\tilde{x}_k(s)|ds + \sum_{k=1}^{m} \frac{\delta\tilde{\phi}_k}{1-\rho_2}\tau_2 e^{\delta\tau_2} \int_{-\tau_2}^{0} |\tilde{y}_k(s)|ds,$$

which only depends on the initial condition $\vartheta(s)$. By (6.7) and (6.16), we have

$$\begin{aligned} \|\omega(t)\| &= \sum_{j=1}^{n}(|x_j| + |y_j|) + \sum_{k=1}^{m}(|\tilde{x}_k| + |\tilde{y}_k|) \\ &\le \frac{\varphi}{\varrho\delta} + \frac{V(0)+\Pi}{\varrho} e^{-\delta t}. \end{aligned} \tag{6.17}$$

According to Definition 6.2, system (6.1) or (6.2) is globally exponentially stable in Lagrange sense and it uniformly exponentially converges to the ball Ω_1 with a rate δ. □

Remark 6.4 In system (6.1) or (6.2), if $f(0) = g(0) = 0$ and $U(t) = \mathcal{V}(t) = 0$, then $\varphi = 0$ in Theorem 6.1. Next, if the inequalities (6.4) still hold, then the zero solution of system (6.1) or (6.2) is globally exponentially stable in Lyapunov sense.

6.3.2 Stability Criterion Dependent on Nonseparable Method

From Sect. 6.3.1, we can know that Theorem 6.1 is a criterion dependent on separable case. If system (6.1) or (6.2) is considered directly in complex domain, what about the stability criterion? In this subsection, we will consider real-imaginary parts as nonseparable case and propose a sufficient criterion based on algebraic structure.

Theorem 6.2 *Assume that Assumptions 6.2 and 6.4 hold, system (6.1) or (6.2) with (6.3) is globally exponentially stable in Lagrange sense, if there exist positive constants γ_j, $\tilde{\gamma}_k$ and scalars ε_{jk}, $\tilde{\varepsilon}_{kj}$, ε'_{jk}, $\tilde{\varepsilon}'_{kj}$, ζ_j and $\tilde{\zeta}_k$ ($j = 1, 2, \ldots, n$, $k = 1, 2, \ldots, m$) such that the following inequalities hold*

$$\begin{aligned} \hat{\kappa}_{1j} &= 2d_{1j} - \sum_{k=1}^{m}(\varepsilon_{jk}^{-1} a_{jk} a_{jk}^* + \varepsilon_{jk}'^{-1} b_{jk} b_{jk}^* + 2\frac{\tilde{\gamma}_k}{\gamma_j}\tilde{\varepsilon}_{kj}\mu_j^2 + \frac{2\tilde{\gamma}_k \tilde{\varepsilon}'_{kj}\mu_j^2}{(1-\rho_1)\gamma_j}) - \zeta_j^{-1} > 0, \\ \hat{\kappa}_{2k} &= 2d_{2k} - \sum_{j=1}^{n}(\tilde{\varepsilon}_{kj}^{-1} \tilde{a}_{kj} \tilde{a}_{kj}^* + \tilde{\varepsilon}_{kj}'^{-1} \tilde{b}_{kj} \tilde{b}_{kj}^* + 2\frac{\gamma_j}{\tilde{\gamma}_k}\varepsilon_{jk} l_k^2 + \frac{2\gamma_j \varepsilon'_{jk} l_k^2}{(1-\rho_2)\tilde{\gamma}_k}) - \tilde{\zeta}_k^{-1} > 0. \end{aligned} \tag{6.18}$$

Moreover, system (6.1) or (6.2) globally exponentially converges to the ball

$$\Omega_2 = \{\Omega(t) \in \mathbb{C}^{n+m} \Big| \|\omega(t)\| = \sqrt{\sum_{j=1}^{n} u_j^*(t)u_j(t) + \sum_{k=1}^{m} v_k^*(t)v_k(t)} \leq \sqrt{\frac{\hat{\varphi}}{\hat{\delta}\hat{\varrho}}}\}, \tag{6.19}$$

where

$$\hat{\varrho} = \min_{1\leq j\leq n, 1\leq k\leq m}\{\gamma_j, \tilde{\gamma}_k\},$$

$$\hat{\varphi} = \sum_{k=1}^{m}\sum_{j=1}^{n}(2\gamma_j\varepsilon_{jk} + 2\gamma_j\varepsilon'_{jk})f_k^*(0)f_k(0) + \sum_{j=1}^{n}\gamma_j\zeta_j\tilde{U}_j^2 + \sum_{j=1}^{n}\sum_{k=1}^{m}(2\tilde{\gamma}_k\tilde{\varepsilon}_{kj} + 2\tilde{\gamma}_k\tilde{\varepsilon}'_{kj})g_j^*(0)g_j(0) + \sum_{k=1}^{m}\tilde{\gamma}_k\tilde{\zeta}_k\tilde{V}_k^2$$

and $\hat{\delta}$ will be determined later.

Proof Firstly, to simplify, we denote $u_j(t)$, $v_k(t)$, $u_j(t-\tau_{1j}(t))$ and $v_k(t-\tau_{2k}(t))$ as u_j, v_k, $u_j^{\tau_{1j}}$ and $v_k^{\tau_{2k}}$, respectively. Construct a candidate Lyapunov functional as follows.

$$V(t) = \sum_{j=1}^{n}\gamma_j u_j^*(t)u_j(t) + \sum_{j=1}^{n}\sum_{k=1}^{m}\frac{2\tilde{\gamma}_k\tilde{\varepsilon}'_{kj}\mu_j^2}{1-\rho_1}\int_{t-\tau_{1j}(t)}^{t} u_j^*(s)u_j(s)ds + \sum_{k=1}^{m}\tilde{\gamma}_k v_k^*(t)v_k(t) + \sum_{k=1}^{m}\sum_{j=1}^{n}\frac{2\gamma_j\varepsilon'_{jk}l_k^2}{1-\rho_2}\int_{t-\tau_{2k}(t)}^{t} v_k^*(s)v_k(s)ds. \tag{6.20}$$

Computing the derivative of $V(t)$ along the solution trajectories of system (6.1) with (6.3), we have

$$\begin{aligned}\dot{V}(t) \leq\ & 2\sum_{j=1}^{n}\gamma_j u_j^*\Big(-d_{1j}u_j + \sum_{k=1}^{m}a_{jk}f_k(v_k) + \sum_{k=1}^{m}b_{jk}f_k(v_k^{\tau_{2k}}) + U_j\Big) \\ & + 2\sum_{k=1}^{m}\tilde{\gamma}_k v_k^*\Big(-d_{2k}v_k + \sum_{j=1}^{n}\tilde{a}_{kj}g_j(u_j) + \sum_{j=1}^{n}\tilde{b}_{kj}g_j(u_j^{\tau_{1j}}) + V_k\Big) \\ & + 2\sum_{j=1}^{n}\sum_{k=1}^{m}\frac{\tilde{\gamma}_k\tilde{\varepsilon}'_{kj}\mu_j^2}{1-\rho_1}\Big(u_j^*u_j - (1-\rho_1)u_j^{*\tau_{1j}}u_j^{\tau_{1j}}\Big) \\ & + 2\sum_{k=1}^{m}\sum_{j=1}^{n}\frac{\gamma_j\varepsilon'_{jk}l_k^2}{1-\rho_2}\Big(v_k^*v_k - (1-\rho_2)v_k^{*\tau_{2k}}v_k^{\tau_{2k}}\Big) \\ \leq\ & -2\sum_{j=1}^{n}\gamma_j d_{1j}u_j^*u_j - 2\sum_{k=1}^{m}\tilde{\gamma}_k d_{2k}v_k^*v_k + \sum_{j=1}^{n}\gamma_j\zeta_j^{-1}u_j^*u_j + \sum_{k=1}^{m}\tilde{\gamma}_k\tilde{\zeta}_k^{-1}v_k^*v_k\end{aligned}$$

$$
\begin{aligned}
&+\sum_{j=1}^{n}\sum_{k=1}^{m}\gamma_j\varepsilon_{jk}^{-1}a_{jk}a_{jk}^{*}u_j^{*}u_j+\sum_{j=1}^{n}\sum_{k=1}^{m}\gamma_j\varepsilon_{jk}f_k^{*}(v_k)f_k(v_k)\\
&+\sum_{j=1}^{n}\sum_{k=1}^{m}\gamma_j\varepsilon_{jk}^{\prime-1}b_{jk}b_{jk}^{*}u_j^{*}u_j+\sum_{j=1}^{n}\sum_{k=1}^{m}\gamma_j\varepsilon'_{jk}f_k^{*}(v_k^{\tau_{2k}})f_k(v_k^{\tau_{2k}})\\
&+\sum_{k=1}^{m}\sum_{j=1}^{n}\tilde{\gamma}_k\tilde{\varepsilon}_{kj}^{-1}\tilde{a}_{kj}\tilde{a}_{kj}^{*}v_k^{*}v_k+\sum_{k=1}^{m}\sum_{j=1}^{n}\tilde{\gamma}_k\tilde{\varepsilon}_{kj}g_j^{*}(u_j)g_j(u_j)\\
&+\sum_{k=1}^{m}\sum_{j=1}^{n}\tilde{\gamma}_k\tilde{\varepsilon}'_{kj}g_j^{*}(u_j^{\tau_{1j}})g_j(u_j^{\tau_{1j}})+\sum_{k=1}^{m}\sum_{j=1}^{n}\tilde{\gamma}_k\tilde{\varepsilon}_{kj}^{\prime-1}\tilde{b}_{kj}\tilde{b}_{kj}^{*}v_k^{*}v_k\\
&+\sum_{j=1}^{n}\gamma_j\zeta_jU_j^{*}U_j+\sum_{k=1}^{m}\tilde{\gamma}_k\tilde{\zeta}_kV_k^{*}V_k\\
&+2\sum_{k=1}^{m}\sum_{j=1}^{n}\frac{\gamma_j\varepsilon'_{jk}l_k^2}{1-\rho_2}v_k^{*}v_k-2\sum_{k=1}^{m}\sum_{j=1}^{n}\gamma_j\varepsilon'_{jk}l_k^2v_k^{*\tau_{2k}}v_k^{\tau_{2k}}\\
&+2\sum_{j=1}^{n}\sum_{k=1}^{m}\frac{\tilde{\gamma}_k\tilde{\varepsilon}'_{kj}\mu_j^2}{1-\rho_1}u_j^{*}u_j-2\sum_{j=1}^{n}\sum_{k=1}^{m}\tilde{\gamma}_k\tilde{\varepsilon}'_{kj}\mu_j^2u_j^{*\tau_{1j}}u_j^{\tau_{1j}}.
\end{aligned}
\tag{6.21}
$$

Based on Assumption 6.2 and (6.18), one has

$$
\begin{aligned}
\dot{V}(t)\le&-\sum_{j=1}^{n}\Big(2d_{1j}-\sum_{k=1}^{m}(\varepsilon_{jk}^{-1}a_{jk}a_{jk}^{*}+\varepsilon_{jk}^{\prime-1}b_{jk}b_{jk}^{*}\\
&+2\frac{\tilde{\gamma}_k}{\gamma_j}\tilde{\varepsilon}_{kj}\mu_j^2+\frac{2\tilde{\gamma}_k\tilde{\varepsilon}'_{kj}\mu_j^2}{(1-\rho_1)\gamma_j})-\zeta_j^{-1}\Big)\gamma_ju_j^{*}u_j\\
&-\sum_{k=1}^{m}\Big(2d_{2k}-\sum_{j=1}^{n}(\tilde{\varepsilon}_{kj}^{-1}\tilde{a}_{kj}\tilde{a}_{kj}^{*}+\tilde{\varepsilon}_{kj}^{\prime-1}\tilde{b}_{kj}\tilde{b}_{kj}^{*}\\
&+2\frac{\gamma_j}{\tilde{\gamma}_k}\varepsilon_{jk}l_k^2+\frac{2\gamma_j\varepsilon'_{jk}l_k^2}{(1-\rho_2)\tilde{\gamma}_k})-\tilde{\zeta}_k^{-1}\Big)\tilde{\gamma}_kv_k^{*}v_k\\
&+\sum_{k=1}^{m}\sum_{j=1}^{n}(2\gamma_j\varepsilon_{jk}+2\gamma_j\varepsilon'_{jk})f_k^{*}(0)f_k(0)+\sum_{j=1}^{n}\gamma_j\zeta_j\tilde{U}_j^2\\
&+\sum_{j=1}^{n}\sum_{k=1}^{m}(2\tilde{\gamma}_k\tilde{\varepsilon}_{kj}+2\tilde{\gamma}_k\tilde{\varepsilon}'_{kj})g_j^{*}(0)g_j(0)+\sum_{k=1}^{m}\tilde{\gamma}_k\tilde{\zeta}_k\tilde{V}_k^2\\
\le&-\hat{\kappa}(\sum_{j=1}^{n}\gamma_ju_j^{*}u_j+\sum_{k=1}^{m}\tilde{\gamma}_kv_k^{*}v_k)+\hat{\varphi},
\end{aligned}
\tag{6.22}
$$

where $\hat{\kappa}=\min_{1\le j\le n,1\le k\le m}\{\hat{\kappa}_{1j},\hat{\kappa}_{2k}\}$.

On the other hand, for $j = 1, 2, \ldots, n, k = 1, 2, \ldots, m$, there must exist a positive constant $\hat{\delta}$ such that

$$
\begin{aligned}
&(-\hat{\kappa} + \hat{\delta})\gamma_j + \frac{\hat{\delta}}{1-\rho_1}\sum_{k=1}^{m} 2\tilde{\gamma}_k \tilde{\varepsilon}'_{kj}\mu_j^2 \tau_1 e^{\hat{\delta}\tau_1} \leq 0, \\
&(-\hat{\kappa} + \hat{\delta})\tilde{\gamma}_k + \frac{\hat{\delta}}{1-\rho_2}\sum_{j=1}^{n} 2\gamma_j \varepsilon'_{jk} l_k^2 \tau_2 e^{\hat{\delta}\tau_2} \leq 0.
\end{aligned} \tag{6.23}
$$

Next, let $V'(t) = (V(t) - \frac{\hat{\varphi}}{\hat{\delta}})e^{\hat{\delta}t}$, and by a derivation process similar to (6.11)–(6.15), we can obtain

$$
V(t) - \frac{\hat{\varphi}}{\hat{\delta}} \leq (V(0) - \frac{\hat{\varphi}}{\hat{\delta}} + \hat{\Pi})e^{-\hat{\delta}t} \leq (V(0) + \hat{\Pi})e^{-\hat{\delta}t}, \tag{6.24}
$$

where

$$
\begin{aligned}
\hat{\Pi} = &\sum_{j=1}^{n} \frac{\hat{\delta}}{1-\rho_1}\sum_{k=1}^{m} 2\tilde{\gamma}_k \tilde{\varepsilon}'_{jk}\mu_j^2 \tau_1 e^{\hat{\delta}\tau_1} \int_{-\tau_1}^{0} u_j^*(s)u_j(s)ds \\
&+ \sum_{k=1}^{m} \frac{\hat{\delta}}{1-\rho_2}\sum_{j=1}^{n} 2\gamma_j \varepsilon'_{jk} l_k^2 \tau_2 e^{\hat{\delta}\tau_2} \int_{-\tau_2}^{0} v_j^*(s)v_j(s)ds,
\end{aligned}
$$

which only depends on the initial condition $\vartheta(s)$. Hence,

$$
\begin{aligned}
\|\omega(t)\| &= \sqrt{\sum_{j=1}^{n} u_j^*(t)u_j(t) + \sum_{k=1}^{m} v_k^*(t)v_k(t)} \\
&\leq \sqrt{\frac{\hat{\varphi}}{\hat{\varrho}\hat{\delta}}} + \sqrt{\frac{V(0) + \hat{\Pi}}{\hat{\varrho}}} e^{-\frac{\hat{\delta}}{2}t}.
\end{aligned} \tag{6.25}
$$

According to Definition 6.2, system (6.1) or (6.2) is globally exponentially stable in Lagrange sense and it uniformly exponentially converges to the ball Ω_2 with a rate $\frac{\hat{\delta}}{2}$. □

6.4 Stability Criterion in Terms of LMI

In the previous section, sufficient criteria based on algebraic structure for the Lagrange exponential stability of system (6.1) or (6.2) were derived. In order to explore this stability problem deeply, we will present a sufficient criterion in terms of LMI dependent on nonseparable case in this section.

Theorem 6.3 *Assume that Assumptions 6.2 and 6.4 hold, system (6.1) or (6.2) with (6.3) is globally exponentially stable in Lagrange sense, if there exist positive definite Hermitian matrices P_1, P_2, and positive diagonal matrices H_1, H_2, H_3, H_4, for any given positive scalars ε_1 and ε_2 such that the following LMI holds:*

$$\Psi = \begin{bmatrix} \Psi_{11} & 0 & P_1A & P_1B & 0 & 0 \\ * & \Psi_{22} & 0 & 0 & P_2\tilde{A} & P_2\tilde{B} \\ * & * & -H_1 & 0 & 0 & 0 \\ * & * & * & -H_2 & 0 & 0 \\ * & * & * & * & -H_3 & 0 \\ * & * & * & * & * & -H_4 \end{bmatrix} < 0, \tag{6.26}$$

$$\Psi_{11} = -P_1D_1 - D_1P_1 + \frac{2KH_4K}{1-\rho_1} + 2KH_3K + (\varepsilon_1+1)P_1,$$

$$\Psi_{22} = -P_2D_2 - D_2P_2 + \frac{2LH_2L}{1-\rho_2} + 2LH_1L + (\varepsilon_2+1)P_2.$$

Moreover, system (6.1) or (6.2) globally exponentially converges to the ball

$$\Omega_3 = \{\omega(t) \in \mathbb{C}^{n+m} \Big| \|\omega(t)\| = \sqrt{u^*(t)u(t) + v^*(t)v(t)} \le \sqrt{\frac{\tilde{\varphi}}{\tilde{\varrho}\tilde{\delta}}}\}, \tag{6.27}$$

where $\tilde{\varrho} = \min\{\lambda_{\min}(P_1), \lambda_{\min}(P_2)\}$, $\tilde{\varphi} = 2f^*(0)(H_1+H_2)f(0) + 2g^*(0)(H_3 + H_4)g(0) + \varepsilon_1^{-1}\lambda_{\max}(P_1)\tilde{U}^T\tilde{U} + \varepsilon_2^{-1}\lambda_{\max}(P_2)\tilde{V}^T\tilde{V}$, *and* $\tilde{\delta}$ *will be determined later.*

Proof Firstly, to simplify, we denote $u(t)$, $v(t)$, $u(t-\tau_1(t))$ and $v(t-\tau_2(t))$ as u, v, u^{τ_1} and v^{τ_2}, respectively. Construct a candidate Lyapunov functional as follows.

$$V(t) = u^*P_1u + \int_{t-\tau_1(t)}^{t} u^*(s)R_1u(s)ds + v^*P_2v + \int_{t-\tau_2(t)}^{t} v^*(s)R_2v(s)ds, \tag{6.28}$$

where $R_1 > 0$ and $R_2 > 0$. The time derivative of $V(t)$ along the trajectories of (6.2) with (6.3) can be bounded as

$$\begin{aligned} \dot{V}(t) \le\ & 2u^*P_1(-D_1u + Af(v) + Bf(v^{\tau_2}) + U) + u^*R_1u \\ & + 2v^*P_2(-D_2v + \tilde{A}g(u) + \tilde{B}g(u^{\tau_1}) + \mathcal{V}) + v^*R_2v \\ & - (1-\rho_1)u^{\tau_1 *}R_1u^{\tau_1} - (1-\rho_2)v^{\tau_2 *}R_2v^{\tau_2}. \end{aligned} \tag{6.29}$$

Moreover, by Assumption 6.4 and Lemma 6.1, one has

$$\begin{aligned} & f^*(v)H_1f(v) \le 2v^*LH_1Lv + 2f^*(0)H_1f(0), \\ & f^*(v^{\tau_2})H_2f(v^{\tau_2}) \le 2v^{\tau_2 *}LH_2Lv^{\tau_2} + 2f^*(0)H_2f(0), \\ & g^*(u)H_3g(u) \le 2u^*KH_3Ku + 2g^*(0)H_3g(0), \end{aligned}$$

$$
\begin{aligned}
&g^*(u^{\mathcal{T}_1})H_4g(u^{\mathcal{T}_1}) \le 2u^{\mathcal{T}_1*}KH_4Ku^{\mathcal{T}_1} + 2g^*(0)H_4f(0),\\
&u^*P_1U + U^*P_1u \le \varepsilon_1 u^*P_1u + \varepsilon_1^{-1}U^*P_1U,\\
&v^*P_2\mathcal{V} + \mathcal{V}^*P_2v \le \varepsilon_2 v^*P_2v + \varepsilon_2^{-1}\mathcal{V}^*P_2\mathcal{V}.
\end{aligned}
\tag{6.30}
$$

Combining (6.29) and (6.30), we have

$$
\begin{aligned}
\dot{V}(t) \le\ & u^*(-P_1D_1 - D_1P_1 + R_1 + 2KH_3K + \varepsilon_1P_1)u\\
&+ v^*(-P_2D_2 - D_2P_2 + R_2 + 2LH_1L + \varepsilon_2P_2)v\\
&+ u^{\mathcal{T}_1*}(2KH_4K - (1-\rho_1)R_1)u^{\mathcal{T}_1} + v^{\mathcal{T}_2*}(2LH_2L - (1-\rho_2)R_2)v^{\mathcal{T}_2}\\
&+ u^*P_1Af(v) + f^*(v)A^*P_1u + u^*P_1Bf(v^{\mathcal{T}_2}) + f^*(v^{\mathcal{T}_2})B^*P_1u\\
&+ v^*P_2\tilde{A}g(u) + g^*(u)\tilde{A}^*P_2v + v^*P_2\tilde{B}g(u^{\mathcal{T}_1}) + g^*(u^{\mathcal{T}_1})\tilde{B}^*P_2v\\
&- g^*(u)H_3g(u) - g^{\mathcal{T}_1*}(u)H_4g(u^{\mathcal{T}_1}) - f^*(v)H_1f(v) - f^{\mathcal{T}_2*}(v)H_2f(v^{\mathcal{T}_2})\\
&+ 2f^*(0)(H_1 + H_2)f(0) + 2g^*(0)(H_3 + H_4)g(0)\\
&+ \varepsilon_1^{-1}U^*P_1U + \varepsilon_2^{-1}\mathcal{V}^*P_2\mathcal{V}\\
\le\ & \xi^*(t)\hat{\Psi}\xi(t) - u^*P_1u - v^*P_2v + \tilde{\varphi},
\end{aligned}
\tag{6.31}
$$

where

$$
\xi(t) = [u^*, u^{\mathcal{T}_1*}, v^*, v^{\mathcal{T}_2*}, f^*(v), f^*(v^{\mathcal{T}_2}), g^*(u), g^*(u^{\mathcal{T}_1})]^*,
$$

$$
\hat{\Psi} = \begin{bmatrix}
\hat{\Psi}_{11} & 0 & 0 & 0 & P_1A & P_1B & 0 & 0\\
* & \hat{\Psi}_{22} & 0 & 0 & 0 & 0 & 0 & 0\\
* & * & \hat{\Psi}_{33} & 0 & 0 & 0 & P_2\tilde{A} & P_2\tilde{B}\\
* & * & * & \hat{\Psi}_{44} & 0 & 0 & 0 & 0\\
* & * & * & * & -H_1 & 0 & 0 & 0\\
* & * & * & * & * & -H_2 & 0 & 0\\
* & * & * & * & * & * & -H_3 & 0\\
* & * & * & * & * & * & * & -H_4
\end{bmatrix},
$$

$$
\begin{aligned}
&\hat{\Psi}_{11} = -P_1D_1 - D_1P_1 + R_1 + 2KH_3K + (\varepsilon_1 + 1)P_1,\\
&\hat{\Psi}_{22} = 2KH_4K - (1-\rho_1)R_1,\\
&\hat{\Psi}_{33} = -P_2D_2 - D_2P_2 + R_2 + 2LH_1L + (\varepsilon_2 + 1)P_2,\\
&\hat{\Psi}_{44} = 2LH_2L - (1-\rho_2)R_2.
\end{aligned}
$$

It is easy to see that (6.26) is equivalent to $\hat{\Psi} < 0$. Hence, it follows from (6.31) that

$$
\begin{aligned}
\dot{V}(t) &\le -u^*P_1u - v^*P_2v + \tilde{\varphi}\\
&\le -\lambda_{\min}(P_1)u^*u - \lambda_{\min}(P_2)v^*v + \tilde{\varphi}.
\end{aligned}
\tag{6.32}
$$

On the other hand, choosing $R_1 > \frac{2KH_4K}{1-\rho_1}$ and $R_2 > \frac{2LH_2L}{1-\rho_2}$, there must exist a positive constant $\tilde{\delta}$ such that

$$\begin{aligned} -\lambda_{\min}(P_1) + \tilde{\delta}\lambda_{\max}(P_1) + \tilde{\delta}\tau_1 e^{\tilde{\delta}\tau_1}\lambda_{\max}(R_1) \leq 0, \\ -\lambda_{\min}(P_2) + \tilde{\delta}\lambda_{\max}(P_2) + \tilde{\delta}\tau_2 e^{\tilde{\delta}\tau_2}\lambda_{\max}(R_2) \leq 0. \end{aligned} \tag{6.33}$$

Next, let $V'(t) = (V(t) - \frac{\tilde{\varphi}}{\tilde{\delta}})e^{\tilde{\delta}t}$, and by a derivation process similar to (6.11)–(6.15), we can obtain

$$V(t) - \frac{\tilde{\varphi}}{\tilde{\delta}} \leq (V(0) - \frac{\tilde{\varphi}}{\tilde{\delta}} + \tilde{\Pi})e^{-\tilde{\delta}t} \leq (V(0) + \tilde{\Pi})e^{-\tilde{\delta}t}, \tag{6.34}$$

where $\tilde{\Pi} = \tilde{\delta}\lambda_{\max}(R_1)\tau_1 e^{\tilde{\delta}\tau_1}\int_{-\tau_1}^{0} u^* u ds + \tilde{\delta}\lambda_{\max}(R_2)\tau_2 e^{\tilde{\delta}\tau_2}\int_{-\tau_2}^{0} v^* v ds$, which only depends on the initial condition $\vartheta(s)$. Therefore,

$$\|\omega(t)\| = \sqrt{u^*(t)u(t) + v^*(t)v(t)} \leq \sqrt{\frac{\tilde{\varphi}}{\tilde{\varrho}\tilde{\delta}}} + \sqrt{\frac{V(0) + \tilde{\Pi}}{\tilde{\varrho}}}e^{-\frac{\tilde{\delta}}{2}t}. \tag{6.35}$$

According to Definition 6.2, system (6.1) or (6.2) is globally exponentially stable in Lagrange sense and it uniformly exponentially converges to the ball Ω_3 with a rate $\frac{\tilde{\delta}}{2}$. □

Remark 6.5 Similar to the discussions in Remark 6.4, if $f(0) = g(0) = 0$ and $U(t) = \mathcal{V}(t) = 0$, then $\hat{\varphi} = 0$ in Theorem 6.2 and $\tilde{\varphi} = 0$ in Theorem 6.3. Now, if the inequalities (6.18) in Theorem 6.2 still hold, then the zero solution of system (6.1) or (6.2) is globally exponentially stable in Lyapunov sense. Likewise, in Theorem 6.3, similar LMI can also be obtained to guarantee the exponential stability of the zero solution in Lyapunov sense. Here, we omit it to save space.

Remark 6.6 As stated in Remark 6.1, the result of Theorem 6.1 is for the real-imaginary separate-type activation functions, but in Theorems 6.2 and 6.3, the considered system is treated as an entirety and the corresponding results are for the case with activation functions which can't be expressed explicitly by real-imaginary parts. Especially, according to Remark 6.2, we know that sufficient conditions of Theorems 6.2 and 6.3 are also applicable for the real-imaginary separate-type case. Therefore, we can choose a suitable criterion for handling problems. Moreover, looked from the forms, the LMI condition in Theorem 6.3 is simple and easy to handle in contrast to the algebraic conditions in Theorems 6.1 and 6.2. Further, for their effectiveness, comparison and superiority-inferiority, we will provide a detailed explanation with illustrative examples in the next section.

Remark 6.7 Unlike the results for globally Lagrange α-exponential stability of fractional-order CVNNs without time delays in Jian and Wan (2017), this chapter studies globally Lagrange exponential stability of CVBAMNNs with time-varying delays. In order to deal with time-delay terms and obtain the desired results, a parameter scalar δ in (6.10) is introduced and a new Lyapunov functional $V'(t)$ is given by a exponential function transform in Theorem 6.1. Then, by light of the proposed techniques in (6.11)–(6.15), our goal is achieved. Similar approaches are presented and our desired results are derived in Theorems 6.2 and 6.3.

Remark 6.8 In order to meet the actual requirement, it becomes urgent to explore multistable dynamical behaviors in the theory development of neural networks models. Naturally, the Lagrange stability different from the common Lyapunov stability appears, which focuses on the boundedness of solutions and the estimation of globally attractive set for the entire system. For various real-valued neural networks, the stability in Lagrange sense is investigated in Liao et al. (2008), Wang et al. (2009), Wu and Zeng (2014a), Wu and Zeng (2014b), Wang et al. (2010), Jian and Wang (2015), Tu et al. (2016), Tu et al. (2011), Xiong et al. (2008). However, For CVNNs, there are few results in corresponding issues. For example, Jian and Wan (2017) considers Lagrange α-exponential stability of fractional-order CVNNs without time delays, and Song et al. (2017), Tu et al. (2016) study Lagrange stability of delayed CVNNs. Moreover, so far, no result on Lagrange stability of delayed CVBAMNNs has been reported. Thus, our work fills the gap in this field.

Remark 6.9 From a practical perspective, complex numbers have great superiority in reflecting real things in comparison with real ones. For example, speed and direction in the wind profile model (Goh et al. 2006). Thus, complex-valued systems have been a hot research issue, especially, the dynamics of various complex-valued neural networks models. However, because the states, connection weights and activation functions of these systems are complex-valued, it becomes more difficult in studying them. Moreover, complex-valued activation functions have great variety. This also leads to some difficulties in the theoretical research. Even so, it is urgent and essential to do this work because of the extensive applications and rich background for these models. In this chapter, we have overcome the obstacle involved by the increase for space dimension and the theory approach on the complex domain and derived some results for the Lagrange exponential stability of delayed complex-valued BAM neural networks with two types of common activation functions.

6.5 Illustrative Examples

In this section, we will provide two illustrative examples to illustrate the effectiveness of the obtained results in Theorems 6.1–6.3 and make some comparisons between their advantages and disadvantages.

Example 6.1 Consider the following system:

$$
\begin{aligned}
\dot{u}_1 &= -7u_1 + (1-i)f_1(v_1) - 0.2if_2(v_2) \\
&\quad + (0.3+i)f_1(v_1^{\tau_{21}}) + (1-0.5i)f_2(v_2^{\tau_{22}}) + U_1(t), \\
\dot{u}_2 &= -6u_2 + 0.2if_1(v_1) + (0.5-i)f_2(v_2) \\
&\quad + (0.1+i)f_1(v_1^{\tau_{21}}) + 0.3if_2(v_2^{\tau_{22}}) + U_2(t), \\
\dot{v}_1 &= -8v_1 - ig_1(u_1) + (0.2+0.3i)g_2(u_2) \\
&\quad + 0.2ig_1(u_1^{\tau_{11}}) + (0.1-i)g_2(u_2^{\tau_{12}}) + V_1(t), \\
\dot{v}_2 &= -9v_2 + (i-0.5)g_1(u_1) + ig_2(u_2) \\
&\quad + (-0.1+i)g_1(u_1^{\tau_{11}}) - 0.2ig_2(u_2^{\tau_{12}}) + V_2(t),
\end{aligned}
\tag{6.36}
$$

where

$$
\begin{aligned}
&\tau_{11}(t) = \tau_{21}(t) = 0.2\sin^2(t) + 0.3, \\
&\tau_{12}(t) = \tau_{22}(t) = 0.2\cos^2(t), \\
&U_1(t) = V_2(t) = 1.3\sin(t) + 1.3i\cos(t), \\
&U_2(t) = V_1(t) = 1.3\cos(t) + 1.3i\sin(t), \\
&f_k(v_k) = \frac{|\tilde{x}_k + 1| - |\tilde{x}_k - 1|}{2} + i\frac{|\tilde{y}_k + 1| - |\tilde{y}_k - 1|}{2}, \\
&g_j(u_j) = \frac{x_j + y_j}{2} + i\frac{x_j - y_j}{2}, \ k = 1, 2, \ j = 1, 2.
\end{aligned}
$$

Obviously, $f_k^R(\tilde{x}_k, \tilde{y}_k) = 0.5(|\tilde{x}_k + 1| - |\tilde{x}_k - 1|)$, $f_k^I(\tilde{x}_k, \tilde{y}_k) = 0.5(|\tilde{y}_k + 1| - |\tilde{y}_k - 1|)$, $g_j^R(x_j, y_j) = 0.5(x_j + y_j)$ and $g_j^I(x_j, y_j) = 0.5(x_j - y_j)$. For $\sigma = R, I$, it is easy to get

$$
\begin{aligned}
&|f_k^R(\tilde{x}_k, \tilde{x}_k) - f_k^R(\tilde{x}_k', \tilde{y}_k')| \le |\tilde{x}_k - \tilde{x}_k'|, \\
&|f_k^I(\tilde{x}_k, \tilde{x}_k) - f_k^I(\tilde{x}_k', \tilde{y}_k')| \le |\tilde{y}_k - \tilde{y}_k'|, \\
&|g_j^\sigma(x_j, y_j) - g_j^\sigma(x_j', y_j')| \le \frac{1}{2}|x_j - x_j'| + \frac{1}{2}|y_j - y_j'|.
\end{aligned}
$$

Moreover,

$$
\begin{aligned}
&|f_k(v_k) - f_k(v_k')| \le \sqrt{2}|v_k - v_k'|, \\
&|g_j(u_j) - g_j(u_j')| \le \frac{\sqrt{2}}{2}|u_j - u_j'|.
\end{aligned}
$$

So, Assumptions 6.1 and 6.2 are satisfied with $\lambda_k^{RR} = \lambda_k^{II} = 1$, $\lambda_k^{RI} = \lambda_k^{IR} = 0$, $\chi_j^{RR} = \chi_j^{RI} = \chi_j^{IR} = \chi_j^{II} = \frac{1}{2}$ and $l_k = \sqrt{2}$, $\mu_j = \frac{\sqrt{2}}{2}$, respectively. Meanwhile, $U_1(t)$, $U_2(t)$, $V_1(t)$ and $V_2(t)$ satisfy Assumptions 6.3 and 6.4 with $\tilde{U}_1^R = \tilde{U}_1^I = \tilde{V}_2^R = \tilde{V}_2^I = \tilde{U}_2^R = \tilde{V}_1^R = \tilde{U}_2^I = \tilde{V}_1^I = 1.3$ and $\tilde{U}_1 = \tilde{V}_2 = \tilde{U}_2 = \tilde{V}_1 = 1.3$, respectively. Next, we will verify the obtained results in Theorems 6.1–6.3 based on this example.

Table 6.1 Computed values B_1 and $\frac{\tilde{\delta}}{2}$ for different ε_1 and ε_2 in Case 3, for Example 6.1

ε_1 and ε_2	$\varepsilon_1=\varepsilon_2=1.2$	$\varepsilon_1=\varepsilon_2=1.5$	$\varepsilon_1=\varepsilon_2=1.8$	$\varepsilon_1=\varepsilon_2=2.1$	$\varepsilon_1=\varepsilon_2=2.4$
B_1	3.5163	3.1389	2.8577	2.6364	2.4721
$\frac{\tilde{\delta}}{2}$	0.1755	0.1765	0.1785	0.181	0.1815

Case 1: In Theorem 6.1, choosing $\alpha_j=\beta_j=2$, $\tilde{\alpha}_k=\tilde{\beta}_k=3$ for $j,k=1,2$. After a simple calculation, it can be easy to verify that (6.4) is true. Then, system (6.36) is globally exponentially stable in Lagrange sense and it globally exponentially converges to the ball Ω_{11} with a rate 0.5101, where

$$\Omega_{11}=\Big\{u(t)=x(t)+y(t)\in\mathbb{C}^n,\ v(t)=\tilde{x}(t)+\tilde{y}(t)$$
$$\in\mathbb{C}^m\Big|\sum_{j=1}^{n}(|x_j|+|y_j|)+\sum_{k=1}^{m}(|\tilde{x}_k|+|\tilde{y}_k|)\le 25.4852\Big\}.$$

Case 2: In Theorem 6.2, choosing $\gamma_j=\tilde{\gamma}_k=1$, $\varepsilon_{jk}=\tilde{\varepsilon}_{kj}=\varepsilon'_{jk}=\tilde{\varepsilon}'_{kj}=0.5$, $\zeta_j=\tilde{\zeta}_k=0.5$ for $j,k=1,2$. After a simple calculation, it can be easy to verify that (6.18) is true. Then, system (6.36) is globally exponentially stable in Lagrange sense and it globally exponentially converges to the ball Ω_{12} with a rate 0.0549, where

$$\Omega_{12}=\{\omega(t)\in\mathbb{C}^{n+m}\Big|\,\|\omega(t)\|=\sqrt{\sum_{j=1}^{n}u_j^*(t)u_j(t)+\sum_{k=1}^{m}v_k^*(t)v_k(t)}\le 5.5483\}.$$

Case 3: In Theorem 6.3, choosing suitable positive scalars ε_1 and ε_2 such that (6.26) holds, then system (6.36) is globally exponentially stable in Lagrange sense and it globally exponentially converges to the ball $\Omega_{13}=\{\omega(t)\in\mathbb{C}^{n+m}\Big|\,\|\omega(t)\|=\sqrt{u^*(t)u(t)+v^*(t)v(t)}\le B_1\}$. Here, we choose multiple sets of constants ε_1 and ε_2 and obtain different B_1 and different convergence rates $\frac{\tilde{\delta}}{2}$ which are shown in Table 6.1. It can be seen from Table 6.1 that B_1 decreases and convergence rate $\frac{\tilde{\delta}}{2}$ increases when scalars ε_1 and ε_2 increases. But, B_1 changes obviously whereas convergence rate $\frac{\tilde{\delta}}{2}$ only changes a little. Anyway, we can obtain a better convergence ball by adjusting ε_1 and ε_2 in (6.26). Moreover, from cases 1–3, it is clear that Ω_{13} is smaller than Ω_{11} and Ω_{12}, which shows that the LMI condition in Theorem 6.3 is less conservative than two other algebraic conditions in Theorems 6.1 and 6.2 in investigating Lagrange exponential stability of system (6.36). Hence, Ω_{13} is a better exponential convergence ball of system (6.36). This conclusion coincides with the one for fractional-order complex-valued neural networks in Jian and Wan (2017).

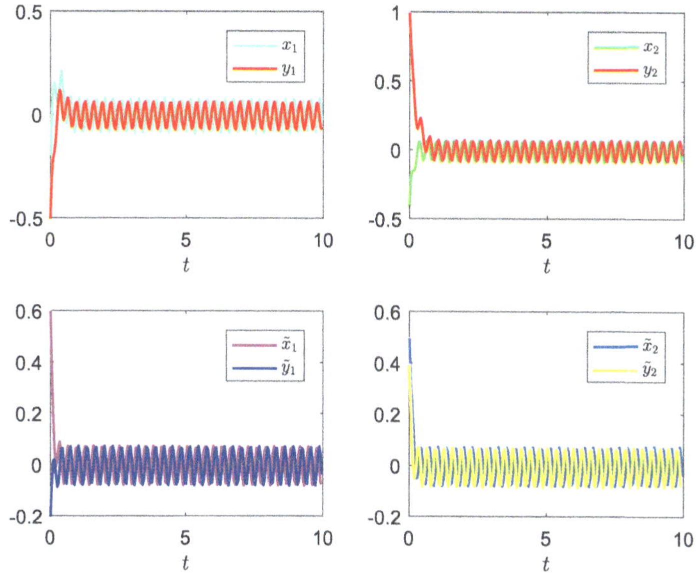

Fig. 6.1 Time responses of real and imaginary parts of the state variables of system (6.36), for Example 6.1

For simulations, the initial conditions are chosen by $\vartheta(s) = \mathrm{col}\{-0.2 - 0.5i, -0.4 + i, 0.6 - 0.2i, 0.5 + 0.4i\}, t \in [-0.5, 0]$. The time responses of real and imaginary parts of system (6.36) with input vectors are shown in Fig. 6.1. The corresponding phase graphs are depicted in Fig. 6.2. The time responses of real and imaginary parts of system (6.36) without inputs are shown in Fig. 6.3. Obviously, Figs. 6.1 and 6.2 show the effectiveness of the obtained results of Theorems 6.1–6.3 and Fig. 6.3 verifies the conclusions of Remarks 6.4 and 6.5.

In order to further make some comparisons and show their effectiveness, different from the activation functions in Example 6.1, the following illustrative example will be presented.

Example 6.2 Consider the same system as in Example 6.1, where

$$f_k(v_k) = \frac{1 - e^{-\tilde{x}_k}}{1 + e^{-\tilde{x}_k}} + i\frac{1}{1 + e^{-\tilde{y}_k}},$$
$$g_j(u_j) = \frac{1 - e^{-x_j}}{1 + e^{-x_j}} + i\frac{1}{1 + e^{-y_j}}, \quad k = 1, 2, \quad j = 1, 2,$$

and the other parameters are as in Example 6.1.

Here, $f_k^R(\tilde{x}_k, \tilde{y}_k) = (1 - e^{-\tilde{x}_k})/(1 + e^{-\tilde{x}_k})$, $f_k^I(\tilde{x}_k, \tilde{y}_k) = 1/(1 + e^{-\tilde{y}_k})$, $g_j^R(x_j, y_j) = (1 - e^{-x_j})/(1 + e^{-x_j})$ and $g_j^I(x_j, y_j) = 1/(1 + e^{-y_j})$. By the Lagrange mean value theorem, we have

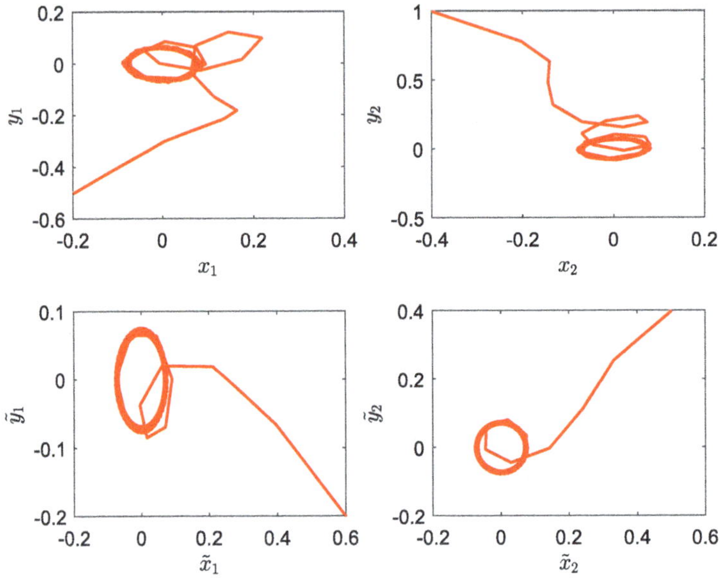

Fig. 6.2 Curve of the state variables of system (6.36) for Example 6.1

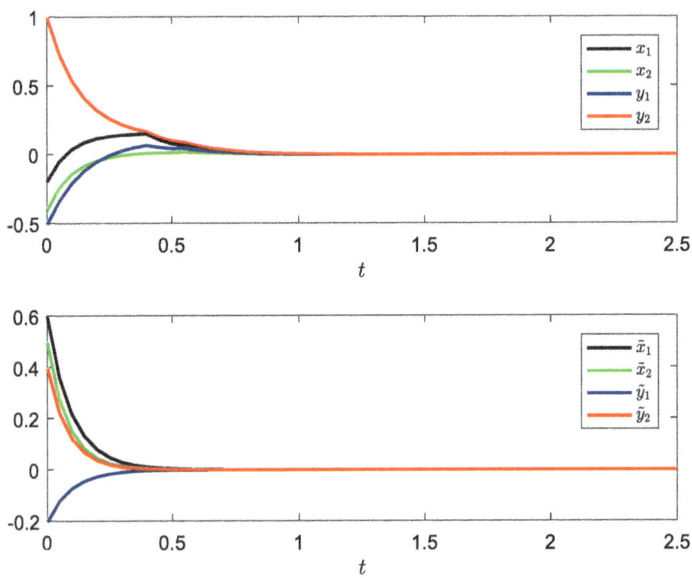

Fig. 6.3 Time responses of real and imaginary parts of the states variables of system (6.36) without inputs, for Example 6.1

$$
\begin{aligned}
|f_k^R(\tilde{x}_k, \tilde{x}_k) - f_k^R(\tilde{x}_k', \tilde{y}_k')| &\le \frac{1}{2}|\tilde{x}_k - \tilde{x}_k'|, \\
|f_k^I(\tilde{x}_k, \tilde{x}_k) - f_k^I(\tilde{x}_k', \tilde{y}_k')| &\le \frac{1}{4}|\tilde{y}_k - \tilde{y}_k'|, \\
|g_j^R(x_j, y_j) - g_j^R(x_j', y_j')| &\le \frac{1}{2}|x_j - x_j'|, \\
|g_j^I(x_j, y_j) - g_j^I(x_j', y_j')| &\le \frac{1}{4}|y_j - y_j'|.
\end{aligned}
$$

On the other hand,

$$
\begin{aligned}
|f_k(v_k) - f_k(v_k')| &= \sqrt{\left(\frac{2}{1+e^{-\tilde{x}_k}} - \frac{2}{1+e^{-\tilde{x}_k'}}\right)^2 + \left(\frac{1}{1+e^{-\tilde{y}_k}} - \frac{1}{1+e^{-\tilde{y}_k'}}\right)^2} \\
&\le \sqrt{\frac{1}{4}(\tilde{x}_k - \tilde{x}_k')^2 + \frac{1}{16}(\tilde{y}_k - \tilde{y}_k')^2} \\
&\le \frac{1}{2}\sqrt{(\tilde{x}_k - \tilde{x}_k')^2 + (\tilde{y}_k - \tilde{y}_k')^2} = \frac{1}{2}|v_k - v_k'|.
\end{aligned}
$$

Similarly,

$$
|g_j(u_j) - g_j(u_j')| \le \frac{1}{2}|u_j - u_j'|.
$$

Thus, Assumptions 6.1 and 6.2 are satisfied with $\lambda_k^{RR} = \chi_j^{RR} = \frac{1}{2}$, $\lambda_k^{II} = \chi_j^{II} = \frac{1}{4}$, $\lambda_k^{RI} = \lambda_k^{IR} = \chi_j^{RI} = \chi_j^{IR} = 0$ and $l_k = \mu_j = \frac{1}{2}$, respectively. Next, we will verify the results of Theorems 6.1–6.3 based on this example.

Case 1: In Theorem 6.1, by choosing $\alpha_j = \beta_j = \tilde{\alpha}_k = \tilde{\beta}_k = 2$ for $j, k = 1, 2$, we can get that (6.4) is true. Thus, this system is globally exponentially stable in Lagrange sense and it globally exponentially converges to the ball Ω_{21} with a rate 1.641, where

$$
\begin{aligned}
\Omega_{21} = \Big\{ & u(t) = x(t) + y(t) \in \mathbb{C}^n, v(t) = \tilde{x}(t) + \tilde{y}(t) \\
& \in \mathbb{C}^m \Big| \sum_{j=1}^{n}(|x_j| + |y_j|) + \sum_{k=1}^{m}(|\tilde{x}_k| + |\tilde{y}_k|) \le 10.8166 \Big\}.
\end{aligned}
$$

Case 2: In Theorem 6.2, by choosing the same scalars as in Example 6.1, we can verify that (6.18) is true. Then, this system is globally exponentially stable in Lagrange sense and it globally exponentially converges to the ball Ω_{22} with a rate 0.659, where

$$
\Omega_{22} = \{\omega(t) \in \mathbb{C}^{n+m} \Big| \|\omega(t)\| = \sqrt{\sum_{j=1}^{n} u_j^*(t)u_j(t) + \sum_{k=1}^{m} v_k^*(t)v_k(t)} \le 2.3663\}.
$$

Table 6.2 Computed values B_2 and $\frac{\tilde{\delta}}{2}$ for different ε_1 and ε_2 in Case 3, for Example 6.2

ε_1 and ε_2	$\varepsilon_1 = \varepsilon_2 = 3.7$	$\varepsilon_1 = \varepsilon_2 = 3.9$	$\varepsilon_1 = \varepsilon_2 = 4.1$	$\varepsilon_1 = \varepsilon_2 = 4.3$
B_2	2.2838	2.2254	2.1674	2.1177
$\frac{\tilde{\delta}}{2}$	0.1315	0.1325	0.134	0.135

Case 3: In Theorem 6.3, by choosing suitable positive scalars ε_1 and ε_2 such that (5.26) holds, then this system is globally exponentially stable in Lagrange sense and it globally exponentially converges to the ball $\Omega_{23} = \{\omega(t) \in \mathbb{C}^{n+m} \big| \|\omega(t)\| = \sqrt{u^*(t)u(t) + v^*(t)v(t)} \leq B_2\}$. Here, we also choose different constants ε_1 and ε_2 and obtain different B_2 and $\frac{\tilde{\delta}}{2}$ which are shown in Table 6.2. Equally, from Table 6.2, similar conclusion can be obtained and it is clear Ω_{23} is a better exponential convergence ball of this system.

Moreover, for simulations, the corresponding figures are shown in Figs. 6.4 and 6.5.

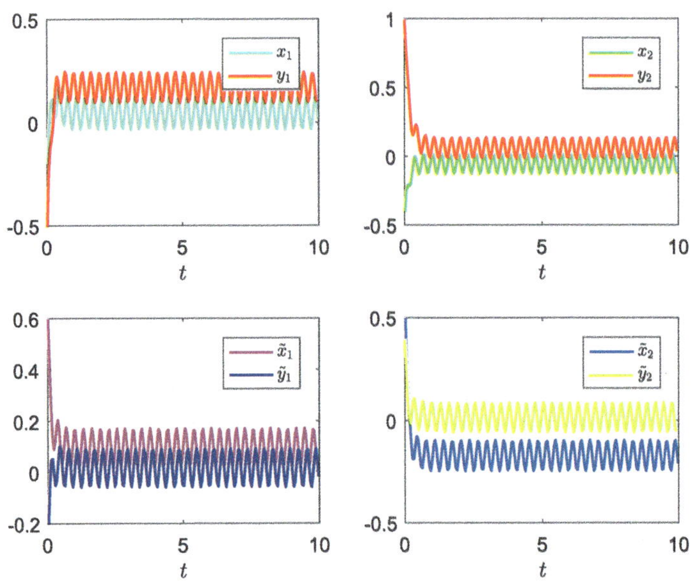

Fig. 6.4 Time responses of real and imaginary parts of the state variables for Example 6.2

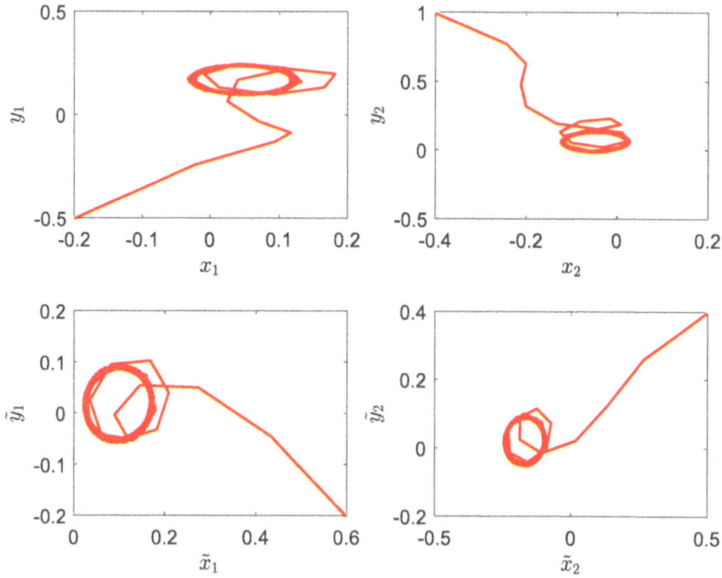

Fig. 6.5 Curve of the state variables for Example 6.2

6.6 Conclusion

In this chapter, we have discussed the Lagrange exponential stability problem for CVBAMNNs with time-varying delays. We have derived different sufficient criteria based on activation functions satisfying different assumption conditions, which guarantee the Lagrange exponential stability of the addressed system. Meanwhile, we have also given the estimations of different globally attractive sets. The effectiveness and superiority-inferiority of these results have been shown by illustrative examples.

References

Hu J, Wang J (2012) Global stability of complex-valued recurrent neural networks with time-delays. IEEE Trans Neural Netw Learn Syst 23(6):853–865

Zhou B, Song Q (2013) Boundedness and complete stability of complex-valued neural networks with time delay. IEEE Trans Neural Netw Learn Syst 24(8):1227–1238

Zhang Z, Lin C, Chen B (2014) Global stability criterion for delayed complex-valued recurrent neural networks. IEEE Trans Neural Netw Learn Syst 25(9):1704–1708

Liu X, Chen T (2016) Global exponential stability for complex-valued recurrent neural networks with asynchronous time delays. IEEE Trans Neural Netw Learn Syst 27(3):593–606

Wang H, Duan S, Huang T, Wang L, Li C (2017) Exponential stability of complex-valued memristive recurrent neural networks. IEEE Trans Neural Netw Learn Syst 28:766–771

Zhang Z, Liu X, Zhou D, Lin C, Chen J, Wang H (2018) Finite-time stabilizability and instabilizability for complex-valued memristive neural networks with time delays. IEEE Trans Syst Man Cybern Syst 48(12):2371–2382
Wang Z, Wang X, Li Y, Huang X (2017) Stability and Hopf bifurcation of fractional-order complex-valued single neuron model with time delay. Int J Bifurc Chaos 27(13):1750209
Guo R, Zhang Z, Gao M (2018) State estimation for complex-valued memristive neural networks with time-varying delays. Adv Differ Equ 118
Fang T, Sun J (2014) Further investigate the stability of complex-valued recurrent neural networks with time-delays. IEEE Trans Neural Netw Learn Syst 25(9):1709–1713
Zhang Z, Liu X, Chen J, Guo R, Zhou S (2017) Further stability analysis for delayed complex-valued recurrent neural networks. Neurocomputing 251:81–89
Song Q, Yan H, Zhao Z, Liu Y (2016) Global exponential stability of complex-valued neural networks with both time-varying delays and impulsive effects. Neural Netw 79:108–116
Song Q, Yan H, Zhao Z, Liu Y (2016) Global exponential stability of impulsive complex-valued neural networks with both asynchronous time-varying and continuously distributed delays. Neural Netw 81:1–10
Zhang Z, Yu S (2016) Global asymptotic stability for a class of complex-valued Cohen-Grossberg neural networks with time delays. Neurocomputing 171:1158–1166
Wang Z, Huang L, Liu Y (2016) Global stability analysis for delayed complex-valued BAM neural networks. Neurocomputing 173:2083–2089
Rajivganthi C, Rihan FA, Lakshmanan S (2019) Dissipativity analysis of complex-valued BAM neural networks with time delay. Neural Comput Appl 31:127–137
Subramanian K, Muthukumar P (2018) Existence, uniqueness, and global asymptotic stability analysis for delayed complex-valued Cohen-Grossberg BAM neural networks. Neural Comput Appl 29:565–584
Xu D, Tan M (2017) Delay-independent stability criteria for complex-valued BAM neutral-type neural networks with time delays. Nonlinear Dyn 89(2):819–832
Guo R, Zhang Z, Liu X, Lin C (2017) Existence, uniqueness, and exponential stability analysis for complex-valued memristor-based BAM neural networks with time delays. Appl Math Comput 311:100–117
Guo R, Zhang Z, Liu X, Lin C, Wang H, Chen J (2018) Exponential input-to-state stability for complex-valued memristor-based BAM neural networks with multiple time-varying delays. Neurocomputing 275:2041–2054
Foss J, Longtin A, Mensour B, Milton J (1996) Multistability and delayed recurrent loops. Phys Rev Lett 76(4):708–711
Huang G, Cao J (2008) Multistability in bidirectional associative memory neural networks. Phys Lett A 372(16):2842–2854
Liao X, Luo Q, Zeng Z, Guo Y (2008) Global exponential stability in Lagrange sense for recurrent neural networks with time delays. Nonlinear Anal-Real 9(4):1535–1557
Wang X, Jiang M, Fang S (2009) Stability analysis in Lagrange sense for a non-autonomous Cohen-Grossberg neural networks with mixed delays. Nonlinear Anal-Theor 70(12):4294–4306
Wu A, Zeng Z (2014) Lagrange stability of memristive neural networks with discrete and distributed delays. IEEE Trans Neural Netw Learn Syst 25(4):690–703
Wu A, Zeng Z (2014) Lagrange stability of neural networks with memristive synapses and multiple delays. Inf Sci 280:135–151
Wang B, Jian J, Jiang M (2010) Stability in lagrange sense for cohen-grossberg neural networks with time-varying delays and finite distributed delays. Nonlinear Anal-Hybri 4(1):65–78
Jian J, Wang B (2015) Stability analysis in lagrange sense for a class of BAM neural networks of neutral type with multiple time-varying delays. Neurocomputing 149:930–939
Tu Z, Cao J, Tasawar H (2016) Global exponential stability in Lagrange sense for inertial neural networks with time-varying delays. Neurocomputing 171:524–531

Tu Z, Jian J, Wang K (2011) Global exponential stability in lagrange sense for recurrent neural networks with both time-varying delays and general activation functions via lmi approach. Nonlinear Anal-Real 12(4):2174–2182

Xiong W, Song L, Cao J (2008) Adaptive robust convergence of neural networks with time-varying delays. Nonlinear Anal-Real 9(4):1283–1291

Jian J, Wan P (2017) Lagrange α-exponential stability and α-exponential convergence for fractional-order complex-valued neural networks. Neural Netw 91:1–10

Song Q, Shu H, Zhao Z, Liu Y, Alsaadi FE (2017) Lagrange stability analysis for complex-valued neural networks with leakage delay and mixed time-varying delays. Neurocomputing 244:33–41

Tu Z, Cao J, Alsaedi A, Alsaadi FE, Hayat T (2016) Global lagrange stability of complex-valued neural networks of neutral type with time-varying delays. Complexity 21(S2):438–450

Goh SL, Chen M, Popovic DH, Aihara K, Obradovic D, Mandic DP (2006) Complex-valued forecasting of wind profile. Renew Energy 31(11):1733–1750

Chapter 7
Anti-synchronization Control for CVBAMNNs with Time-Varying Delays

Abstract This chapter aims to investigate the exponential anti-synchronization control of complex-valued bidirectional associative memory neural networks (CVBAMNNs) with time-varying delays. We will use a suitable Lyapunov functional and some inequalities techniques to solve the exponential anti-synchronization problem. We will propose a sufficient criterion based on algebraic structure for the exponential anti-synchronization control design.

7.1 Introduction

As everyone knows, when analyzing dynamic behavior of systems, synchronization and anti-synchronization are very hot topic (Xing et al. 2018; Wang et al. 2019a, b) since they have been successfully applied in engineering applications and hardware implementations, such as image processing, information science and so on. As a result, it becomes very important and necessary to study the synchronization and anti-synchronization control problems in theoretical work. Moreover, for complex-valued neural networks (CVNNs), some results involving these problems have been reported, such as synchronization of fractional-order delayed CVNNs (Bao et al. 2016; Zhang et al. 2019), finite-time synchronization for delayed CVNNs (Shen and Cao 2011; Zhou et al. 2017; Zhang et al. 2018a, b) and anti-synchronization of memristor-based delayed CVNNs (Liu et al. 2019, 2018). At the same time, for complex-valued bidirectional associative memory neural networks (CVBAMNNs), many researchers are interested in the synchronization problem of CVBAMNNs and have achieved some results. For example, a sufficient condition on global asymptotic periodic synchronization of CVBAMNNs is established by using novel LMI method in Zhang et al. (2018). However, for the anti-synchronization control problem, there has been still no information.

In this chapter, we will study the anti-synchronization control problem for CVBAMNNs with time-varying delays. The main contributions of our work can be shown in the following points: (1) Compared with the previous results, it is the first time that the exponential anti-synchronization control problem of CVBAMNNs with time-varying delays is investigated; (2) Via a suitable Lyapunov functional and the

Z. Zhang et al., *Complex-Valued Neural Networks Systems with Time Delay*, Intelligent Control and Learning Systems 4,
https://doi.org/10.1007/978-981-19-5450-4_7

inequality techniques, a sufficient condition is established to ensure the exponential anti-synchronization of the considered system; (3) According to Hölder inequality, the right inequalities different from those in the existing references are used to derive the main result.

7.2 Problem Formulation and Preliminaries

Consider the following CVBAMNNs model with time-varying delays:

$$\begin{cases} \dot{z}_k(t) = -d_{1k}z_k(t) + \sum\limits_{l=1}^{m} a_{1lk} f_l(h_l(t)) + \sum\limits_{l=1}^{m} b_{1lk} f_l(h_l(t-\tau_{2kl}(t))), \\ \dot{h}_l(t) = -d_{2l}h_l(t) + \sum\limits_{k=1}^{n} a_{2kl} g_k(z_k(t)) + \sum\limits_{k=1}^{n} b_{2kl} g_k(z_k(t-\tau_{1lk}(t))) \end{cases} \tag{7.1}$$

with the initial conditions

$$z_k(s) = \psi_{1k}(s),\ h_l(s) = \psi_{2l}(s),\ s \in [-\tau, 0], \tag{7.2}$$

where $k = 1, 2, ..., n$ and $l = 1, 2, ..., m$; $z_k(t)$ and $h_l(t)$ denote the complex-valued state variables; $d_{1k} > 0$ and $d_{2l} > 0$ are constants; $a_{1lk}, b_{1lk}, a_{2kl}$ and b_{2kl} are complex-valued connection weights; $f_l(\cdot)$ and $g_k(\cdot)$ denote the complex-valued activation functions; $\tau_{2kl}(t)$ and $\tau_{1lk}(t)$ are time-varying delays with $0 \le \tau_{2kl}(t) \le \tau_1, \dot{\tau}_{2kl}(t) \le \rho_1 \le 1, 0 \le \tau_{1lk}(t) \le \tau_2, \dot{\tau}_{1lk}(t) \le \rho_2 \le 1, \tau_1, \tau_2, \rho_1, \rho_2$ are positive constants; $\tau = \max\{\tau_1, \tau_2\}$, $\psi_{1k}(s)$ and $\psi_{2l}(s)$ ($\forall s \in [-\tau, 0]$) are continuous.

Moreover, complex-valued activation functions $f_l(h_l(t))$ and $g_k(z_k(t))$ can be separated into the real and imaginary parts as

$$f_l(h_l) = f_l^R(x_{2l}, y_{2l}) + i f_l^I(x_{2l}, y_{2l}),\ g_k(z_k) = g_k^R(x_{1k}, y_{1k}) + i g_k^I(x_{1k}, y_{1k}),$$

where $h_l(t) = x_{2l}(t) + iy_{2l}(t)$, $z_k(t) = x_{1k}(t) + iy_{1k}(t)$, and $h_l(t)$, $z_k(t)$, $x_{2l}(t)$, $y_{2l}(t)$, $x_{1k}(t)$, $y_{1k}(t)$ are simplified as h_l, z_k, x_{2l}, y_{2l}, x_{1k}, y_{1k}, respectively. They satisfy the following assumption.

Assumption 7.1 $f_l(-h_l) = -f_l(h_l)$, $g_k(-z_k) = -g_k(z_k)$, and for any $x_{1k}, y_{1k}, x'_{1k}, y'_{1k}, x_{2l}, y_{2l}, x'_{2l}$ and $y'_{2l} \in \mathbb{R}$, there exist positive constants $\lambda_{1l}^{RR}, \lambda_{1l}^{RI}, \lambda_{1l}^{IR}, \lambda_{1l}^{II}, \lambda_{2k}^{RR}, \lambda_{2k}^{RI}, \lambda_{2k}^{IR}$ and λ_{2k}^{II} such that

$$\begin{aligned} |f_l^R(x_{2l}, y_{2l}) - f_l^R(x'_{2l}, y'_{2l})| &\le \lambda_{1l}^{RR}|x_{2l} - x'_{2l}| + \lambda_{1l}^{RI}|y_{2l} - y'_{2l}|, \\ |f_l^I(x_{2l}, y_{2l}) - f_l^I(x'_{2l}, y'_{2l})| &\le \lambda_{1l}^{IR}|x_{2l} - x'_{2l}| + \lambda_{1l}^{II}|y_{2l} - y'_{2l}|, \\ |g_k^R(x_{1k}, y_{1k}) - g_k^R(x'_{1k}, y'_{1k})| &\le \lambda_{2k}^{RR}|x_{1k} - x'_{1k}| + \lambda_{2k}^{RI}|y_{1k} - y'_{1k}|, \\ |g_k^I(x_{1k}, y_{1k}) - g_k^I(x'_{1k}, y'_{1k})| &\le \lambda_{2k}^{IR}|x_{1k} - x'_{1k}| + \lambda_{2k}^{II}|y_{1k} - y'_{1k}|. \end{aligned} \tag{7.3}$$

According to the drive system (7.1), the response system is described as follows:

$$\begin{cases} \dot{z}_k^*(t) = -d_{1k}z_k^*(t) + \sum_{l=1}^{m} a_{1lk} f_l(h_l^*(t)) + \sum_{l=1}^{m} b_{1lk} f_l(h_l^*(t-\tau_{2kl}(t))) + u_{1k}(t), \\ \dot{h}_l^*(t) = -d_{2l}h_l^*(t) + \sum_{k=1}^{n} a_{2kl} g_k(z_k^*(t)) + \sum_{k=1}^{n} b_{2kl} g_k(z_k^*(t-\tau_{1lk}(t))) + u_{2l}(t) \end{cases} \tag{7.4}$$

with the initial conditions

$$z_k^*(s) = \theta_{1k}(s),\ h_l^*(s) = \theta_{2l}(s),\ s \in [-\tau, 0], \tag{7.5}$$

where $u_{1k}(t)$ and $u_{2l}(t)$ are to be designed to achieve a certain control objective. $\theta_{1k}(s)$ and $\theta_{2l}(s)$ ($\forall s \in [-\tau, 0]$) are continuous.

Let $e_{1k}(t) = z_k^*(t) + z_k(t)$ and $e_{2l}(t) = h_l^*(t) + h_l(t)$, the anti-synchronization error system between (7.1) and (7.4) can be given by

$$\begin{cases} \dot{e}_{1k}(t) = -d_{1k}e_{1k}(t) + \sum_{l=1}^{m} a_{1lk}\mathcal{F}_{1l}(e_{2l}(t)) + \sum_{l=1}^{m} b_{1lk}\mathcal{F}_{1l}(e_{2l}(t-\tau_{2kl}(t))) + u_{1k}(t), \\ \dot{e}_{2l}(t) = -d_{2l}e_{2l}(t) + \sum_{k=1}^{n} a_{2kl}\mathcal{F}_{2k}(e_{1k}(t)) + \sum_{k=1}^{n} b_{2kl}\mathcal{F}_{2k}(e_{1k}(t-\tau_{1lk}(t))) + u_{2l}(t) \end{cases} \tag{7.6}$$

with the initial conditions

$$e_{1k}(s) = \phi_{1k}(s),\ e_{2l}(s) = \phi_{2l}(s),\ s \in [-\tau, 0], \tag{7.7}$$

where $\mathcal{F}_{1l}(e_{2l}(t)) = f_l(h_l^*(t)) + f_l(h_l(t))$, $\mathcal{F}_{1l}(e_{2l}(t-\tau_{2kl}(t))) = f_l(h_l^*(t-\tau_{2kl}(t))) + f_l(h_l(t-\tau_{2kl}(t)))$, $\mathcal{F}_{2k}(e_{1k}(t)) = g_k(z_k^*(t)) + g_k(z_k(t))$, $\mathcal{F}_{2k}(e_{1k}(t-\tau_{1lk}(t))) = g_k(z_k^*(t-\tau_{1lk}(t))) + g_k(z_k(t-\tau_{1lk}(t)))$, $\phi_{1k}(s) = \psi_{1k}(s) + \theta_{1k}(s)$ and $\phi_{2l}(s) = \psi_{2l}(s) + \theta_{2l}(s)$.

Define $e_{1k}(t) = e_{1k}^R(t) + ie_{1k}^I(t)$, $e_{2l}(t) = e_{2l}^R(t) + ie_{2l}^I(t)$, $u_{1k}(t) = u_{1k}^R(t) + iu_{1k}^I(t)$, $u_{2l}(t) = u_{2l}^R(t) + iu_{2l}^I(t)$, $a_{1lk} = a_{1lk}^R + ia_{1lk}^I$, $b_{1lk} = b_{1lk}^R + ib_{1lk}^I$, $a_{2kl} = a_{2kl}^R + ia_{2kl}^I$ and $b_{2kl} = b_{2kl}^R + ib_{2kl}^I$. Obviously, $e_{1k}^R(t) = x_{1k}(t) + x_{1k}^*(t)$, $e_{1k}^I(t) = y_{1k}(t) + y_{1k}^*(t)$, $e_{2l}^R(t) = x_{2l}(t) + x_{2l}^*(t)$ and $e_{2l}^I(t) = y_{2l}(t) + y_{2l}^*(t)$. For brevity, $e_{1k}(t)$, $e_{2l}(t)$, $e_{1k}(t-\tau_{1lk}(t))$, $e_{2l}(t-\tau_{2kl}(t))$, $e_{1k}^R(t)$, $e_{1k}^I(t)$, $e_{2l}^R(t)$, $e_{2l}^I(t)$, $e_{1k}^R(t-\tau_{1lk}(t))$, $e_{1k}^I(t-\tau_{1lk}(t))$, $e_{2l}^R(t-\tau_{2kl}(t))$, $e_{2l}^I(t-\tau_{2kl}(t))$, $u_{1k}^R(t)$, $u_{1k}^I(t)$, $u_{2l}^R(t)$, $u_{2l}^I(t)$, $x_{2l}(t-\tau_{2kl}(t))$, $y_{2l}(t-\tau_{2kl}(t))$, $x_{1k}(t-\tau_{1lk}(t))$ and $y_{1k}(t-\tau_{1lk}(t))$ are simplified as e_{1k}, e_{2l}, e_{1k}^τ, e_{2l}^τ, e_{1k}^R, e_{1k}^I, e_{2l}^R, e_{2l}^I, $e_{1k}^{R\tau}$, $e_{1k}^{I\tau}$, $e_{2l}^{R\tau}$, $e_{2l}^{I\tau}$, u_{1k}^R, u_{1k}^I, u_{2l}^R, u_{2l}^I, x_{2l}^τ, y_{2l}^τ, x_{1k}^τ and y_{1k}^τ. Then, system (7.6) can be separated into real and imaginary parts as

$$\dot{e}_{1k}^R = -d_{1k}e_{1k}^R + \sum_{l=1}^{m} a_{1lk}^R \mathcal{F}_{1l}^R(e_{2l}^R, e_{2l}^I) - \sum_{l=1}^{m} a_{1lk}^I \mathcal{F}_{1l}^I(e_{2l}^R, e_{2l}^I)$$
$$+ \sum_{l=1}^{m} b_{1lk}^R \mathcal{F}_{1l}^R(e_{2l}^{R\tau}, e_{2l}^{I\tau}) - \sum_{l=1}^{m} b_{1lk}^I \mathcal{F}_{1l}^I(e_{2l}^{R\tau}, e_{2l}^{I\tau}) + u_{1k}^R,$$
$$\dot{e}_{1k}^I = -d_{1k}e_{1k}^I + \sum_{l=1}^{m} a_{1lk}^R \mathcal{F}_{1l}^I(e_{2l}^R, e_{2l}^I) + \sum_{l=1}^{m} a_{1lk}^I \mathcal{F}_{1l}^R(e_{2l}^R, e_{2l}^I)$$
$$+ \sum_{l=1}^{m} b_{1lk}^R \mathcal{F}_{1l}^I(e_{2l}^{R\tau}, e_{2l}^{I\tau}) + \sum_{l=1}^{m} b_{1lk}^I \mathcal{F}_{1l}^R(e_{2l}^{R\tau}, e_{2l}^{I\tau}) + u_{1k}^I,$$
$$\dot{e}_{2l}^R = -d_{2l}e_{2l}^R + \sum_{k=1}^{n} a_{2kl}^R \mathcal{F}_{2k}^R(e_{1k}^R, e_{1k}^I) - \sum_{k=1}^{n} a_{2kl}^I \mathcal{F}_{2k}^I(e_{1k}^R, e_{1k}^I)$$
$$+ \sum_{k=1}^{n} b_{2kl}^R \mathcal{F}_{2k}^R(e_{1k}^{R\tau}, e_{1k}^{I\tau}) - \sum_{k=1}^{n} b_{2kl}^I \mathcal{F}_{2k}^I(e_{1k}^{R\tau}, e_{1k}^{I\tau}) + u_{2l}^R,$$
$$\dot{e}_{2l}^I = -d_{2l}e_{2l}^I + \sum_{k=1}^{n} a_{2kl}^R \mathcal{F}_{2k}^I(e_{1k}^R, e_{1k}^I) + \sum_{k=1}^{n} a_{2kl}^I \mathcal{F}_{2k}^R(e_{1k}^R, e_{1k}^I)$$
$$+ \sum_{k=1}^{n} b_{2kl}^R \mathcal{F}_{2k}^I(e_{1k}^{R\tau}, e_{1k}^{I\tau}) + \sum_{k=1}^{n} b_{2kl}^I \mathcal{F}_{2k}^R(e_{1k}^{R\tau}, e_{1k}^{I\tau}) + u_{2l}^I, \tag{7.8}$$

where $\mathcal{F}_{1l}^\sigma(e_{2l}^R, e_{2l}^I) = f_l^\sigma(x_{2l}^*, y_{2l}^*) + f_l^\sigma(x_{2l}, y_{2l})$, $\mathcal{F}_{1l}^\sigma(e_{2l}^{\sigma\tau}, e_{2l}^{\sigma\tau}) = f_l^\sigma(x_{2l}^{*\tau}, y_{2l}^{*\tau}) + f_l^\sigma(x_{2l}^\tau, y_{2l}^\tau)$, $\mathcal{F}_{2k}^\sigma(e_{1k}^R, e_{1k}^I) = g_k^\sigma(x_{1k}^*, y_{1k}^*) + g_k^\sigma(x_{1k}, y_{1k})$, $\mathcal{F}_{2k}^\sigma(e_{1k}^{R\tau}, e_{1k}^{I\tau}) = g_k^\sigma(x_{1k}^{*\tau}, y_{1k}^{*\tau}) + g_k^\sigma(x_{1k}^\tau, y_{1k}^\tau)$, for $\sigma = R, I$.

Next, we introduce the following definition and lemmas to get the desired result. Let

$$w_{1k}(t) = \begin{cases} e_{1k}^R(t), & k = 1, 2, ..., n, \\ e_{1k}^I(t), & k = n+1, n+2, ..., 2n, \end{cases}$$
$$w_{2l}(t) = \begin{cases} e_{2l}^R(t), & l = 1, 2, ..., m, \\ e_{2l}^I(t), & l = m+1, m+2, ..., 2m. \end{cases}$$

Moreover, for continuous function $\vartheta(s) = (\vartheta_1(s), \vartheta_2(s), ..., \vartheta_n(s))^T$ $(s \in [-\tau, 0])$, the norm $\| \vartheta \| = \sup_{-\tau \le s \le 0}\{\sum_{k=1}^n |\vartheta_k(s)|^\varrho\}^{\frac{1}{\varrho}}$ $(\varrho > 1)$ is equipped.

Definition 7.1 Systems (7.1) and (7.4) can achieve the exponential anti-synchronization if for arbitrary initial condition $\Phi(s) \in \mathbb{R}^{2n+2m}$, properly designed feedback controllers, and $\varrho \ge 1$, there exist constants $\alpha \ge 1$ and $\epsilon > 0$ such that

$$\left\{\sum_{k=1}^{2n} |w_{1k}(t)|^\varrho + \sum_{l=1}^{2m} |w_{2l}(t)|^\varrho\right\}^{\frac{1}{\varrho}} \le \alpha e^{-\epsilon t} \| \Phi \|, \text{ for } \forall\, t \ge 0, \tag{7.9}$$

where ϵ is the estimated rate of exponential anti-synchronization.

Lemma 7.1 (Young Inequality). *Let $a \geq 0$, $b \geq 0$, $p > 1$, $\frac{1}{p} + \frac{1}{q} = 1$, then the following inequality hold:*

$$ab \leq \frac{a^p}{p} + \frac{b^q}{q}. \tag{7.10}$$

Lemma 7.2 (Hölder Inequality). *Let $a_i \geq 0$, $b_i \geq 0 (i = 1, 2, ..., n)$, $p > 1$, $\frac{1}{p} + \frac{1}{q} = 1$, then the following inequality hold:*

$$\sum_{i=1}^{n} a_i b_i \leq \Big(\sum_{i=1}^{n} a_i^p\Big)^{1/p} \Big(\sum_{i=1}^{n} b_i^q\Big)^{1/q}. \tag{7.11}$$

Especially, let $b_i = 1$ and $n = 2$, one has

$$(a_1 + a_2)^p \leq 2^{p-1}(a_1^p + a_2^p). \tag{7.12}$$

Moreover, it is obvious that the inequality (7.12) also hold when $p = 1$.

Remark 7.1 Based on Assumption 7.1 and Lemma 7.2, for $\mathcal{F}_{1l}^R(e_{2l}^R, e_{2l}^I)$, $\mathcal{F}_{1l}^I(e_{2l}^R, e_{2l}^I)$, $\mathcal{F}_{2k}^R(e_{1k}^R, e_{1k}^I)$, $\mathcal{F}_{2k}^I(e_{1k}^R, e_{1k}^I)$ and $p \geq 1$, the following inequalities hold:

$$\begin{aligned}
|\mathcal{F}_{1l}^R(e_{2l}^R, e_{2l}^I)|^p &\leq 2^{p-1}[(\lambda_{1l}^{RR})^p |e_{2l}^R|^p + (\lambda_{1l}^{RI})^p |e_{2l}^I|^p],\\
|\mathcal{F}_{1l}^I(e_{2l}^R, e_{2l}^I)|^p &\leq 2^{p-1}[(\lambda_{1l}^{IR})^p |e_{2l}^R|^p + (\lambda_{1l}^{II})^p |e_{2l}^I|^p],\\
|\mathcal{F}_{2k}^R(e_{1k}^R, e_{1k}^I)|^p &\leq 2^{p-1}[(\lambda_{2k}^{RR})^p |e_{1k}^R|^p + (\lambda_{2k}^{RI})^p |e_{1k}^I|^p],\\
|\mathcal{F}_{2k}^I(e_{1k}^R, e_{1k}^I)|^p &\leq 2^{p-1}[(\lambda_{2k}^{IR})^p |e_{1k}^R|^p + (\lambda_{2k}^{II})^p |e_{1k}^I|^p].
\end{aligned} \tag{7.13}$$

Here, in fact, since $f_l^R(x_{2l}, y_{2l})$ is odd function, $|\mathcal{F}_{1l}^R(e_{2l}^R, e_{2l}^I)| = |f_l^R(x_{2l}^*, y_{2l}^*) + f_l^R(x_{2l}, y_{2l})| = |f_l^R(x_{2l}^*, y_{2l}^*) - f_l^R(-x_{2l}, -y_{2l})|$. Then, according to Assumption 7.1 and the inequality (7.12), we can obtain

$$\begin{aligned}
|\mathcal{F}_{1l}^R(e_{2l}^R, e_{2l}^I)|^p &\leq (\lambda_{1l}^{RR}|e_{2l}^R| + \lambda_{1l}^{RI}|e_{2l}^I|)^p\\
&\leq 2^{p-1}[(\lambda_{1l}^{RR})^p |e_{2l}^R|^p + (\lambda_{1l}^{RI})^p |e_{2l}^I|^p].
\end{aligned} \tag{7.14}$$

Similarly, we can also obtain the other inequalities in (7.13). Moreover, $\mathcal{F}_{1l}^R(0, 0) = f_l^R(x_{2l}, y_{2l}) + f_l^R(-x_{2l}, -y_{2l}) = 0$, $\mathcal{F}_{1l}^I(0, 0) = 0$, $\mathcal{F}_{2k}^R(0, 0) = 0$ and $\mathcal{F}_{2k}^I(0, 0) = 0$.

Remark 7.2 It should be pointed out that the inequalities in Remark 1 of Liu et al. (2019) are problematic. In Remark 1 of Liu et al. (2019), $|\mathcal{F}_l^R(e_l^R(t), e_l^I(t))|^r \leq (\lambda_l^{RR})^r |e_l^R(t)|^r + (\lambda_l^{RI})^r |e_l^I(t)|^r$ and $|\mathcal{F}_l^I(e_l^R(t), e_l^I(t))|^r \leq (\lambda_l^{IR})^r |e_l^R(t)|^r + (\lambda_l^{II})^r |e_l^I(t)|^r$ $(r \geq 1)$, where the mathematical notations are the same as those in Liu et al. (2019) as a matter of convenience. In fact, $|\mathcal{F}_l^R(e_l^R(t), e_l^I(t))|^r \leq (\lambda_l^{RR}|e_l^R(t)| + \lambda_l^{RI}|e_l^I(t)|)^r$ whereas $(\lambda_l^{RR}|e_l^R(t)| + \lambda_l^{RI}|e_l^I(t)|)^r \geq (\lambda_l^{RR})^r |e_{2l}^R(t)|^r + (\lambda_l^{RI})^r |e_l^I(t)|^r$, so $|\mathcal{F}_l^R(e_l^R(t), e_l^I(t))|^r \leq$

$(\lambda_l^{RR})^r|e_l^R(t)|^r + (\lambda_l^{RI})^r|e_l^I(t)|^r$ may not be true. Here, we rectify it and give the right inequalities (7.13) by using Hölder inequality.

7.3 Anti-Synchronization Control Criterion

In this section, we will give a sufficient condition to ensure the anti-synchronization for system (7.1). The state feedback control laws are designed as

$$\begin{aligned} u_{1k}^R(t) &= -\pi_{1k}^R(x_{1k}(t) + x_{1k}^*(t)),\ u_{1k}^I(t) = -\pi_{1k}^I(y_{1k}(t) + y_{1k}^*(t)), \\ u_{2l}^R(t) &= -\pi_{2l}^R(x_{2l}(t) + x_{2l}^*(t)),\ u_{2l}^I(t) = -\pi_{2l}^I(y_{2l}(t) + y_{2l}^*(t)), \end{aligned} \tag{7.15}$$

where π_{1k}^R, π_{1k}^I, π_{2l}^R and π_{2l}^I are the control gains to be determined.

Remark 7.3 So as to obtain the proposed result, we designed the control laws as shown above, which are simple and effective. Moreover, our work is carried out in theory. Hence, from the theoretical viewpoint, it seems good. In practice, it also has obvious advantages because of its simplicity. So, our designed control laws are suitable.

Theorem 7.1 *Suppose that Assumption 7.1 holds, systems (7.1) and (7.4) can be exponentially anti-synchronized under control inputs (7.15) if there exist positive constants ξ_{1k}, ν_{1k}, ξ_{2l}, ν_{2l} and $r \geq 1$ such that*

$$\begin{aligned}
&-r(d_{1k} + \pi_{1k}^R) + \sum_{l=1}^{m}\Big\{\Big[\Big(\frac{\xi_{2l}}{\xi_{1k}}|a_{2kl}^R| + \frac{\nu_{2l}}{\xi_{1k}}|a_{2kl}^I| + \frac{\xi_{2l}\delta_{2l}}{\xi_{1k}(1-\rho_2)}\Big)(\lambda_{2k}^{RR})^r \\
&+\Big(\frac{\xi_{2l}}{\xi_{1k}}|a_{2kl}^I| + \frac{\nu_{2l}}{\xi_{1k}}|a_{2kl}^R| + \frac{\nu_{2l}\eta_{2l}}{\xi_{1k}(1-\rho_2)}(\lambda_{2k}^{IR})^r\Big)\Big]2^{r-1} + L_1(r-1)\Big\} \leq 0, \\
&-r(d_{1k} + \pi_{1k}^I) + \sum_{l=1}^{m}\Big\{\Big[\Big(\frac{\xi_{2l}}{\nu_{1k}}|a_{2kl}^R| + \frac{\nu_{2l}}{\nu_{1k}}|a_{2kl}^I| + \frac{\xi_{2l}\delta_{2l}}{\nu_{1k}(1-\rho_2)}\Big)(\lambda_{2k}^{RI})^r \\
&+\Big(\frac{\xi_{2l}}{\nu_{1k}}|a_{2kl}^I| + \frac{\nu_{2l}}{\nu_{1k}}|a_{2kl}^R| + \frac{\nu_{2l}\eta_{2l}}{\nu_{1k}(1-\rho_2)}\Big)(\lambda_{2k}^{II})^r\Big]2^{r-1} + L_1(r-1)\Big\} \leq 0, \\
&-r(d_{2l} + \pi_{2l}^R) + \sum_{k=1}^{n}\Big\{\Big[\Big(\frac{\xi_{1k}}{\xi_{2l}}|a_{1lk}^R| + \frac{\nu_{1k}}{\xi_{2l}}|a_{1lk}^I| + \frac{\xi_{1k}\delta_{1k}}{\xi_{2l}(1-\rho_1)}\Big)(\lambda_{1l}^{RR})^r \\
&+\Big(\frac{\xi_{1k}}{\xi_{2l}}|a_{1lk}^I| + \frac{\nu_{1k}}{\xi_{2l}}|a_{1lk}^R| + \frac{\nu_{1k}\eta_{1k}}{\xi_{2l}(1-\rho_1)}\Big)(\lambda_{1l}^{IR})^r\Big]2^{r-1} + L_2(r-1)\Big\} \leq 0, \\
&-r(d_{2l} + \pi_{2l}^I) + \sum_{k=1}^{n}\Big\{\Big[\Big(\frac{\xi_{1k}}{\nu_{2l}}|a_{1lk}^R| + \frac{\nu_{1k}}{\nu_{2l}}|a_{1lk}^I| + \frac{\xi_{1k}\delta_{1k}}{\nu_{2l}(1-\rho_1)}\Big)(\lambda_{1l}^{RI})^r \\
&+\Big(\frac{\xi_{1k}}{\nu_{2l}}|a_{1lk}^I| + \frac{\nu_{1k}}{\nu_{2l}}|a_{1lk}^R| + \frac{\nu_{1k}\eta_{1k}}{\nu_{2l}(1-\rho_1)}\Big)(\lambda_{1l}^{II})^r\Big]2^{r-1} + L_2(r-1)\Big\} \leq 0,
\end{aligned} \tag{7.16}$$

where $\delta_{1k} = |b^R_{1lk}| + (\nu_{1k}/\xi_{1k})|b^I_{1lk}|$, $\eta_{1k} = |b^R_{1lk}| + (\xi_{1k}/\nu_{1k})|b^I_{1lk}|$, $\delta_{2l} = |b^R_{2kl}| + (\nu_{2l}/\xi_{2l})|b^I_{2kl}|$, $\eta_{2l} = |b^R_{2kl}| + (\xi_{2l}/\nu_{2l})|b^I_{2kl}|$, $L_1 = |a^R_{1lk}| + |a^I_{1lk}| + |b^R_{1lk}| + |b^I_{1lk}|$, $L_2 = |a^R_{2kl}| + |a^I_{2kl}| + |b^R_{2kl}| + |b^I_{2kl}|$.

Proof Firstly, for the inequalities (7.16), we choose a sufficient small constant $\epsilon > 0$ such that

$$\begin{aligned}
&r(\epsilon - d_{1k} - \pi^R_{1k}) + \sum_{l=1}^{m}\Big\{\Big[\Big(\frac{\xi_{2l}}{\xi_{1k}}|a^R_{2kl}| + \frac{\nu_{2l}}{\xi_{1k}}|a^I_{2kl}| + \frac{\xi_{2l}e^{r\epsilon\tau_2}}{\xi_{1k}(1-\rho_2)}\delta_{2l}\Big)(\lambda^{RR}_{2k})^r \\
&+\Big(\frac{\xi_{2l}}{\xi_{1k}}|a^I_{2kl}| + \frac{\nu_{2l}}{\xi_{1k}}|a^R_{2kl}| + \frac{\nu_{2l}e^{r\epsilon\tau_2}}{\xi_{1k}(1-\rho_2)}\eta_{2l}\Big)(\lambda^{IR}_{2k})^r\Big]2^{r-1} + L_1(r-1)\Big\} \le 0, \\
&r(\epsilon - d_{1k} - \pi^I_{1k}) + \sum_{l=1}^{m}\Big\{\Big[\Big(\frac{\xi_{2l}}{\nu_{1k}}|a^R_{2kl}| + \frac{\nu_{2l}}{\nu_{1k}}|a^I_{2kl}| + \frac{\xi_{2l}e^{r\epsilon\tau_2}}{\nu_{1k}(1-\rho_2)}\delta_{2l}\Big)(\lambda^{RI}_{2k})^r \\
&+\Big(\frac{\xi_{2l}}{\nu_{1k}}|a^I_{2kl}| + \frac{\nu_{2l}}{\nu_{1k}}|a^R_{2kl}| + \frac{\nu_{2l}e^{r\epsilon\tau_2}}{\nu_{1k}(1-\rho_2)}\eta_{2l}\Big)(\lambda^{II}_{2k})^r\Big]2^{r-1} + L_1(r-1)\Big\} \le 0, \\
&r(\epsilon - d_{2l} - \pi^R_{2l}) + \sum_{k=1}^{n}\Big\{\Big[\Big(\frac{\xi_{1k}}{\xi_{2l}}|a^R_{1lk}| + \frac{\nu_{1k}}{\xi_{2l}}|a^I_{1lk}| + \frac{\xi_{1k}e^{r\epsilon\tau_1}}{\xi_{2l}(1-\rho_1)}\delta_{1k}\Big)(\lambda^{RR}_{1l})^r \\
&+\Big(\frac{\xi_{1k}}{\xi_{2l}}|a^I_{1lk}| + \frac{\nu_{1k}}{\xi_{2l}}|a^R_{1lk}| + \frac{\nu_{1k}e^{r\epsilon\tau_1}}{\xi_{2l}(1-\rho_1)}\eta_{1k}\Big)(\lambda^{IR}_{1l})^r\Big]2^{r-1} + L_2(r-1)\Big\} \le 0, \\
&r(\epsilon - d_{2l} - \pi^I_{2l}) + \sum_{k=1}^{n}\Big\{\Big[\Big(\frac{\xi_{1k}}{\nu_{2l}}|a^R_{1lk}| + \frac{\nu_{1k}}{\nu_{2l}}|a^I_{1lk}| + \frac{\xi_{1k}e^{r\epsilon\tau_1}}{\nu_{2l}(1-\rho_1)}\delta_{1k}\Big)(\lambda^{RI}_{1l})^r \\
&+\Big(\frac{\xi_{1k}}{\nu_{2l}}|a^I_{1lk}| + \frac{\nu_{1k}}{\nu_{2l}}|a^R_{1lk}| + \frac{\nu_{1k}e^{r\epsilon\tau_1}}{\nu_{2l}(1-\rho_1)}\eta_{1k}\Big)(\lambda^{II}_{1l})^r\Big]2^{r-1} + L_2(r-1)\Big\} \le 0.
\end{aligned} \tag{7.17}$$

Next, we construct a Lyapunov functional as follows:

$$\begin{aligned}
V(t) = &\sum_{k=1}^{2n}\zeta_{1k}\Big\{|w_{1k}(t)|^r e^{r\epsilon t} + \frac{1}{1-\rho_1}\sum_{l=1}^{2m}\beta_{1k} \\
&\times \int_{t-\tau_{2kl}(t)}^{t} |\mathcal{F}_{1l}(w_{2l}(s))|^r e^{r\epsilon(s+\tau_{2kl}(s))}ds\Big\} \\
&+\sum_{l=1}^{2m}\zeta_{2l}\Big\{|w_{2l}(t)|^r e^{r\epsilon t} + \frac{1}{1-\rho_2}\sum_{k=1}^{2n}\beta_{2l} \\
&\times \int_{t-\tau_{1lk}(t)}^{t} |\mathcal{F}_{2k}(w_{1k}(s))|^r e^{r\epsilon(s+\tau_{1lk}(s))}ds\Big\},
\end{aligned} \tag{7.18}$$

where

$$\mathcal{F}_{1l}(w_{2l}(t)) = \begin{cases} \mathcal{F}_{1l}^R(e_{2l}^R(t), e_{2l}^I(t)), l = 1, 2, ..., m, \\ \mathcal{F}_{1l}^I(e_{2l}^R(t), e_{2l}^I(t)), l = m+1, ..., 2m, \end{cases}$$

$$\mathcal{F}_{2k}(w_{1k}(t)) = \begin{cases} \mathcal{F}_{2k}^R(e_{1k}^R(t), e_{1k}^I(t)), k = 1, 2, ..., n, \\ \mathcal{F}_{2k}^I(e_{1k}^R(t), e_{1k}^I(t)), k = n+1, ..., 2n, \end{cases}$$

$$\zeta_{1k} = \begin{cases} \xi_{1k}, & k = 1, 2, ..., n, \\ \nu_{1k}, & k = n+1, n+2, ..., 2n, \end{cases}$$

$$\zeta_{2l} = \begin{cases} \xi_{2l}, & l = 1, 2, ..., m, \\ \nu_{2l}, & l = m+1, m+2, ..., 2m, \end{cases}$$

$$\beta_{1k} = \begin{cases} \delta_{1k}, & k = 1, 2, ..., n, \\ \eta_{1k}, & k = n+1, n+2, ..., 2n, \end{cases}$$

$$\beta_{2l} = \begin{cases} \delta_{2l}, & l = 1, 2, ..., m, \\ \eta_{2l}, & l = m+1, m+2, ..., 2m. \end{cases}$$

Then,

$$\begin{aligned} V(t) = & \sum_{k=1}^{n} \xi_{1k} |e_{1k}^R(t)|^r e^{r\epsilon t} + \frac{1}{1-\rho_1} \sum_{k=1}^{n} \sum_{l=1}^{m} \xi_{1k} \delta_{1k} \\ & \times \int_{t-\tau_{2kl}(t)}^{t} |\mathcal{F}_{1l}^R(e_{2l}^R(s), e_{2l}^I(s))|^r e^{r\epsilon(s+\tau_{2kl}(s))} ds \\ & + \sum_{k=1}^{n} \nu_{1k} |e_{1k}^I(t)|^r e^{r\epsilon t} + \frac{1}{1-\rho_1} \sum_{k=1}^{n} \sum_{l=1}^{m} \nu_{1k} \eta_{1k} \\ & \times \int_{t-\tau_{2kl}(t)}^{t} |\mathcal{F}_{1l}^I(e_{2l}^R(s), e_{2l}^I(s))|^r e^{r\epsilon(s+\tau_{2kl}(s))} ds \\ & + \sum_{l=1}^{m} \xi_{2l} |e_{2l}^R(t)|^r e^{r\epsilon t} + \frac{1}{1-\rho_2} \sum_{l=1}^{m} \sum_{k=1}^{n} \xi_{2l} \delta_{2l} \\ & \times \int_{t-\tau_{1lk}(t)}^{t} |\mathcal{F}_{2k}^R(e_{1k}^R(s), e_{1k}^I(s))|^r e^{r\epsilon(s+\tau_{1lk}(s))} ds \\ & + \sum_{l=1}^{m} \nu_{2l} |e_{2l}^I(t)|^r e^{r\epsilon t} + \frac{1}{1-\rho_2} \sum_{l=1}^{m} \sum_{k=1}^{n} \nu_{2l} \eta_{2l} \\ & \times \int_{t-\tau_{1lk}(t)}^{t} |\mathcal{F}_{2k}^I(e_{1k}^R(s), e_{1k}^I(s))|^r e^{r\epsilon(s+\tau_{1lk}(s))} ds. \end{aligned} \tag{7.19}$$

By taking the upper Dini-derivative of $V(t)$ along the solution trajectories of system (7.4), we have

$$
\begin{aligned}
D^{+}V(t) &= e^{r\epsilon t}\sum_{k=1}^{n} r\xi_{1k}\Big\{\epsilon|e_{1k}^{R}|^{r} + |e_{1k}^{R}|^{r-1}\operatorname{sgn}(e_{1k}^{R})\Big[-(d_{1k}+\pi_{1k}^{R})e_{1k}^{R} \\
&+ \sum_{l=1}^{m} a_{1lk}^{R}\mathcal{F}_{1l}^{R}(e_{2l}^{R},e_{2l}^{I}) - \sum_{l=1}^{m} a_{1lk}^{I}\mathcal{F}_{1l}^{I}(e_{2l}^{R},e_{2l}^{I}) + \sum_{l=1}^{m} b_{1lk}^{R}\mathcal{F}_{1l}^{R}(e_{2l}^{R\tau},e_{2l}^{I\tau}) \\
&- \sum_{l=1}^{m} b_{1lk}^{I}\mathcal{F}_{1l}^{I}(e_{2l}^{R\tau},e_{2l}^{I\tau})\Big]\Big\} + e^{r\epsilon t}\sum_{k=1}^{n} r\nu_{1k}\Big\{\epsilon|e_{1k}^{I}|^{r} + |e_{1k}^{I}|^{r-1}\operatorname{sgn}(e_{1k}^{I})\Big[-(d_{1k} \\
&+\pi_{1k}^{I})e_{1k}^{I} + \sum_{l=1}^{m} a_{1lk}^{R}\mathcal{F}_{1l}^{I}(e_{2l}^{R},e_{2l}^{I}) + \sum_{l=1}^{m} a_{1lk}^{I}\mathcal{F}_{1l}^{R}(e_{2l}^{R},e_{2l}^{I}) + \sum_{l=1}^{m} b_{1lk}^{R}\mathcal{F}_{1l}^{I}(e_{2l}^{R\tau},e_{2l}^{I\tau}) \\
&+ \sum_{l=1}^{m} b_{1lk}^{I}\mathcal{F}_{1l}^{R}(e_{2l}^{R\tau},e_{2l}^{I\tau})\Big]\Big\} + \frac{1}{1-\rho_1}\sum_{k=1}^{n}\sum_{l=1}^{m} e^{r\epsilon t}\xi_{1k}\delta_{1k}\Big\{|\mathcal{F}_{1l}^{R}(e_{2l}^{R},e_{2l}^{I})|^{r} e^{r\epsilon\tau_{2kl}(t)} \\
&-(1-\dot{\tau}_{2kl}(t))|\mathcal{F}_{1l}^{R}(e_{2l}^{R\tau},e_{2l}^{I\tau})|^{r}\Big\} + \frac{1}{1-\rho_1}\sum_{k=1}^{n}\sum_{l=1}^{m} e^{r\epsilon t}\nu_{1k}\eta_{1k}\Big\{|\mathcal{F}_{1l}^{I}(e_{2l}^{R},e_{2l}^{I})|^{r} e^{r\epsilon\tau_{2kl}(t)} \\
&-(1-\dot{\tau}_{2kl}(t))|\mathcal{F}_{1l}^{I}(e_{2l}^{R\tau},e_{2l}^{I\tau})|^{r}\Big\} + e^{r\epsilon t}\sum_{l=1}^{m} r\xi_{2l}\Big\{\epsilon|e_{2l}^{R}|^{r} + |e_{2l}^{R}|^{r-1}\operatorname{sgn}(e_{2l}^{R})\Big[-(d_{2l} \\
&+\pi_{2l}^{R})e_{2l}^{R} + \sum_{k=1}^{n} a_{2kl}^{R}\mathcal{F}_{2k}^{R}(e_{1k}^{R},e_{1k}^{I}) - \sum_{k=1}^{n} a_{2kl}^{I}\mathcal{F}_{2k}^{I}(e_{1k}^{R},e_{1k}^{I}) + \sum_{k=1}^{n} b_{2kl}^{R}\mathcal{F}_{2k}^{R}(e_{1k}^{R\tau},e_{1k}^{I\tau}) \\
&- \sum_{k=1}^{n} b_{2kl}^{I}\mathcal{F}_{2k}^{I}(e_{1k}^{R\tau},e_{1k}^{I\tau})\Big]\Big\} + e^{r\epsilon t}\sum_{l=1}^{m} r\nu_{2l}\Big\{\epsilon|e_{2l}^{I}|^{r} + |e_{2l}^{I}|^{r-1}\operatorname{sgn}(e_{2l}^{I})\Big[-(d_{2l} \\
&+\pi_{2l}^{I})e_{2l}^{I} + \sum_{k=1}^{n} a_{2kl}^{R}\mathcal{F}_{2k}^{I}(e_{1k}^{R},e_{1k}^{I}) + \sum_{k=1}^{n} a_{2kl}^{I}\mathcal{F}_{2k}^{R}(e_{1k}^{R},e_{1k}^{I}) + \sum_{k=1}^{n} b_{2kl}^{R}\mathcal{F}_{2k}^{I}(e_{1k}^{R\tau},e_{1k}^{I\tau}) \\
&+ \sum_{k=1}^{n} b_{2kl}^{I}\mathcal{F}_{2k}^{R}(e_{1k}^{R\tau},e_{1k}^{I\tau})\Big]\Big\} + \frac{1}{1-\rho_2}\sum_{l=1}^{m}\sum_{k=1}^{n} e^{r\epsilon t}\xi_{2l}\delta_{2l}\Big\{|\mathcal{F}_{2k}^{R}(e_{1k}^{R},e_{1k}^{I})|^{r} e^{r\epsilon\tau_{1lk}(t)} \\
&-(1-\dot{\tau}_{1lk}(t))|\mathcal{F}_{2k}^{R}(e_{1k}^{R\tau},e_{1k}^{I\tau})|^{r}\Big\} + \frac{1}{1-\rho_2}\sum_{l=1}^{m}\sum_{k=1}^{n} e^{r\epsilon t}\nu_{2l}\eta_{2l}\Big\{|\mathcal{F}_{2k}^{I}(e_{1k}^{R},e_{1k}^{I})|^{r} e^{r\epsilon\tau_{1lk}(t)} \\
&-(1-\dot{\tau}_{1lk}(t))|\mathcal{F}_{2k}^{I}(e_{1k}^{R\tau},e_{1k}^{I\tau})|^{r}\Big\} \\
&\le e^{r\epsilon t}\sum_{k=1}^{n}\Big\{r\xi_{1k}(\epsilon-d_{1k}-\pi_{1k}^{R})|e_{1k}^{R}|^{r} + \sum_{l=1}^{m}|a_{1lk}^{R}|r\xi_{1k}|\mathcal{F}_{1l}^{R}(e_{2l}^{R},e_{2l}^{I})||e_{1k}^{R}|^{r-1} \\
&+ \sum_{l=1}^{m}|a_{1lk}^{I}|r\xi_{1k}|\mathcal{F}_{1l}^{I}(e_{2l}^{R},e_{2l}^{I})||e_{1k}^{R}|^{r-1} + \sum_{l=1}^{m}|b_{1lk}^{R}|r\xi_{1k}|\mathcal{F}_{1l}^{R}(e_{2l}^{R\tau},e_{2l}^{I\tau})||e_{1k}^{R}|^{r-1} \\
&+ \sum_{l=1}^{m}|b_{1lk}^{I}|r\xi_{1k}|\mathcal{F}_{1l}^{I}(e_{2l}^{R\tau},e_{2l}^{I\tau})||e_{1k}^{R}|^{r-1} + \frac{e^{r\epsilon\tau_1}}{1-\rho_1}\sum_{l=1}^{m}\xi_{1k}\delta_{1k}|\mathcal{F}_{1l}^{R}(e_{2l}^{R},e_{2l}^{I})|^{r}
\end{aligned}
$$

$$
\begin{aligned}
&-\sum_{l=1}^{m}\xi_{1k}\delta_{1k}|\mathcal{F}_{1l}^{R}(e_{2l}^{R\tau},e_{2l}^{I\tau})|^{r}\Big\}+e^{r\epsilon t}\sum_{k=1}^{n}\Big\{r\nu_{1k}(\epsilon-d_{1k}-\pi_{1k}^{I})|e_{1k}^{I}|^{r}\\
&+\sum_{l=1}^{m}|a_{1lk}^{R}|r\nu_{1k}|\mathcal{F}_{1l}^{I}(e_{2l}^{R},e_{2l}^{I})||e_{1k}^{I}|^{r-1}+\sum_{l=1}^{m}|a_{1lk}^{I}|r\nu_{1k}|\mathcal{F}_{1l}^{R}(e_{2l}^{R},e_{2l}^{I})||e_{1k}^{I}|^{r-1}\\
&+\sum_{l=1}^{m}|b_{1lk}^{R}|r\nu_{1k}|\mathcal{F}_{1l}^{I}(e_{2l}^{R\tau},e_{2l}^{I\tau})||e_{1k}^{I}|^{r-1}+\sum_{l=1}^{m}|b_{1lk}^{I}|r\nu_{1k}|\mathcal{F}_{1l}^{R}(e_{2l}^{R\tau},e_{2l}^{I\tau})||e_{1k}^{I}|^{r-1}\\
&+\frac{e^{r\epsilon\tau_{1}}}{1-\rho_{1}}\sum_{l=1}^{m}\nu_{1k}\eta_{1k}|\mathcal{F}_{1l}^{I}(e_{2l}^{R},e_{2l}^{I})|^{r}-\sum_{l=1}^{m}\nu_{1k}\eta_{1k}|\mathcal{F}_{1l}^{I}(e_{2l}^{R\tau},e_{2l}^{I\tau})|^{r}\Big\}\\
&+e^{r\epsilon t}\sum_{l=1}^{m}\Big\{r\xi_{2l}(\epsilon-d_{2l}-\pi_{2l}^{R})|e_{2l}^{R}|^{r}+\sum_{k=1}^{n}|a_{2kl}^{R}|r\xi_{2l}|\mathcal{F}_{2k}^{R}(e_{1k}^{R},e_{1k}^{I})||e_{2l}^{R}|^{r-1}\\
&+\sum_{k=1}^{n}|a_{2kl}^{I}|r\xi_{2l}|\mathcal{F}_{2k}^{I}(e_{1k}^{R},e_{1k}^{I})||e_{2l}^{R}|^{r-1}+\sum_{k=1}^{n}|b_{2kl}^{R}|r\xi_{2l}|\mathcal{F}_{2k}^{R}(e_{1k}^{R\tau},e_{1k}^{I\tau})||e_{2l}^{R}|^{r-1}\\
&+\sum_{k=1}^{n}|b_{2kl}^{I}|r\xi_{2l}|\mathcal{F}_{2k}^{I}(e_{1k}^{R\tau},e_{1k}^{I\tau})||e_{2l}^{R}|^{r-1}+\frac{e^{r\epsilon\tau_{2}}}{1-\rho_{2}}\sum_{k=1}^{n}\xi_{2l}\delta_{2l}|\mathcal{F}_{2k}^{R}(e_{1k}^{R},e_{1k}^{I})|^{r}\\
&-\sum_{k=1}^{n}\xi_{2l}\delta_{2l}|\mathcal{F}_{2k}^{R}(e_{1k}^{R\tau},e_{1k}^{I\tau})|^{r}\Big\}+e^{r\epsilon t}\sum_{l=1}^{m}\Big\{r\nu_{2l}(\epsilon-d_{2l}-\pi_{2l}^{I})|e_{2l}^{I}|^{r}\\
&+\sum_{k=1}^{n}|a_{2kl}^{R}|r\nu_{2l}|\mathcal{F}_{2k}^{I}(e_{1k}^{R},e_{1k}^{I})||e_{2l}^{I}|^{r-1}+\sum_{k=1}^{n}|a_{2kl}^{I}|r\nu_{2l}|\mathcal{F}_{2k}^{R}(e_{1k}^{R},e_{1k}^{I})||e_{2l}^{I}|^{r-1}\\
&+\sum_{k=1}^{n}|b_{2kl}^{R}|r\nu_{2l}|\mathcal{F}_{2k}^{I}(e_{1k}^{R\tau},e_{1k}^{I\tau})||e_{2l}^{I}|^{r-1}+\sum_{k=1}^{n}|b_{2kl}^{I}|r\nu_{2l}|\mathcal{F}_{2k}^{R}(e_{1k}^{R\tau},e_{1k}^{I\tau})||e_{2l}^{I}|^{r-1}\\
&+\frac{e^{r\epsilon\tau_{2}}}{1-\rho_{2}}\sum_{k=1}^{n}\nu_{2l}\eta_{2l}|\mathcal{F}_{2k}^{I}(e_{1k}^{R},e_{1k}^{I})|^{r}-\sum_{k=1}^{n}\nu_{2l}\eta_{2l}|\mathcal{F}_{2k}^{I}(e_{1k}^{R\tau},e_{1k}^{I\tau})|^{r}\Big\}. \qquad (7.20)
\end{aligned}
$$

By using Lemma 7.1, one has

$$
\begin{aligned}
&D^{+}V(t)\le e^{r\epsilon t}\sum_{k=1}^{n}\Big\{r\xi_{1k}(\epsilon-d_{1k}-\pi_{1k}^{R})|e_{1k}^{R}|^{r}\\
&+\sum_{l=1}^{m}|a_{1lk}^{R}|r\xi_{1k}\Big[\frac{1}{r}|\mathcal{F}_{1l}^{R}(e_{2l}^{R},e_{2l}^{I})|^{r}+\frac{r-1}{r}(|e_{1k}^{R}|^{r-1})^{\frac{r}{r-1}}\Big]\\
&+\sum_{l=1}^{m}|a_{1lk}^{I}|r\xi_{1k}\Big[\frac{1}{r}|\mathcal{F}_{1l}^{I}(e_{2l}^{R},e_{2l}^{I})|^{r}+\frac{r-1}{r}(|e_{1k}^{R}|^{r-1})^{\frac{r}{r-1}}\Big]\\
&+\sum_{l=1}^{m}|b_{1lk}^{R}|r\xi_{1k}\Big[\frac{1}{r}|\mathcal{F}_{1l}^{R}(e_{2l}^{R\tau},e_{2l}^{I\tau})|^{r}+\frac{r-1}{r}(|e_{1k}^{R}|^{r-1})^{\frac{r}{r-1}}\Big]
\end{aligned}
$$

$$+\sum_{l=1}^{m}|b_{1lk}^{I}|r\xi_{1k}\Big[\frac{1}{r}|\mathcal{F}_{1l}^{I}(e_{2l}^{R\tau},e_{2l}^{I\tau})|^{r}+\frac{r-1}{r}(|e_{1k}^{R}|^{r-1})^{\frac{r}{r-1}}\Big]$$

$$+\frac{e^{r\epsilon\tau_1}}{1-\rho_1}\sum_{l=1}^{m}\xi_{1k}\delta_{1k}|\mathcal{F}_{1l}^{R}(e_{2l}^{R},e_{2l}^{I})|^{r}-\sum_{l=1}^{m}\xi_{1k}\delta_{1k}|\mathcal{F}_{1l}^{R}(e_{2l}^{R\tau},e_{2l}^{I\tau})|^{r}\Big\}$$

$$+e^{r\epsilon t}\sum_{k=1}^{n}\Big\{r\nu_{1k}(\epsilon-d_{1k}-\pi_{1k}^{I})|e_{1k}^{I}|^{r}$$

$$+\sum_{l=1}^{m}|a_{1lk}^{R}|r\nu_{1k}\Big[\frac{1}{r}|\mathcal{F}_{1l}^{I}(e_{2l}^{R},e_{2l}^{I})|^{r}+\frac{r-1}{r}(|e_{1k}^{I}|^{r-1})^{\frac{r}{r-1}}\Big]$$

$$+\sum_{l=1}^{m}|a_{1lk}^{I}|r\nu_{1k}\Big[\frac{1}{r}|\mathcal{F}_{1l}^{R}(e_{2l}^{R},e_{2l}^{I})|^{r}+\frac{r-1}{r}(|e_{1k}^{I}|^{r-1})^{\frac{r}{r-1}}\Big]$$

$$+\sum_{l=1}^{m}|b_{1lk}^{R}|r\nu_{1k}\Big[\frac{1}{r}|\mathcal{F}_{1l}^{I}(e_{2l}^{R\tau},e_{2l}^{I\tau})|^{r}+\frac{r-1}{r}(|e_{1k}^{I}|^{r-1})^{\frac{r}{r-1}}\Big]$$

$$+\sum_{l=1}^{m}|b_{1lk}^{I}|r\nu_{1k}\Big[\frac{1}{r}|\mathcal{F}_{1l}^{R}(e_{2l}^{R\tau},e_{2l}^{I\tau})|^{r}+\frac{r-1}{r}(|e_{1k}^{I}|^{r-1})^{\frac{r}{r-1}}\Big]$$

$$+\frac{e^{r\epsilon\tau_1}}{1-\rho_1}\sum_{l=1}^{m}\nu_{1k}\eta_{1k}|\mathcal{F}_{1l}^{I}(e_{2l}^{R},e_{2l}^{I})|^{r}-\sum_{l=1}^{m}\nu_{1k}\eta_{1k}|\mathcal{F}_{1l}^{I}(e_{2l}^{R\tau},e_{2l}^{I\tau})|^{r}\Big\}$$

$$+e^{r\epsilon t}\sum_{l=1}^{m}\Big\{r\xi_{2l}(\epsilon-d_{2l}-\pi_{2l}^{R})|e_{2l}^{R}|^{r}$$

$$+\sum_{k=1}^{n}|a_{2kl}^{R}|r\xi_{2l}\Big[\frac{1}{r}|\mathcal{F}_{2k}^{R}(e_{1k}^{R},e_{1k}^{I})|^{r}+\frac{r-1}{r}(|e_{2l}^{R}|^{r-1})^{\frac{r}{r-1}}\Big]$$

$$+\sum_{k=1}^{n}|a_{2kl}^{I}|r\xi_{2l}\Big[\frac{1}{r}|\mathcal{F}_{2k}^{I}(e_{1k}^{R},e_{1k}^{I})|^{r}+\frac{r-1}{r}(|e_{2l}^{R}|^{r-1})^{\frac{r}{r-1}}\Big]$$

$$+\sum_{k=1}^{n}|b_{2kl}^{R}|r\xi_{2l}\Big[\frac{1}{r}|\mathcal{F}_{2k}^{R}(e_{1k}^{R\tau},e_{1k}^{R\tau})|^{r}+\frac{r-1}{r}(|e_{2l}^{R}|^{r-1})^{\frac{r}{r-1}}\Big]$$

$$+\sum_{k=1}^{n}|b_{2kl}^{I}|r\xi_{2l}\Big[\frac{1}{r}|\mathcal{F}_{2k}^{I}(e_{1k}^{R\tau},e_{1k}^{R\tau})|^{r}+\frac{r-1}{r}(|e_{2l}^{R}|^{r-1})^{\frac{r}{r-1}}\Big]$$

$$+\frac{e^{r\epsilon\tau_2}}{1-\rho_2}\sum_{k=1}^{n}\xi_{2l}\delta_{2l}|\mathcal{F}_{2k}^{R}(e_{1k}^{R},e_{1k}^{I})|^{r}-\sum_{k=1}^{n}\xi_{2l}\delta_{2l}|\mathcal{F}_{2k}^{R}(e_{1k}^{R\tau},e_{1k}^{I\tau})|^{r}\Big\}$$

$$+e^{r\epsilon t}\sum_{l=1}^{m}\Big\{r\nu_{2l}(\epsilon-d_{2l}-\pi_{2l}^{I})|e_{2l}^{I}|^{r}$$

$$+\sum_{k=1}^{n}|a_{2kl}^{R}|r\nu_{2l}\Big[\frac{1}{r}|\mathcal{F}_{2k}^{I}(e_{1k}^{R},e_{1k}^{I})|^{r}+\frac{r-1}{r}(|e_{2l}^{I}|^{r-1})^{\frac{r}{r-1}}\Big]$$

$$+\sum_{k=1}^{n}|a_{2kl}^{I}|r\nu_{2l}\Big[\frac{1}{r}|\mathcal{F}_{2k}^{R}(e_{1k}^{R},e_{1k}^{I})|^{r}+\frac{r-1}{r}(|e_{2l}^{I}|^{r-1})^{\frac{r}{r-1}}\Big]$$

$$+\sum_{k=1}^{n}|b_{2kl}^{R}|r\nu_{2l}\Big[\frac{1}{r}|\mathcal{F}_{2k}^{I}(e_{1k}^{R\tau},e_{1k}^{I\tau})|^{r}+\frac{r-1}{r}(|e_{2l}^{I}|^{r-1})^{\frac{r}{r-1}}\Big]$$

$$+\sum_{k=1}^{n}|b_{2kl}^{I}|r\nu_{2l}\Big[\frac{1}{r}|\mathcal{F}_{2k}^{R}(e_{1k}^{R\tau},e_{1k}^{I\tau})|^{r}+\frac{r-1}{r}(|e_{2l}^{I}|^{r-1})^{\frac{r}{r-1}}\Big]$$

$$+\frac{e^{r\epsilon\tau_2}}{1-\rho_2}\sum_{k=1}^{n}\nu_{2l}\eta_{2l}|\mathcal{F}_{2k}^{I}(e_{1k}^{R},e_{1k}^{I})|^{r}-\sum_{k=1}^{n}\nu_{2l}\eta_{2l}|\mathcal{F}_{2k}^{I}(e_{1k}^{R\tau},e_{1k}^{I\tau})|^{r}\Big\}$$

$$=e^{r\epsilon t}\sum_{k=1}^{n}\Big\{r\xi_{1k}(\epsilon-d_{1k}-\pi_{1k}^{R})|e_{1k}^{R}|^{r}+r\nu_{1k}(\epsilon-d_{1k}-\pi_{1k}^{I})|e_{1k}^{I}|^{r}$$

$$+\sum_{l=1}^{m}\Big[|a_{1lk}^{R}|\xi_{1k}+|a_{1lk}^{I}|\nu_{1k}+\frac{e^{r\epsilon\tau_1}}{1-\rho_1}\xi_{1k}\delta_{1k}\Big]|\mathcal{F}_{1l}^{R}(e_{2l}^{R},e_{2l}^{I})|^{r}$$

$$+\sum_{l=1}^{m}\Big[|a_{1lk}^{I}|\xi_{1k}+|a_{1lk}^{R}|\nu_{1k}+\frac{e^{r\epsilon\tau_1}}{1-\rho_1}\nu_{1k}\eta_{1k}\Big]|\mathcal{F}_{1l}^{I}(e_{2l}^{R},e_{2l}^{I})|^{r}$$

$$+\sum_{l=1}^{m}\xi_{1k}\Big[|a_{1lk}^{R}|+|a_{1lk}^{I}|+|b_{1lk}^{R}|+|b_{1lk}^{I}|\Big](r-1)|e_{1k}^{R}|^{r}$$

$$+\sum_{l=1}^{m}\nu_{1k}\Big[|a_{1lk}^{R}|+|a_{1lk}^{I}|+|b_{1lk}^{R}|+|b_{1lk}^{I}|\Big](r-1)|e_{1k}^{I}|^{r}$$

$$+\sum_{l=1}^{m}\Big[|b_{1lk}^{R}|\xi_{1k}+|b_{1lk}^{I}|\nu_{1k}\Big]|\mathcal{F}_{1l}^{R}(e_{2l}^{R\tau},e_{2l}^{I\tau})|^{r}-\sum_{l=1}^{m}\xi_{1k}\delta_{1k}|\mathcal{F}_{1l}^{R}(e_{2l}^{R\tau},e_{2l}^{I\tau})|^{r}$$

$$+\sum_{l=1}^{m}\Big[|b_{1lk}^{I}|\xi_{1k}+|b_{1lk}^{R}|\nu_{1k}\Big]|\mathcal{F}_{1l}^{I}(e_{2l}^{R\tau},e_{2l}^{I\tau})|^{r}-\sum_{l=1}^{m}\nu_{1k}\eta_{1k}|\mathcal{F}_{1l}^{I}(e_{2l}^{R\tau},e_{2l}^{I\tau})|^{r}\Big\}$$

$$+e^{r\epsilon t}\sum_{l=1}^{m}\Big\{r\xi_{2l}(\epsilon-d_{2l}-\pi_{2l}^{R})|e_{2l}^{R}|^{r}+r\nu_{2l}(\epsilon-d_{2l}-\pi_{2l}^{I})|e_{2l}^{I}|^{r}$$

$$+\sum_{k=1}^{n}\Big[|a_{2kl}^{R}|\xi_{2l}+|a_{2lk}^{I}|\nu_{2l}+\frac{e^{r\epsilon\tau_2}}{1-\rho_2}\xi_{2l}\delta_{2l}\Big]|\mathcal{F}_{2k}^{R}(e_{1k}^{R},e_{1k}^{I})|^{r}$$

$$+\sum_{k=1}^{n}\Big[|a_{2kl}^{I}|\xi_{2l}+|a_{2lk}^{R}|\nu_{2l}+\frac{e^{r\epsilon\tau_2}}{1-\rho_2}\nu_{2l}\eta_{2l}\Big]|\mathcal{F}_{2k}^{I}(e_{1k}^{R},e_{1k}^{I})|^{r}$$

$$
\begin{aligned}
&+\sum_{k=1}^{n}\xi_{2l}\Big[|a_{2kl}^{R}|+|a_{2kl}^{I}|+|b_{2kl}^{R}|+|b_{2kl}^{I}|\Big](r-1)|e_{2l}^{R}|^{r}\\
&+\sum_{k=1}^{n}\nu_{2l}\Big[|a_{2kl}^{R}|+|a_{2kl}^{I}|+|b_{2kl}^{R}|+|b_{2kl}^{I}|\Big](r-1)|e_{2l}^{I}|^{r}\\
&+\sum_{k=1}^{n}\Big[|b_{2kl}^{R}|\xi_{2l}+|b_{2kl}^{I}|\nu_{2l}\Big]|\mathcal{F}_{2k}^{R}(e_{1k}^{R\tau},e_{1k}^{I\tau})|^{r}-\sum_{k=1}^{n}\xi_{2l}\delta_{2l}|\mathcal{F}_{2k}^{R}(e_{1k}^{R\tau},e_{1k}^{I\tau})|^{r}\\
&+\sum_{k=1}^{n}\Big[|b_{2kl}^{I}|\xi_{2l}+|b_{2kl}^{R}|\nu_{2l}\Big]|\mathcal{F}_{2k}^{I}(e_{1k}^{R\tau},e_{1k}^{I\tau})|^{r}-\sum_{k=1}^{n}\nu_{2l}\eta_{2l}|\mathcal{F}_{2k}^{I}(e_{1k}^{R\tau},e_{1k}^{I\tau})|^{r}\Big\}.
\end{aligned}
\tag{7.21}
$$

Based on Assumption 7.1, $\delta_{1k}=|b_{1lk}^{R}|+(\nu_{1k}/\xi_{1k})|b_{1lk}^{I}|$, $\eta_{1k}=|b_{1lk}^{R}|+(\xi_{1k}/\nu_{1k})|b_{1lk}^{I}|$, $\delta_{2l}=|b_{2kl}^{R}|+(\nu_{2l}/\xi_{2l})|b_{2kl}^{I}|$ and $\eta_{2l}=|b_{2kl}^{R}|+(\xi_{2l}/\nu_{2l})|b_{2kl}^{I}|$, we can further get that

$$
\begin{aligned}
&D^{+}V(t)\\
&\leq e^{r\epsilon t}\sum_{k=1}^{n}\Big\{r\xi_{1k}(\epsilon-d_{1k}-\pi_{1k}^{R})|e_{1k}^{R}|^{r}+r\nu_{1k}(\epsilon-d_{1k}-\pi_{1k}^{I})|e_{1k}^{I}|^{r}\\
&+\sum_{l=1}^{m}2^{r-1}\Big[|a_{1lk}^{R}|\xi_{1k}+\frac{e^{r\epsilon\tau_{1}}}{1-\rho_{1}}\xi_{1k}\delta_{1k}+|a_{1lk}^{I}|\nu_{1k}\Big]\Big[(\lambda_{1l}^{RR})^{r}|e_{2l}^{R}|^{r}+(\lambda_{1l}^{RI})^{r}|e_{2l}^{I}|^{r}\Big]\\
&+\sum_{l=1}^{m}2^{r-1}\Big[|a_{1lk}^{I}|\xi_{1k}+\frac{e^{r\epsilon\tau_{1}}}{1-\rho_{1}}\nu_{1k}\eta_{1k}+|a_{1lk}^{R}|\nu_{1k}\Big]\Big[(\lambda_{1l}^{IR})^{r}|e_{2l}^{R}|^{r}+(\lambda_{1l}^{II})^{r}|e_{2l}^{I}|^{r}\Big]\\
&+\sum_{l=1}^{m}\xi_{1k}\Big[|a_{1lk}^{R}|+|a_{1lk}^{I}|+|b_{1lk}^{R}|+|b_{1lk}^{I}|\Big](r-1)|e_{1k}^{R}|^{r}\\
&+\sum_{l=1}^{m}\nu_{1k}\Big[|a_{1lk}^{R}|+|a_{1lk}^{I}|+|b_{1lk}^{R}|+|b_{1lk}^{I}|\Big](r-1)|e_{1k}^{I}|^{r}\Big\}\\
&+e^{r\epsilon t}\sum_{l=1}^{m}\Big\{r\xi_{2l}(\epsilon-d_{2l}-\pi_{2l}^{R})|e_{2l}^{R}|^{r}+r\nu_{2l}(\epsilon-d_{2l}-\pi_{2l}^{I})|e_{2l}^{I}|^{r}\\
&+\sum_{k=1}^{n}2^{r-1}\Big[|a_{2kl}^{R}|\xi_{2l}+\frac{e^{r\epsilon\tau_{2}}}{1-\rho_{2}}\xi_{2l}\delta_{2l}+|a_{2kl}^{I}|\nu_{2l}\Big]\Big[(\lambda_{2k}^{RR})^{r}|e_{1k}^{R}|^{r}+(\lambda_{2k}^{RI})^{r}|e_{1k}^{I}|^{r}\Big]\\
&+\sum_{k=1}^{n}2^{r-1}\Big[|a_{2kl}^{I}|\xi_{2l}+\frac{e^{r\epsilon\tau_{2}}}{1-\rho_{2}}\nu_{2l}\eta_{2l}+|a_{2kl}^{R}|\nu_{2l}\Big]\Big[(\lambda_{2k}^{IR})^{r}|e_{1k}^{R}|^{r}+(\lambda_{2k}^{II})^{r}|e_{1k}^{I}|^{r}\Big]\\
&+\sum_{k=1}^{n}\xi_{2l}\Big[|a_{2kl}^{R}|+|a_{2kl}^{I}|+|b_{2kl}^{R}|+|b_{2kl}^{I}|\Big](r-1)|e_{2l}^{R}|^{r}
\end{aligned}
$$

$$
\begin{aligned}
&+\sum_{k=1}^{n}\nu_{2l}\Big[|a^R_{2kl}|+|a^I_{2kl}|+|b^R_{2kl}|+|b^I_{2kl}|\Big](r-1)|e^I_{2l}|^r\Big\}\\
&=e^{r\epsilon t}\sum_{k=1}^{n}\Big\{r\xi_{1k}(\epsilon-d_{1k}-\pi^R_{1k})+\sum_{l=1}^{m}\Big[\Big(|a^R_{2kl}|\xi_{2l}+|a^I_{2kl}|\nu_{2l}\\
&+\frac{e^{r\epsilon\tau_2}}{1-\rho_2}\xi_{2l}\delta_{2l})(\lambda^{RR}_{2k})^r+(|a^I_{2kl}|\xi_{2l}+|a^R_{2kl}|+\frac{e^{r\epsilon\tau_2}}{1-\rho_2}\nu_{2l}\eta_{2l})(\lambda^{IR}_{2k})^r\Big)2^{r-1}\\
&+\xi_{1k}(|a^R_{1lk}|+|a^I_{1lk}|+|b^R_{1lk}|+|b^I_{1lk}|)(r-1)\Big]\Big\}|e^R_{1k}|^r\\
&+e^{r\epsilon t}\sum_{k=1}^{n}\Big\{r\nu_{1k}(\epsilon-d_{1k}-\pi^I_{1k})+\sum_{l=1}^{m}\Big[\Big((|a^R_{2kl}|\xi_{2l}+|a^I_{2kl}|\nu_{2l}\\
&+\frac{e^{r\epsilon\tau_2}}{1-\rho_2}\xi_{2l}\delta_{2l})(\lambda^{RI}_{2k})^r+(|a^I_{2kl}|\xi_{2l}+|a^R_{2kl}|\nu_{2l}+\frac{e^{r\epsilon\tau_2}}{1-\rho_2}\nu_{2l}\eta_{2l})(\lambda^{II}_{2k})^r\Big)2^{r-1}\\
&+\nu_{1k}(|a^R_{1lk}|+|a^I_{1lk}|+|b^R_{1lk}|+|b^I_{1lk}|)(r-1)\Big]\Big\}|e^I_{1k}|^r\\
&+e^{r\epsilon t}\sum_{l=1}^{m}\Big\{r\xi_{2l}(\epsilon-d_{2l}-\pi^R_{2l})+\sum_{k=1}^{n}\Big[\Big((|a^R_{1lk}|\xi_{1k}+|a^I_{1lk}|\nu_{1k}\\
&+\frac{e^{r\epsilon\tau_1}}{1-\rho_1}\xi_{1k}\delta_{1k})(\lambda^{RR}_{1l})^r+(|a^I_{1lk}|\xi_{1k}+|a^R_{1lk}|\nu_{1k}+\frac{e^{r\epsilon\tau_1}}{1-\rho_1}\nu_{1k}\eta_{1k})(\lambda^{IR}_{1l})^r\Big)2^{r-1}\\
&+\xi_{2l}(|a^R_{2kl}|+|a^I_{2kl}|+|b^R_{2kl}|+|b^I_{2kl}|)(r-1)\Big]\Big\}|e^R_{2l}|^r\\
&+e^{r\epsilon t}\sum_{l=1}^{m}\Big\{r\nu_{2l}(\epsilon-d_{2l}-\pi^I_{2l})+\sum_{k=1}^{n}\Big[\Big((|a^R_{1lk}|\xi_{1k}+|a^I_{1lk}|\nu_{1k}\\
&+\frac{e^{r\epsilon\tau_1}}{1-\rho_1}\xi_{1k}\delta_{1k})(\lambda^{RI}_{1l})^r+(|a^I_{1lk}|\xi_{1k}+|a^R_{1lk}|\nu_{1k}+\frac{e^{r\epsilon\tau_1}}{1-\rho_1}\nu_{1k}\eta_{1k})(\lambda^{II}_{1l})^r\Big)2^{r-1}\\
&+\nu_{2l}(|a^R_{2kl}|+|a^I_{2kl}|+|b^R_{2kl}|+|b^I_{2kl}|)(r-1)\Big]\Big\}|e^I_{2l}|^r.
\end{aligned}
\tag{7.22}
$$

Then, according to the inequalities (7.17), we have

$$D^+V(t)\le 0. \tag{7.23}$$

Hence,

$$V(t)\le V(0),\ t\ge 0. \tag{7.24}$$

Moreover,

$$
\begin{aligned}
V(0) \leq \Big\{ & \max_{1\leq k\leq 2n} \zeta_{1k} + \frac{\tau_1 e^{r\epsilon\tau_1}}{1-\rho_1}\lambda_1^r \sum_{l=1}^{2m} \zeta_{1l} \max_{1\leq l\leq 2m} \beta_{1l} \\
& + \max_{1\leq l\leq 2m} \zeta_{2l} + \frac{\tau_2 e^{r\epsilon\tau_2}}{1-\rho_2}\lambda_2^r \sum_{k=1}^{2n} \zeta_{2k} \max_{1\leq k\leq 2n} \beta_{2k} \Big\} \|\Phi\|^r \triangleq \Pi \|\Phi\|^r, \\
V(t) \geq & \sum_{k=1}^{2n} \zeta_{1k} |w_{1k}(t)|^r e^{r\epsilon t} + \sum_{l=1}^{2m} \zeta_{2l} |w_{2l}(t)|^r e^{r\epsilon t} \\
\geq & (\min_{1\leq k\leq 2n} \zeta_{1k}) e^{r\epsilon t} \sum_{k=1}^{2n} |w_{1k}(t)|^r + (\min_{1\leq l\leq 2m} \zeta_{2l}) e^{r\epsilon t} \sum_{l=1}^{2m} |w_{2l}(t)|^r \\
\geq & \gamma e^{r\epsilon t} \Big\{ \sum_{k=1}^{2n} |w_{1k}(t)|^r + \sum_{l=1}^{2m} |w_{2l}(t)|^r \Big\},
\end{aligned}
\tag{7.25}
$$

where $\lambda_1 = \max\{\lambda_{1l}^{RR}, \lambda_{1l}^{RI}, \lambda_{1l}^{IR}, \lambda_{1l}^{II}\}$, $\lambda_2 = \max\{\lambda_{2k}^{RR}, \lambda_{2k}^{RI}, \lambda_{2k}^{IR}, \lambda_{2k}^{II}\}$, and $\gamma = \min\{\min_{1\leq k\leq 2n} \zeta_{1k}, \min_{1\leq l\leq 2m} \zeta_{2l}\}$.

From (7.24)–(7.25), we have

$$
\gamma e^{r\epsilon t} \Big\{ \sum_{k=1}^{2n} |w_{1k}(t)|^r + \sum_{l=1}^{2m} |w_{2l}(t)|^r \Big\} \leq \Pi \|\Phi\|^r,
\tag{7.26}
$$

that is,

$$
\Big\{ \sum_{k=1}^{2n} |w_{1k}(t)|^r + \sum_{l=1}^{2m} |w_{2l}(t)|^r \Big\}^{\frac{1}{r}} \leq \alpha e^{-\epsilon t} \parallel \Phi \parallel,
\tag{7.27}
$$

where

$$
\alpha = \Big(\frac{\Pi}{\gamma}\Big)^{\frac{1}{r}} \geq 1.
\tag{7.28}
$$

Thus, by Definition 7.1, systems (7.1) and (7.4) can achieve the exponential anti-synchronization. □

Remark 7.4 In recent years, great achievements in dynamic behavior analysis for CVNNs models have developed, in which some results are related to CVBAMNNs. But there has been no information for the anti-synchronization problem of CVBAMNNs up to now. In this chapter, by applying the inequalities techniques, the anti-synchronization of CVBAMNNs with time-varying delays is studied and the corresponding criterion is shown in Theorem 7.1. Hence, our work fills the gap in this respect and supplements the existing results.

Remark 7.5 Compared with real-valued neural networks, the states, connection weights and activation functions of CVNNs are defined in complex domain, which can provide a simple and natural way to maintain the physical characteristics of the original problems. For instance, the XOR problem in real numbers cannot be solved with a single real-valued neuron, which can be solved by complex-valued neurons. Therefore, CVNNs models have more complex properties and wider applications than real-valued ones, and it is very necessary and meaningful to study the dynamical behaviors of these models.

7.4 Illustrative Example

In this section, we will provide an example to illustrate the availability of our result.

Example 7.1 Consider system (7.1) with the following parameters:

$$
\begin{aligned}
f_l(h_l) &= \frac{e^{x_{2l}} - e^{-x_{2l}}}{e^{x_{2l}} + e^{-x_{2l}}} + i\frac{e^{y_{2l}} - e^{-y_{2l}}}{e^{y_{2l}} + e^{-y_{2l}}}, (l = 1, 2), d_{11} = 2.5,\\
g_k(z_k) &= \frac{e^{x_{1k}} - e^{-x_{1k}}}{e^{x_{1k}} + e^{-x_{1k}}} + i\frac{e^{y_{1k}} - e^{-y_{1k}}}{e^{y_{1k}} + e^{-y_{1k}}}, (k = 1, 2), d_{12} = 2.5,\\
\tau_{2kl}(t) &= \tau_{1lk}(t) = \frac{1}{1 + e^{-t}}, \rho_1 = \rho_2 = 0.25, d_{21} = 2.5, d_{22} = 2.5,\\
a_{111} &= -2 + 3i, a_{112} = 0.5 + 3i, a_{121} = 1.2 + 2i, a_{122} = 0.9 - 0.3i,\\
b_{111} &= -2 - 0.5i, b_{112} = 1.5 + 0.2i, b_{121} = 0.6 + 2i, b_{122} = 1.8 + 2.5i,\\
a_{211} &= 1.6 + 2.5i, a_{212} = -1 + 2.8i, a_{221} = 0.6 + 2i, a_{222} = 1.6 - 1.9i,\\
b_{211} &= 1.5 + 3i, b_{212} = -2 + 1.9i, b_{221} = 1.5 + 2.5i, b_{222} = 2 - i.
\end{aligned}
$$

It is easy to get that $\lambda_{11}^{RR} = \lambda_{11}^{II} = \lambda_{12}^{RR} = \lambda_{12}^{II} = \lambda_{21}^{RR} = \lambda_{21}^{II} = \lambda_{22}^{RR} = \lambda_{22}^{II} = 1$ and $\lambda_{11}^{IR} = \lambda_{11}^{RI} = \lambda_{12}^{IR} = \lambda_{12}^{RI} = \lambda_{21}^{IR} = \lambda_{21}^{RI} = \lambda_{22}^{IR} = \lambda_{22}^{RI} = 0$. Then, let $\pi_{11}^{R} = 26$, $\pi_{12}^{R} = 18.5$, $\pi_{11}^{I} = 23.5$, $\pi_{12}^{I} = 18.5$, $\pi_{21}^{R} = 19.5$, $\pi_{22}^{R} = 16$, $\pi_{21}^{I} = 19.5$, $\pi_{22}^{I} = 16$, $r = 2$ and $\xi_{2l} = \nu_{2l} = \xi_{1k} = \nu_{1k} = 1$, via a simple calculation, the conditions of Theorem 7.1 are satisfied. Hence, the drive system (7.1) and the response system (7.4) with the above parameters can achieve the exponential anti-synchronization. For simulations, the corresponding response curves are depicted in Figs. 7.1, 7.2, 7.3 and 7.4. Figures 7.1 and 7.2 show the time responses of real and imaginary parts of variables $z_1, z_2, h_1, h_2, z_1^*, z_2^*, h_1^*$ and h_2^* for the drive-response system without external control inputs. Figures 7.3 and 7.4 display the time responses of anti-synchronization errors $e_{11}^R, e_{12}^R, e_{11}^I, e_{12}^I, e_{21}^R, e_{22}^R, e_{21}^I$ and e_{22}^I under external control inputs with 15 different initial conditions. Figures 7.1, 7.2, 7.3 and 7.4 further show the effectiveness of the proposed result.

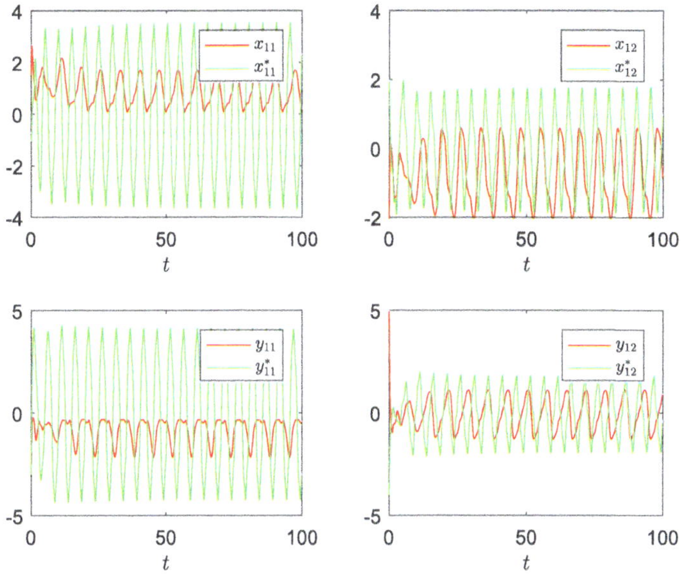

Fig. 7.1 The trajectories of real and imaginary parts of variables z_1, h_1, z_1^*, h_1^* for the drive-response system without external control inputs in Example 7.1

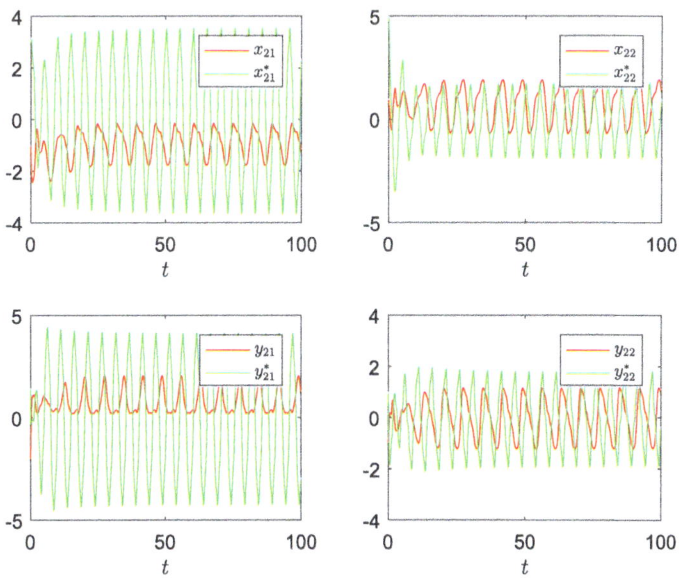

Fig. 7.2 The trajectories of real and imaginary parts of variables z_2, h_2, z_2^*, h_2^* for the drive-response system without external control inputs in Example 7.1

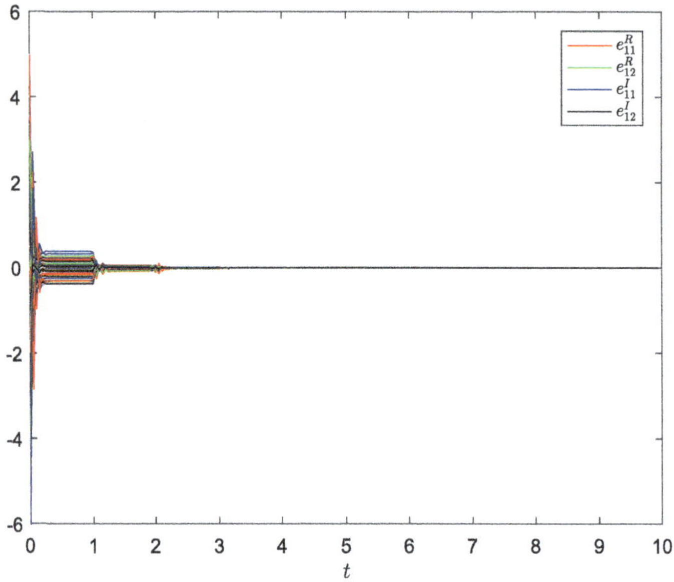

Fig. 7.3 The trajectories of anti-synchronization errors e_{11}^R, e_{11}^I, e_{12}^R and e_{12}^I under external control inputs with 15 random initial conditions in Example 7.1

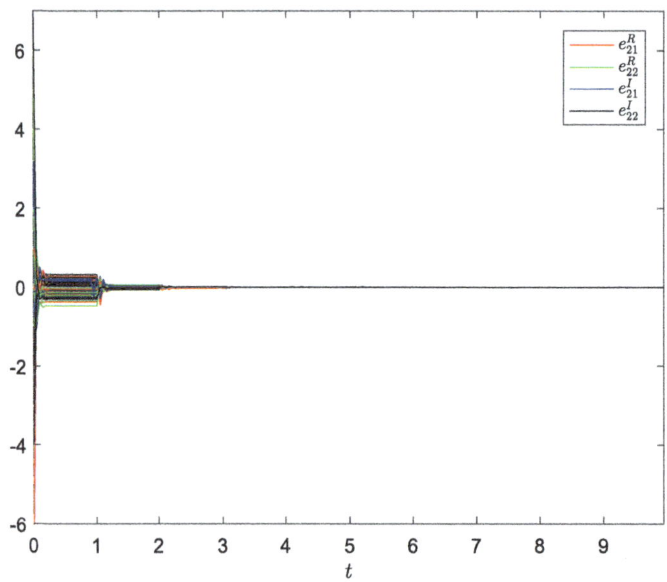

Fig. 7.4 The trajectories of anti-synchronization errors e_{21}^R, e_{21}^I, e_{22}^R and e_{22}^I under external control inputs with 15 random initial conditions in Example 7.1

7.5 Conclusion

In this chapter, we have investigated the exponential anti-synchronization control problem of CVBAMNNs with time-varying delays. Via constructing a suitable Lyapunov functional and using some inequalities techniques, we have obtained a sufficient condition in terms of algebraic structure to ensure the anti-synchronization of the considered system. A numerical example has been given to show the effectiveness of our result. Moreover, the proposed result is the first one for delayed CVBAMNNs, so our work fills the gap in this field and complements the previous results.

References

Xing M, Shen H, Wang Z (2018) H_∞ synchronization of semi-Markovian jump neural networks with randomly occurring time-varying delays. Complexity 8094292

Wang J, Ru T, Xia J, Wei Y, Wang Z (2019) Finite-time synchronization for complex dynamic networks with semi-Markov switching topologies: An H-infinity event-triggered control scheme. Appl Math Comput 356:235–251

Wang J, Hu X, Wei Y, Wang Z (2019) Sampled-data synchronization of semi-Markov jump complex dynamical networks subject to generalized dissipativity property. Appl Math Comput 346:853–864

Bao H, Park JH, Cao J (2016) Synchronization of fractional-order complex-valued neural networks with time delay. Neural Netw 81:16–28

Zhang W, Zhang H, Cao J, Alsaadi FE, Chen D (2019) Synchronization in uncertain fractional-order memristive complex-valued neural networks with multiple time delays. Neural Netw 110:186–198

Shen J, Cao J (2011) Finite-time synchronization of coupled neural networks via discontinuous controllers. Cogn Neurodyn 5:373–385

Zhou C, Zhang W, Yang X, Xu C, Feng J (2017) Finite-time synchronization of complex-valued neural networks with mixed delays and uncertain perturbations. Neural Process Lett 46:271–291

Zhang Z, Li A, Yu S (2018a) Finite-time synchronization for delayed complex-valued neural networks via integrating inequality method. Neurocomputing 318:248–260

Zhang Z, Liu X, Lin C, Chen B (2018b) Finite-time synchronization for complex-valued recurrent neural networks with time delays. Complexity 8456737

Liu D, Zhu S, Sun K (2019) Global anti-synchronization of complex-valued memristive neural networks with time delays. IEEE Trans Cybern 49(5):1735–1747

Liu D, Zhu S, Sun K (2018) Anti-synchronization of complex-valued memristor-based delayed neural networks. Neural Netw 105:1–13

Zhang Z, Li A, Yang L (2018) Global asymptotic periodic synchronization for delayed complex-valued BAM neural networks via vector-valued inequality techniques. Neural Process Lett 48:1019–1041

Chapter 8
Anti-synchronization Control for CVNNs with Mixed Delays

Abstract This chapter discusses the anti-synchronization control of complex-valued neural networks (CVNNs) with leakage delay and time-varying delays. By applying some inequalities techniques and the Lyapunov function approach, and choosing the suitable controllers, we will derive two sufficient criteria for the anti-synchronization problem of the addressed systems on the basis of the separable method and nonseparable method, respectively.

8.1 Introduction

Time delays widely exists in the actual systems such as physical, chemical, engineering and biomedical. They can cause undesirable dynamical behaviors such as instability and chaos. In the existing references, common delays types mainly include constant delays, time-varying delays, additive delays, leakage delays, etc. Among them, it is necessary to consider the influences of the leakage delay because it has an important effect on the dynamic characteristics of the models and can affect the stability of the models to some extent. Many results for systems with leakage delay have been widely developed (Wang et al. 2019; Suresh et al. 2016; Balasubramaniam et al. 2011; Maharajan et al. 2018; Zhu et al. 2018; Aouiti and Assali 2019; Ali et al. 2020). For instance, the passivity and passification problem of memristive neural networks with leakage term is discussed in Wang et al. (2019) and the stabilization problem of neural networks with leakage delay is studied via the delayed state-feedback control scheme in Zhu et al. (2018). When involving complex-valued neural networks (CVNNs), it is worth further studying the dynamic behaviors for CVNNs with leakage delay. For example, asymptotic stability and exponential stability analysis (Chen and Song 2013; Song and Zhao 2016; Samidurai et al. 2019), μ-stability analysis (Gong et al. 2015; Chen et al. 2016), Lagrange stability analysis (Song et al. 2017) and so on.

On the other hand, as a typical dynamical behavior, synchronization occupies very important position because of its wide applications in engineering practice and software supplements such as image processing and information sciences. Without considering leakage delay, there have been many achievements for various

Z. Zhang et al., *Complex-Valued Neural Networks Systems with Time Delay*, Intelligent Control and Learning Systems 4,
https://doi.org/10.1007/978-981-19-5450-4_8

synchronization phenomena of CVNNs, including asymptotic synchronization (Ali et al. 2019), exponential synchronization (Kan et al. 2019), finite-time synchronization (Wang et al. 2019; Liu et al. 2019) and anti-synchronization (Xu et al. 2019; Liu et al. 2019), etc. When leakage delay involved in CVNNs, many results for the synchronization control problem have been obtained. For example, in Hu et al. (2018) synchronization is investigated for a class of impulsive CVNNs with discrete and distributed time-varying delays as well as leakage delay. In Wang et al. (2019), synchronization of CVNNs with leakage delay and time-varying delays is studied by using the integral sliding mode control method and inequalities techniques. However, according to what the authors know, the information on the issue of the anti-synchronization control of CVNNs with leakage delays has not been reported.

In this chapter, we will be committed to analyzing the anti-synchronization of CVNNs with time-varying delays and leakage delay. The main advantages include the following: (1) For the first time, the anti-synchronization of CVNNs with leakage delay and time-varying delays is analyzed; (2) Based on the different assumption conditions satisfied by different types of activation functions, different criteria are established to guarantee the anti-synchronization of the considered system via choosing the suitable controllers; (3) Compared with the existing results, our work fills and supplements this aspect.

8.2 Problem Formulation and Preliminaries

Considering the following CVNNs model with leakage delay and time-varying delays as the drive system:

$$\dot{z}_\kappa(t) = -d_\kappa z_\kappa(t-\delta) + \sum_{\iota=1}^{n} a_{\kappa\iota} f_\iota(z_\iota(t)) + \sum_{\iota=1}^{n} b_{\kappa\iota} f_\iota(z_\iota(t-\tau_{\kappa\iota}(t))) \quad (8.1)$$

with the initial condition

$$z_\kappa(s) = \psi_\kappa(s), \ s \in [-\tau, 0], \quad (8.2)$$

where $z_\kappa(t)$ denotes the complex-valued state variables defined; d_κ is a positive constant; $a_{\kappa\iota}$ and $b_{\kappa\iota}$ are the complex-valued connection weights; $f_\iota(z_\iota(t))$ and $f_\iota(z_\iota(t-\tau_{\kappa\iota}(t)))$ are the complex-valued activation functions; $\delta > 0$ is the leakage delay; $\tau_{\kappa\iota}(t)$ is the time-varying delays with $0 \le \tau_{\kappa\iota}(t) \le \tau$, $\dot{\tau}_{\kappa\iota}(t) \le \rho < 1$; $\psi_\kappa(s)$ is continuous. In addition, $f_\iota(-z_\iota(t)) = -f_\iota(z_\iota(t))$.

Assumption 8.1 For $z_\iota(t) = x_\iota(t) + iy_\iota(t)$, $f_\iota(z_\iota(t))$ is separated as

$$f_\iota(z_\iota(t)) = f_\iota^R(x_\iota(t), y_\iota(t)) + i f_\iota^I(x_\iota(t), y_\iota(t))$$

and there exist constants $\chi_\iota^{RR} > 0$, $\chi_\iota^{RI} > 0$, $\chi_\iota^{IR} > 0$, $\chi_\iota^{II} > 0$, such that

$$|f_\iota^R(x_\iota, y_\iota) - f_\iota^R(x_\iota', y_\iota')| \le \chi_\iota^{RR}|x_\iota - x_\iota'| + \chi_\iota^{RI}|y_\iota - y_\iota'|,$$
$$|f_\iota^I(x_\iota, y_\iota) - f_\iota^I(x_\iota', y_\iota')| \le \chi_\iota^{IR}|x_\iota - x_\iota'| + \chi_\iota^{II}|y_\iota - y_\iota'|.$$

Assumption 8.2 For $\iota = 1, 2, ..., n$ and $z_\iota, z_\iota' \in \mathbb{C}$, there exists $\lambda_\iota > 0$, such that

$$|f_\iota(z_\iota) - f_\iota(z_\iota')| \le \lambda_\iota |z_\iota - z_\iota'|.$$

Let $\lambda = \text{diag}\{\lambda_1, \lambda_2, ..., \lambda_n\}$.

The response system should be provided by

$$\dot{\tilde{z}}_\kappa(t) = -d_\kappa \tilde{z}_\kappa(t-\delta) + \sum_{\iota=1}^{n} a_{\kappa\iota} f_\iota(\tilde{z}_\iota(t)) + \sum_{\iota=1}^{n} b_{\kappa\iota} f_\iota(\tilde{z}_\iota(t - \tau_{\kappa\iota}(t))) + u_\kappa(t) \tag{8.3}$$

with the initial condition

$$\tilde{z}_\kappa(s) = \varphi_\kappa(s),\ s \in [-\tau, 0], \tag{8.4}$$

where $u_\kappa(t)$ is a controller which will be designed; $\varphi_\kappa(s)$ is continuous.

Let $e_\kappa(t) = \tilde{z}_\kappa(t) + z_\kappa(t)$, on the basis of (8.1) and (8.3), the anti-synchronization error system is written below

$$\dot{e}_\kappa(t) = -d_\kappa e_\kappa(t-\delta) + \sum_{\iota=1}^{n} a_{\kappa\iota} \widetilde{\mathcal{F}}_\iota(e_\iota(t)) + \sum_{\iota=1}^{n} b_{\kappa\iota} \widetilde{\mathcal{F}}_\iota(e_\iota(t - \tau_{\kappa\iota}(t))) + u_\kappa(t) \tag{8.5}$$

with the initial condition

$$e_\kappa(s) = \vartheta_\kappa(s),\ s \in [-\tau, 0], \tag{8.6}$$

where $\widetilde{\mathcal{F}}_\iota(e_\iota(t)) = f_\iota(\tilde{z}_\iota(t)) + f_\iota(z_\iota(t))$, $\widetilde{\mathcal{F}}_\iota(e_\iota(t - \tau_{\kappa\iota}(t))) = f_\iota(\tilde{z}_\iota(t - \tau_{\kappa\iota}(t))) + f_\iota(z_\iota(t - \tau_{\kappa\iota}(t)))$ and $\vartheta_\kappa(s) = \psi_\kappa(s) + \varphi_\kappa(s)$. For brevity, $e_\kappa(t-\delta)$, $e_\iota(t)$, $e_\iota(t - \tau_{\kappa\iota}(t))$ and $u_\kappa(t)$ are simplified as e_κ^δ, e_ι, e_ι^τ and u_κ, respectively.

Let $e(t) = \text{col}\{e_1(t), e_2(t), ..., e_n(t)\}$ and the initial condition $e(s) = \vartheta(s) = \text{col}\{\vartheta_1(s), \vartheta_2(s), ..., \vartheta_n(s)\}, s \in [-\tau, 0]$. Here, the norms $\| e \| = \sum_{\kappa=1}^{n}(|e_\kappa^R| + |e_\kappa^I|)$ or $\| e \| = \sqrt{\sum_{\kappa=1}^{n}(((e_\kappa^R)^2 + (e_\kappa^I)^2)}$ and $\| \vartheta \| = \sup_{-\tau \le s \le 0} \sum_{\kappa=1}^{n}(|\vartheta_\kappa^R(s)| + |\vartheta_\kappa^I(s)|)$ or $\| \vartheta \| = \sup_{-\tau \le s \le 0} \sqrt{\sum_{\kappa=1}^{n}(((\vartheta_\kappa^R(s))^2 + (\vartheta_\kappa^I(s))^2)}$ are equipped.

Definition 8.1 If for arbitrary initial condition $\vartheta(s)$ and the properly designed feedback controller, there exists a constant $\alpha \ge 1$, such that

$$\| e \| \le \alpha \| \vartheta \|, \text{ for } \forall\, t \ge 0, \tag{8.7}$$

then systems (8.1) and (8.3) can be said to be anti-synchronized.

Lemma 8.1 (Zhang et al. 2020) *For any vectors* $Z, \tilde{Z} \in \mathbb{C}^n$, *a constant* $\varepsilon > 0$ *and a Hermitian matrix* $Q \in \mathbb{C}^{n\times n}$ *with* $Q > 0$, *one has*

$$Z^*\tilde{Z} + \tilde{Z}^*Z \leq \varepsilon Z^*QZ + \frac{1}{\varepsilon}\tilde{Z}^*Q^{-1}\tilde{Z}. \tag{8.8}$$

8.3 Anti-synchronization Criterion Based on Separable Method

In this section, we will consider the separable real-imaginary activation functions in system (8.1) and propose a sufficient condition to guarantee the anti-synchronization of the drive-response systems.

Firstly, in system (8.5), define $e_\kappa(t) = e_\kappa^R(t) + ie_\kappa^I(t)$, $u_\kappa(t) = u_\kappa^R(t) + iu_\kappa^I(t)$, $a_{\kappa\iota} = a_{\kappa\iota}^R + ia_{\kappa\iota}^I$ and $b_{\kappa\iota} = b_{\kappa\iota}^R + ib_{\kappa\iota}^I$. Evidently, $e_\kappa^R(t) = x_\kappa(t) + \tilde{x}_\kappa(t)$, $e_\kappa^I(t) = y_\kappa(t) + \tilde{y}_\kappa(t)$ with $x_\kappa(t) = \mathrm{Re}(z_\kappa(t))$, $y_\kappa(t) = \mathrm{Im}(z_\kappa(t))$, $\tilde{x}_\kappa(t) = \mathrm{Re}(\tilde{z}_\kappa(t))$ and $\tilde{y}_\kappa(t) = \mathrm{Im}(\tilde{z}_\kappa(t))$. For brevity, $e_\kappa^R(t-\delta)$, $e_\kappa^I(t-\delta)$, $e_\kappa^R(t)$, $e_\kappa^I(t)$, $e_\iota^R(t-\tau_{\kappa\iota}(t))$, $e_\iota^I(t-\tau_{\kappa\iota}(t))$, $u_\kappa^R(t)$, $u_\kappa^I(t)$, $z_\kappa(t)$, $x_\kappa(t)$, $y_\kappa(t)$, $\tilde{x}_\kappa(t)$, $\tilde{y}_\kappa(t)$, $x_\iota(t-\tau_{\kappa\iota}(t))$, $y_\iota(t-\tau_{\kappa\iota}(t))$, $\tilde{x}_\iota(t-\tau_{\kappa\iota}(t))$ and $\tilde{y}_\iota(t-\tau_{\kappa\iota}(t))$ are simplified as $e_\kappa^{R\delta}$, $e_\kappa^{I\delta}$, e_κ^R, e_κ^I, $e_\iota^{R\tau}$, $e_\iota^{I\tau}$, u_κ^R, u_κ^I, z_κ, x_κ, y_κ, $\tilde{x}_\kappa$, $\tilde{y}_\kappa$, x_ι^τ, y_ι^τ, $\tilde{x}_\iota^\tau$ and $\tilde{y}_\iota^\tau$, respectively. Then, by separating the real and imaginary parts, system (8.5) can be shown as

$$\begin{cases} \dot{e}_\kappa^R = -d_\kappa e_\kappa^{R\delta} + \sum\limits_{\iota=1}^n a_{\kappa\iota}^R \tilde{\mathcal{F}}_\iota^R(e_\iota^R, e_\iota^I) - \sum\limits_{\iota=1}^n a_{\kappa\iota}^I \tilde{\mathcal{F}}_\iota^I(e_\iota^R, e_\iota^I) \\ \qquad + \sum\limits_{\iota=1}^n b_{\kappa\iota}^R \tilde{\mathcal{F}}_\iota^R(e_\iota^{R\tau}, e_\iota^{I\tau}) - \sum\limits_{\iota=1}^n b_{\kappa\iota}^I \tilde{\mathcal{F}}_\iota^I(e_\iota^{R\tau}, e_\iota^{I\tau}) + u_\kappa^R, \\ \dot{e}_\kappa^I = -d_\kappa e_\kappa^{I\delta} + \sum\limits_{\iota=1}^n a_{\kappa\iota}^R \tilde{\mathcal{F}}_\iota^I(e_\iota^R, e_\iota^I) + \sum\limits_{\iota=1}^n a_{\kappa\iota}^I \tilde{\mathcal{F}}_\iota^R(e_\iota^R, e_\iota^I) \\ \qquad + \sum\limits_{\iota=1}^n b_{\kappa\iota}^R \tilde{\mathcal{F}}_\iota^I(e_\iota^{R\tau}, e_\iota^{I\tau}) + \sum\limits_{\iota=1}^n b_{\kappa\iota}^I \tilde{\mathcal{F}}_\iota^R(e_\iota^{R\tau}, e_\iota^{I\tau}) + u_\kappa^I, \end{cases} \tag{8.9}$$

where $\tilde{\mathcal{F}}_\iota^R(e_\iota^R, e_\iota^I) = f_\iota^R(\tilde{x}_\iota, \tilde{y}_\iota) + f_\iota^R(x_\iota, y_\iota)$, $\tilde{\mathcal{F}}_\iota^I(e_\iota^R, e_\iota^I) = f_\iota^I(\tilde{x}_\iota, \tilde{y}_\iota) + f_\iota^I(x_\iota, y_\iota)$, $\tilde{\mathcal{F}}_\iota^R(e_\iota^{R\tau}, e_\iota^{I\tau}) = f_\iota^R(\tilde{x}_\iota^\tau, \tilde{y}_\iota^\tau) + f_\iota^R(x_\iota^\tau, y_\iota^\tau)$, and $\tilde{\mathcal{F}}_\iota^I(e_\iota^{R\tau}, e_\iota^{I\tau}) = f_\iota^I(\tilde{x}_\iota^\tau, \tilde{y}_\iota^\tau) + f_\iota^I(x_\iota^\tau, y_\iota^\tau)$.

Here, the controllers $u_\kappa^R(t)$ and $u_\kappa^I(t)$ are designed as

$$\begin{aligned} u_\kappa^R(t) &= -\pi_\kappa^R(x_\kappa(t) + \tilde{x}_\kappa(t)) + d_\kappa(x_\kappa(t-\delta) + \tilde{x}_\kappa(t-\delta)), \\ u_\kappa^I(t) &= -\pi_\kappa^I(y_\kappa(t) + \tilde{y}_\kappa(t)) + d_\kappa(y_\kappa(t-\delta) + \tilde{y}_\kappa(t-\delta)), \end{aligned} \tag{8.10}$$

where π_κ^R and π_κ^I are the gains coefficients.

Theorem 8.1 *For the drive-response systems (8.1) and (8.3) satisfying the Assumption 8.1, the anti-synchronization between them can be achieved under the controller (8.10) if there are constants $\beta_\kappa > 0$ and $\nu_\kappa > 0$, such that*

$$\begin{aligned}
&-\pi_\kappa^R\beta_\kappa + \sum_{\iota=1}^{n}\Big[(|a_{\iota\kappa}^R|\beta_\iota + |a_{\iota\kappa}^I|\nu_\iota + \frac{1}{1-\rho}\beta_\iota\sigma_\iota)\chi_\kappa^{RR} \\
&+(|a_{\iota\kappa}^I|\beta_\iota + |a_{\iota\kappa}^R|\nu_\iota + \frac{1}{1-\rho}\nu_\iota\eta_\iota)\chi_\kappa^{IR}\Big] \le 0, \\
&-\pi_\kappa^I\nu_\kappa + \sum_{\iota=1}^{n}\Big[(|a_{\iota\kappa}^R|\beta_\iota + |a_{\iota\kappa}^I|\nu_\iota + \frac{1}{1-\rho}\beta_\iota\sigma_\iota)\chi_\kappa^{RI} \\
&+(|a_{\iota\kappa}^I|\beta_\iota + |a_{\iota\kappa}^R|\nu_\iota + \frac{1}{1-\rho}\nu_\iota\eta_\iota)\chi_\kappa^{II}\Big] \le 0,
\end{aligned} \tag{8.11}$$

where $\sigma_\iota = |b_{\iota\kappa}^R| + (\nu_\iota/\beta_\iota)|b_{\iota\kappa}^I|$ *and* $\eta_\iota = |b_{\iota\kappa}^R| + (\beta_\iota/\nu_\iota)|b_{\iota\kappa}^I|$, $\kappa = 1, 2, ..., n$.

Proof Construct a Lyapunov functional as follows:

$$\begin{aligned}
V(t) &= \sum_{\kappa=1}^{n}\beta_\kappa|e_\kappa^R(t)| + \frac{1}{1-\rho}\sum_{\kappa=1}^{n}\sum_{\iota=1}^{n}\beta_\kappa\sigma_\kappa\int_{t-\tau_{\kappa\iota}(t)}^{t}|\widetilde{\mathcal{F}}_\iota^R(e_\iota^R(s), e_\iota^I(s))|ds \\
&+\sum_{\kappa=1}^{n}\nu_\kappa|e_\kappa^I(t)| + \frac{1}{1-\rho}\sum_{\kappa=1}^{n}\sum_{\iota=1}^{n}\nu_\kappa\eta_\kappa\int_{t-\tau_{\kappa\iota}(t)}^{t}|\widetilde{\mathcal{F}}_\iota^I(e_\iota^R(s), e_\iota^I(s))|ds.
\end{aligned} \tag{8.12}$$

Along with the solution trajectories of system (8.9), we can obtain

$$\begin{aligned}
D^+V(t) &= \sum_{\kappa=1}^{n}\beta_\kappa\mathrm{sgn}(e_\kappa^R)\dot{e}_\kappa^R + \sum_{\kappa=1}^{n}\nu_\kappa\mathrm{sgn}(e_\kappa^I)\dot{e}_\kappa^I \\
&+\frac{1}{1-\rho}\sum_{\kappa=1}^{n}\sum_{\iota=1}^{n}\beta_\kappa\sigma_\kappa\Big\{|\widetilde{\mathcal{F}}_\iota^R(e_\iota^R, e_\iota^I)| - (1-\dot{\tau}_{\kappa\iota}(t))|\widetilde{\mathcal{F}}_\iota^R(e_\iota^{R\tau}, e_\iota^{I\tau})|\Big\} \\
&+\frac{1}{1-\rho}\sum_{\kappa=1}^{n}\sum_{\iota=1}^{n}\nu_\kappa\eta_\kappa\Big\{|\widetilde{\mathcal{F}}_\iota^I(e_\iota^R, e_\iota^I)| - (1-\dot{\tau}_{\kappa\iota}(t))|\widetilde{\mathcal{F}}_\iota^I(e_\iota^{R\tau}, e_\iota^{I\tau})|\Big\} \\
&= \sum_{\kappa=1}^{n}\beta_\kappa\Big\{\mathrm{sgn}(e_\kappa^R)\Big[-d_\kappa e_\kappa^{R\delta} + \sum_{\iota=1}^{n}a_{\kappa\iota}^R\widetilde{\mathcal{F}}_\iota^R(e_\iota^R, e_\iota^I) - \sum_{\iota=1}^{n}a_{\kappa\iota}^I\widetilde{\mathcal{F}}_\iota^I(e_\iota^R, e_\iota^I) \\
&+\sum_{\iota=1}^{n}b_{\kappa\iota}^R\widetilde{\mathcal{F}}_\iota^R(e_\iota^{R\tau}, e_\iota^{I\tau}) - \sum_{\iota=1}^{n}b_{\kappa\iota}^I\widetilde{\mathcal{F}}_\iota^I(e_\iota^{R\tau}, e_\iota^{I\tau}) - \pi_\kappa^R e_\kappa^R + d_\kappa e_\kappa^{R\delta}\Big]\Big\} \\
&+\sum_{\kappa=1}^{n}\nu_\kappa\Big\{\mathrm{sgn}(e_\kappa^I)\Big[-d_\kappa e_\kappa^{I\delta} + \sum_{\iota=1}^{n}a_{\kappa\iota}^R\widetilde{\mathcal{F}}_\iota^I(e_\iota^R, e_\iota^I) + \sum_{\iota=1}^{n}a_{\kappa\iota}^I\widetilde{\mathcal{F}}_\iota^R(e_\iota^R, e_\iota^I)
\end{aligned}$$

$$
\begin{aligned}
&+\sum_{\iota=1}^{n} b_{\kappa\iota}^{R}\widetilde{\mathcal{F}}_{\iota}^{I}(e_{\iota}^{R\tau},e_{\iota}^{I\tau})+\sum_{\iota=1}^{n} b_{\kappa\iota}^{I}\widetilde{\mathcal{F}}_{\iota}^{R}(e_{\iota}^{R\tau},e_{\iota}^{I\tau})-\pi_{\kappa}^{I}e_{\kappa}^{I}+d_{\kappa}e_{\kappa}^{I\delta}\Big]\Big\}\\
&+\frac{1}{1-\rho}\sum_{\kappa=1}^{n}\sum_{\iota=1}^{n}\beta_{\kappa}\sigma_{\kappa}\Big\{|\widetilde{\mathcal{F}}_{\iota}^{R}(e_{\iota}^{R},e_{\iota}^{I})|-(1-\dot{\tau}_{\kappa\iota}(t))|\widetilde{\mathcal{F}}_{\iota}^{R}(e_{\iota}^{R\tau},e_{\iota}^{I\tau})|\Big\}\\
&+\frac{1}{1-\rho}\sum_{\kappa=1}^{n}\sum_{\iota=1}^{n}\nu_{\kappa}\eta_{\kappa}\Big\{|\widetilde{\mathcal{F}}_{\iota}^{I}(e_{\iota}^{R},e_{\iota}^{I})|-(1-\dot{\tau}_{\kappa\iota}(t))|\widetilde{\mathcal{F}}_{\iota}^{I}(e_{\iota}^{R\tau},e_{\iota}^{I\tau})|\Big\}\\
&\le\sum_{\kappa=1}^{n}\Big\{-\pi_{\kappa}^{R}\beta_{\kappa}|e_{\kappa}^{R}|+\sum_{\iota=1}^{n}|a_{\kappa\iota}^{R}|\beta_{\kappa}|\widetilde{\mathcal{F}}_{\iota}^{R}(e_{\iota}^{R},e_{\iota}^{I})|+\sum_{\iota=1}^{n}|a_{\kappa\iota}^{I}|\beta_{\kappa}|\widetilde{\mathcal{F}}_{\iota}^{I}(e_{\iota}^{R},e_{\iota}^{I})|\\
&+\sum_{\iota=1}^{n}|b_{\kappa\iota}^{R}|\beta_{\kappa}|\widetilde{\mathcal{F}}_{\iota}^{R}(e_{\iota}^{R\tau},e_{\iota}^{I\tau})|+\sum_{\iota=1}^{n}|b_{\kappa\iota}^{I}|\beta_{\kappa}|\widetilde{\mathcal{F}}_{\iota}^{I}(e_{\iota}^{R\tau},e_{\iota}^{I\tau})|\Big\}\\
&+\frac{1}{1-\rho}\sum_{\kappa=1}^{n}\sum_{\iota=1}^{n}\beta_{\kappa}\sigma_{\kappa}|\widetilde{\mathcal{F}}_{\iota}^{R}(e_{\iota}^{R},e_{\iota}^{I})|-\sum_{\kappa=1}^{n}\sum_{\iota=1}^{n}\beta_{\kappa}\sigma_{\kappa}|\widetilde{\mathcal{F}}_{\iota}^{R}(e_{\iota}^{R\tau},e_{\iota}^{I\tau})|\\
&+\sum_{\kappa=1}^{n}\Big\{-\pi_{\kappa}^{I}\nu_{\kappa}|e_{\kappa}^{I}|+\sum_{\iota=1}^{n}|a_{\kappa\iota}^{R}|\nu_{\kappa}|\widetilde{\mathcal{F}}_{\iota}^{I}(e_{\iota}^{R},e_{\iota}^{I})|+\sum_{\iota=1}^{n}|a_{\kappa\iota}^{I}|\nu_{\kappa}|\widetilde{\mathcal{F}}_{\iota}^{R}(e_{\iota}^{R},e_{\iota}^{I})|\\
&+\sum_{\iota=1}^{n}|b_{\kappa\iota}^{R}|\nu_{\kappa}|\widetilde{\mathcal{F}}_{\iota}^{I}(e_{\iota}^{R\tau},e_{\iota}^{I\tau})|+\sum_{\iota=1}^{n}|b_{\kappa\iota}^{I}|\nu_{\kappa}|\widetilde{\mathcal{F}}_{\iota}^{R}(e_{\iota}^{R\tau},e_{\iota}^{I\tau})|\Big\}\\
&+\frac{1}{1-\rho}\sum_{\kappa=1}^{n}\sum_{\iota=1}^{n}\nu_{\kappa}\eta_{\kappa}|\widetilde{\mathcal{F}}_{\iota}^{I}(e_{\iota}^{R},e_{\iota}^{I})|-\sum_{\kappa=1}^{n}\sum_{\iota=1}^{n}\nu_{\kappa}\eta_{\kappa}|\widetilde{\mathcal{F}}_{\iota}^{I}(e_{\iota}^{R\tau},e_{\iota}^{I\tau})|\\
&=\sum_{\kappa=1}^{n}\Big\{-\pi_{\kappa}^{R}\beta_{\kappa}|e_{\kappa}^{R}|-\pi_{\kappa}^{I}\nu_{\kappa}|e_{\kappa}^{I}|+\sum_{\iota=1}^{n}(|a_{\kappa\iota}^{R}|\beta_{\kappa}+|a_{\kappa\iota}^{I}|\nu_{\kappa}+\frac{1}{1-\rho}\beta_{\kappa}\sigma_{\kappa})\\
&\times|\widetilde{\mathcal{F}}_{\iota}^{R}(e_{\iota}^{R},e_{\iota}^{I})|+\sum_{\iota=1}^{n}(|a_{\kappa\iota}^{R}|\nu_{\kappa}+|a_{\kappa\iota}^{I}|\beta_{\kappa}+\frac{1}{1-\rho}\nu_{\kappa}\eta_{\kappa})|\widetilde{\mathcal{F}}_{\iota}^{I}(e_{\iota}^{R},e_{\iota}^{I})|\Big\}.
\end{aligned}
\tag{8.13}
$$

By using Assumption 8.1, one has

$$
\begin{aligned}
D^{+}V(t)\le\sum_{\kappa=1}^{n}\Big\{&-\pi_{\kappa}^{R}\beta_{\kappa}|e_{\kappa}^{R}|-\pi_{\kappa}^{I}\nu_{\kappa}|e_{\kappa}^{I}|\\
&+\sum_{\iota=1}^{n}\Big[|a_{\kappa\iota}^{R}|\beta_{\kappa}+|a_{\kappa\iota}^{I}|\nu_{\kappa}+\frac{1}{1-\rho}\beta_{\kappa}\sigma_{\kappa}\Big]\Big[\chi_{\iota}^{RR}|e_{\iota}^{R}|+\chi_{\iota}^{RI}|e_{\iota}^{I}|\Big]\\
&+\sum_{\iota=1}^{n}\Big[|a_{\kappa\iota}^{I}|\beta_{\kappa}+|a_{\kappa\iota}^{R}|\nu_{\kappa}+\frac{1}{1-\rho}\nu_{\kappa}\eta_{\kappa}\Big]\Big[\chi_{\iota}^{IR}|e_{\iota}^{R}|+\chi_{\iota}^{II}|e_{\iota}^{I}|\Big]\Big\}
\end{aligned}
$$

$$
\begin{aligned}
&= \sum_{\kappa=1}^{n} \Big\{ -\pi_{\kappa}^{R}\beta_{\kappa} + \sum_{\iota=1}^{n} \Big[(|a_{\iota\kappa}^{R}|\beta_{\iota} + |a_{\iota\kappa}^{I}|\nu_{\iota} + \frac{1}{1-\rho}\beta_{\iota}\sigma_{\iota})\chi_{\kappa}^{RR} \\
&+ (|a_{\iota\kappa}^{I}|\beta_{\iota} + |a_{\iota\kappa}^{R}|\nu_{\iota} + \frac{1}{1-\rho}\nu_{\iota}\eta_{\iota})\chi_{\kappa}^{IR} \Big] \Big\} |e_{\kappa}^{R}| \\
&+ \sum_{\kappa=1}^{n} \Big\{ -\pi_{\kappa}^{I}\nu_{\kappa} + \sum_{\iota=1}^{n} \Big[(|a_{\iota\kappa}^{R}|\beta_{\iota} + |a_{\iota\kappa}^{I}|\nu_{\iota} + \frac{1}{1-\rho}\beta_{\iota}\sigma_{\iota})\chi_{\kappa}^{RI} \\
&+ (|a_{\iota\kappa}^{I}|\beta_{\iota} + |a_{\iota\kappa}^{R}|\nu_{\iota} + \frac{1}{1-\rho}\nu_{\iota}\eta_{\iota})\chi_{\kappa}^{II} \Big] \Big\} |e_{\kappa}^{I}|.
\end{aligned}
\tag{8.14}
$$

According to (8.11), we have

$$
D^{+}V(t) \leq 0. \tag{8.15}
$$

Thus,

$$
V(t) \leq V(0), \ t \geq 0. \tag{8.16}
$$

Moreover,

$$
\begin{aligned}
V(0) &\leq \Big\{ \max_{1\leq\kappa\leq n}\{\beta_{\kappa}, \nu_{\kappa}\} + \frac{\tau\chi}{1-\rho} \sum_{\iota=1}^{n} (\beta_{\iota}\sigma_{\iota} + \nu_{\iota}\eta_{\iota}) \Big\} \|\vartheta\|, \\
V(t) &\geq \sum_{\kappa=1}^{n} \beta_{\kappa}|e_{\kappa}^{R}(t)| + \sum_{\kappa=1}^{n} \nu_{\kappa}|e_{\kappa}^{I}(t)| \geq \min_{1\leq\kappa\leq n}\{\beta_{\kappa}, \nu_{\kappa}\} \sum_{\kappa=1}^{n} \Big(|e_{\kappa}^{R}(t)| + |e_{\kappa}^{I}(t)| \Big) \\
&= \min_{1\leq\kappa\leq n}\{\beta_{\kappa}, \nu_{\kappa}\} \parallel e \parallel,
\end{aligned}
\tag{8.17}
$$

where $\chi = \max\{\chi_{\iota}^{RR}, \chi_{\iota}^{RI}, \chi_{\iota}^{IR}, \chi_{\iota}^{II}\}$. From (8.17), we have

$$
\parallel e \parallel \leq \alpha \parallel \vartheta \parallel, \tag{8.18}
$$

where

$$
\alpha = \frac{\max_{1\leq\kappa\leq n}\{\beta_{\kappa}, \nu_{\kappa}\} + \frac{\tau\chi}{1-\rho}\sum_{\iota=1}^{n}(\beta_{\iota}\sigma_{\iota} + \nu_{\iota}\eta_{\iota})}{\min_{1\leq\kappa\leq n}\{\beta_{\kappa}, \nu_{\kappa}\}} \geq 1. \tag{8.19}
$$

According to Definition 8.1, the drive-response systems (8.1) and (8.3) can be anti-synchronized. □

8.4 Anti-synchronization Criterion Based on Nonseparable Method

As a continuation of the previous section, in this section, for system (8.1), via the nonseparable method, we will further consider the anti-synchronization control problem. Here, the controller $u_\kappa(t)$ is designed as

$$u_\kappa(t) = -\pi_\kappa(z_\kappa(t) + \tilde{z}_\kappa(t)) + d_\kappa(z_\kappa(t-\delta) + \tilde{z}_\kappa(t-\delta)), \tag{8.20}$$

where π_κ is the gain coefficient.

Theorem 8.2 *For the drive-response systems (8.1) and (8.3) that satisfy the Assumption 8.2, the anti-synchronization between them can be achieved under the controller (8.20) if there are constants $\xi_\kappa > 0$, $\acute{\varepsilon}_\kappa > 0$ and $\tilde{\varepsilon}_\kappa > 0$, such that*

$$-2\pi_\kappa\xi_\kappa + \sum_{\iota=1}^{n}\xi_\kappa(\acute{\varepsilon}_\kappa a_{\kappa\iota}a_{\kappa\iota}^* + \tilde{\varepsilon}_\kappa b_{\kappa\iota}b_{\kappa\iota}^*) + \sum_{\iota=1}^{n}\xi_\iota(\acute{\varepsilon}_\iota^{-1} + \frac{1}{1-\rho}\tilde{\varepsilon}_\iota^{-1})\lambda_\kappa^2 \le 0, \tag{8.21}$$

for $\kappa = 1, 2, ..., n$.

Proof Construct a Lyapunov functional as follows:

$$V(t) = \sum_{\kappa=1}^{n}\xi_\kappa e_\kappa^*(t)e_\kappa(t) + \frac{1}{1-\rho}\sum_{\kappa=1}^{n}\sum_{\iota=1}^{n}\xi_\kappa\tilde{\varepsilon}_\kappa^{-1}\int_{t-\tau_{\kappa\iota}(t)}^{t}\widetilde{\mathcal{F}}_\iota^*(e_\iota(s))\widetilde{\mathcal{F}}_\iota(e_\iota(s))ds. \tag{8.22}$$

Along with the solution trajectories of system (8.5), we can obtain

$$\begin{aligned}
\dot{V}(t) &= 2\sum_{\kappa=1}^{n}\xi_\kappa e_\kappa^*\dot{e}_\kappa + \frac{1}{1-\rho}\sum_{\kappa=1}^{n}\sum_{\iota=1}^{n}\xi_\kappa\tilde{\varepsilon}_\kappa^{-1}[\widetilde{\mathcal{F}}_\iota^*(e_\iota)\widetilde{\mathcal{F}}_\iota(e_\iota) \\
&\quad - (1-\dot{\tau}_{\kappa\iota}(t))\widetilde{\mathcal{F}}_\iota^*(e_\iota^\tau)\widetilde{\mathcal{F}}_\iota(e_\iota^\tau)] \\
&= 2\sum_{\kappa=1}^{n}\xi_\kappa e_\kappa^*\Big\{-d_\kappa e_\kappa^\delta + \sum_{\iota=1}^{n}a_{\kappa\iota}\widetilde{\mathcal{F}}_\iota(e_\iota) + \sum_{\iota=1}^{n}b_{\kappa\iota}\widetilde{\mathcal{F}}_\iota(e_\iota^\tau) - \pi_\kappa e_\kappa + d_\kappa e_\kappa^\delta\Big\} \\
&\quad + \frac{1}{1-\rho}\sum_{\kappa=1}^{n}\sum_{\iota=1}^{n}\xi_\kappa\tilde{\varepsilon}_\kappa^{-1}[\widetilde{\mathcal{F}}_\iota^*(e_\iota)\widetilde{\mathcal{F}}_\iota(e_\iota) - (1-\dot{\tau}_{\kappa\iota}(t))\widetilde{\mathcal{F}}_\iota^*(e_\iota^\tau)\widetilde{\mathcal{F}}_\iota(e_\iota^\tau)] \\
&\le \sum_{\kappa=1}^{n}\xi_\kappa\Big\{-2\pi_\kappa e_\kappa^*e_\kappa + \sum_{\iota=1}^{n}\acute{\varepsilon}_\kappa a_{\kappa\iota}a_{\kappa\iota}^*e_\kappa^*e_\kappa + \sum_{\iota=1}^{n}\acute{\varepsilon}_\kappa^{-1}\widetilde{\mathcal{F}}_\iota^*(e_\iota)\widetilde{\mathcal{F}}_\iota(e_\iota) \\
&\quad + \sum_{\kappa=1}^{n}\tilde{\varepsilon}_\kappa b_{\kappa\iota}b_{\kappa\iota}^*e_\kappa^*e_\kappa + \sum_{\iota=1}^{n}\tilde{\varepsilon}_\kappa^{-1}\widetilde{\mathcal{F}}_\iota^*(e_\iota^\tau)\widetilde{\mathcal{F}}_\iota(e_\iota^\tau)\Big\}
\end{aligned}$$

$$+\frac{1}{1-\rho}\sum_{\kappa=1}^{n}\sum_{\iota=1}^{n}\xi_\kappa\tilde{\varepsilon}_\kappa^{-1}\widetilde{\mathcal{F}}_\iota^*(e_\iota)\widetilde{\mathcal{F}}_\iota(e_\iota)-\sum_{\kappa=1}^{n}\sum_{\iota=1}^{n}\xi_\kappa\tilde{\varepsilon}_\kappa^{-1}\widetilde{\mathcal{F}}_\iota^*(e_\iota^\tau)\widetilde{\mathcal{F}}_\iota(e_\iota^\tau)$$
$$=\sum_{\kappa=1}^{n}\xi_\kappa\Big\{-2\pi_\kappa+\sum_{\iota=1}^{n}\acute{\varepsilon}_\kappa a_{\kappa\iota}a_{\kappa\iota}^*+\sum_{\iota=1}^{n}\tilde{\varepsilon}_\kappa b_{\kappa\iota}b_{\kappa\iota}^*\Big\}e_\kappa^*e_\kappa$$
$$+\sum_{\kappa=1}^{n}\sum_{\iota=1}^{n}\xi_\kappa\Big\{\acute{\varepsilon}_\kappa^{-1}+\frac{1}{1-\rho}\tilde{\varepsilon}_\kappa^{-1}\Big\}\widetilde{\mathcal{F}}_\iota^*(e_\iota)\widetilde{\mathcal{F}}_\iota(e_\iota). \tag{8.23}$$

According to Assumption 8.2, we can have

$$\dot{V}(t)\le\sum_{\kappa=1}^{n}\Big\{-2\pi_\kappa\xi_\kappa+\sum_{\iota=1}^{n}\acute{\varepsilon}_\kappa\xi_\kappa a_{\kappa\iota}^*a_{\kappa\iota}+\sum_{\iota=1}^{n}\tilde{\varepsilon}_\kappa\xi_\kappa b_{\kappa\iota}^*b_{\kappa\iota}\Big\}e_\kappa^*e_\kappa$$
$$+\sum_{\kappa=1}^{n}\sum_{\iota=1}^{n}\xi_\kappa\acute{\varepsilon}_\kappa^{-1}\lambda_\iota^2e_\iota^*e_\iota+\frac{1}{1-\rho}\sum_{\kappa=1}^{n}\sum_{\iota=1}^{n}\xi_\kappa\tilde{\varepsilon}_\kappa^{-1}\lambda_\iota^2e_\iota^*e_\iota$$
$$=\sum_{\kappa=1}^{n}\Big\{-2\pi_\kappa\xi_\kappa+\sum_{\iota=1}^{n}\xi_\kappa(\acute{\varepsilon}_\kappa a_{\kappa\iota}a_{\kappa\iota}^*+\tilde{\varepsilon}_\kappa b_{\kappa\iota}b_{\kappa\iota}^*)$$
$$+\sum_{\iota=1}^{n}\xi_\iota(\acute{\varepsilon}_\iota^{-1}+\frac{1}{1-\rho}\tilde{\varepsilon}_\iota^{-1})\lambda_\kappa^2\Big\}e_\kappa^*e_\kappa. \tag{8.24}$$

By (8.21), one has

$$\dot{V}(t)\le 0. \tag{8.25}$$

Thus,

$$V(t)\le V(0),\ t\ge 0. \tag{8.26}$$

Moreover,

$$V(0)\le\Big\{\max_{1\le\kappa\le n}\xi_\kappa+\frac{\tau\lambda^2}{1-\rho}\sum_{\iota=1}^{n}\xi_\iota\tilde{\varepsilon}_\iota^{-1}\Big\}\|\vartheta\|^2,$$
$$V(t)\ge\sum_{\kappa=1}^{n}\xi_\kappa e_\kappa^*(t)e_\kappa(t)\ge\{\min_{1\le\kappa\le n}\xi_\kappa\}\sum_{\kappa=1}^{n}((e_\kappa^R(t))^2+(e_\kappa^I(t))^2)$$
$$=\{\min_{1\le\kappa\le n}\xi_\kappa\}\parallel e\parallel^2. \tag{8.27}$$

From (8.26)–(8.27), we have

$$\parallel e\parallel\le\alpha\parallel\vartheta\parallel, \tag{8.28}$$

where

$$\alpha = \Big(\frac{\max\limits_{1\le\kappa\le n} \xi_\kappa + \frac{\tau\lambda^2}{1-\rho}\sum\limits_{\iota=1}^{n} \xi_\iota \tilde{\varepsilon}_\iota^{-1}}{\min\limits_{1\le\kappa\le n} \xi_\kappa}\Big)^{\frac{1}{2}} \ge 1.$$

Thus, the drive-response systems (8.1) and (8.3) can be anti-synchronized by Definition 8.1. □

When the leakage delay $\delta = 0$, systems (8.1) and (8.3) will be reduced to the following forms:

$$\dot{z}_\kappa(t) = -d_\kappa z_\kappa(t) + \sum_{\iota=1}^{n} a_{\kappa\iota} f_\iota(z_\iota(t)) + \sum_{\iota=1}^{n} b_{\kappa\iota} f_\iota(z_\iota(t - \tau_{\kappa\iota}(t))), \tag{8.29}$$

$$\dot{\tilde{z}}_\kappa(t) = -d_\kappa \tilde{z}_\kappa(t) + \sum_{\iota=1}^{n} a_{\kappa\iota} f_\iota(\tilde{z}_\iota(t)) + \sum_{\iota=1}^{n} b_{\kappa\iota} f_\iota(\tilde{z}_\iota(t - \tau_{\kappa\iota}(t))) + u_\kappa(t). \tag{8.30}$$

Accordingly, the state-feedback control law are reduced to

$$u_\kappa(t) = -\pi_\kappa(z_\kappa(t) + \tilde{z}_\kappa(t)). \tag{8.31}$$

Corollary 8.1 *The drive-response systems (8.29) and (8.30) can be anti-synchronized with the control input (8.31) if*

$$-2(d_\kappa + \pi_\kappa)\xi_\kappa + \sum_{\iota=1}^{n} \xi_\kappa(\acute{\varepsilon}_\kappa a_{\kappa\iota} a_{\kappa\iota}^* + \tilde{\varepsilon}_\kappa b_{\kappa\iota} b_{\kappa\iota}^*)$$
$$+ \sum_{\iota=1}^{n} \xi_\iota(\acute{\varepsilon}_\iota^{-1} + \frac{1}{1-\rho}\tilde{\varepsilon}_\iota^{-1})\lambda_\kappa^2 \le 0. \tag{8.32}$$

Remark 8.1 As we know, up to now, no information on the issue of the anti-synchronization with respect to CVNNs with leakage delay and time-varying delays has been reported. Here, this topic is discussed by means of some inequalities techniques in the framework of the Lyapunov function approach, and several criteria have been presented to guarantee the anti-synchronization control of the considered system. Hence, our work supplements the existing results.

Remark 8.2 In this chapter, so as to analyze the anti-synchronization problem comprehensively, two different kinds of activation functions are involved. One type is real-imaginary separate type and the corresponding criterion is derived, as stated in Theorem 8.1. The other type is based on the functions which cannot be explicitly separated into their real and imaginary parts. At this moment, the state variables are regarded as a whole and the corresponding criterion is derived, as stated in Theorem 8.2.

Remark 8.3 When the leakage delay $\delta = 0$, Theorem 8.2 reduces to Corollary 8.1. Different from the result in Liu et al. (2019) which only deals with the real-imaginary separate-type activation functions case, the one in Corollary 8.1 considers the entire variables. Thus, our work owns more universality and applicability.

Remark 8.4 In Wang et al. (2019), Suresh et al. (2016), Balasubramaniam et al. (2011), Maharajan et al. (2018), Zhu et al. (2018), Aouiti and Assali (2019), Ali et al. (2020), the stability of real-valued neural networks with leakage delay is studied. For CVNNs, Xu et al. (2019) and Liu et al. (2019) study the anti-synchronization problem of CVNNs with time-varying delay, but they don't involve the leakage delay. Wang et al. (2019) considers the synchronization problem for CVNNs with leakage delay and time-varying delays. As we know, so far there has been no report on the anti-synchronization result of CVNNs with leakage delay and time-varying delays, as stated in Remark 8.1. Therefore, our work have made up for the gap in this field. In addition, we consider two kinds of activation functions. This makes our results more applicable.

Remark 8.5 As is known to all, time delay is inherent in the actual implementation of neural networks models and it can lead to bad behaviors such as oscillations and chaos. Consequently, it is necessary to consider the involvement of time delay in dynamical behaviors of neural networks models. In fact, in line with different forms, there are different types of time delays. As one of them, the leakage delay is a typical type of the negative feedback term of the models and has an important effect on the dynamics, widely introduced in real problems such as population dynamics, automatic control of AIDS epidemic and so on. As a result, it is important to consider the influence of the leakage delay in models.

8.5 Illustrative Examples

In this section, we will provide two illustrative examples to illustrate the effectiveness of the proposed results.

Example 8.1 Consider system (8.1) with

$$
\begin{aligned}
& f_\iota(z_\iota) = \frac{e^{x_\iota} - e^{-x_\iota}}{e^{x_\iota} + e^{-x_\iota}} + i\frac{e^{y_\iota} - e^{-y_\iota}}{e^{y_\iota} + e^{-y_\iota}}, \ (\iota = 1, 2), \ \tau_{\kappa\iota}(t) = \frac{1}{1 + e^{-t}}, \ \delta = 0.005, \\
& d_1 = 0.5, \ d_2 = 0.5, \ a_{11} = -0.5 - 0.41i, \ a_{12} = 0.5 + 1.8i, \\
& a_{21} = -1.2 + 2i, \ a_{22} = -0.89 + 0.3i, \ b_{11} = -1 + 0.5i, \\
& b_{12} = 0.76 + 1.3i, \ b_{21} = -1 + 0.58i, \ b_{22} = -0.9 - 0.25i.
\end{aligned}
$$

Via a simple calculation, one can get that $\tau = 1$, $\rho = 0.25$, $\chi_1^{RR} = \chi_2^{RR} = \chi_1^{II} = \chi_2^{II} = 1$ and $\chi_1^{IR} = \chi_2^{IR} = \chi_1^{RI} = \chi_2^{RI} = 0$. Let $\pi_1^R = \pi_2^R = \pi_1^I = \pi_2^I = 10$ and choose $\chi_\kappa = \nu_\kappa = 1$, then, it can verify that Theorem 8.1 holds. Thus, the anti-synchronization of the drive-response systems (8.1) and (8.3) can be achieved.

For simulations, see Figs. 8.1 and 8.2. Among them, Fig. 8.1 displays the trajectories of variables x_1 and $\tilde{x}_1$, x_2 and $\tilde{x}_2$, y_1 and $\tilde{y}_1$, y_2 and $\tilde{y}_2$ without the controller and Fig. 8.2 displays the curves of the errors e_1^R, e_1^I, e_2^R and e_2^I under the controller, which indicates that systems (8.1) and (8.3) can be anti-synchronized under the controller (8.10).

Example 8.2 Consider system (8.1) with

$$f_\iota(z_\iota) = \frac{1 - e^{-\bar{z}_\iota}}{1 + e^{-\bar{z}_\iota}}, \ (\iota = 1, 2), \ \tau_{\kappa\iota}(t) = 0.2\cos^2(t), \ \delta = 0.009, \ d_1 = d_2 = 0.2,$$
$$a_{11} = -1.1 - 0.41i, \ a_{12} = 0.5 + 1.8i, \ a_{21} = 1.2 + 2i, \ a_{22} = 0.89 - 0.3i,$$
$$b_{11} = 1 - 0.5i, b_{12} = 0.76 + 1.3i, \ b_{21} = 1 - 0.58i, \ b_{22} = -0.9 + 0.25i.$$

Here, activation functions $f_\iota(z_\iota)$ can't be expressed explicitly by separating real and imaginary parts, which can be regarded as an entire. By computing, we can get that $\tau = 0.2$, $\rho = 0.2$ and $\lambda_1 = \lambda_2 = \frac{2}{3}$. Let $\pi_1 = \pi_2 = 10 + 10i$ and choose $\xi_\kappa = \acute{\varepsilon}_\kappa = \tilde{\varepsilon}_\kappa = 1$, it can verify that Theorem 8.2 holds. Thus, the anti-synchronization of the drive-response systems (8.1) and (8.3) can be achieved.

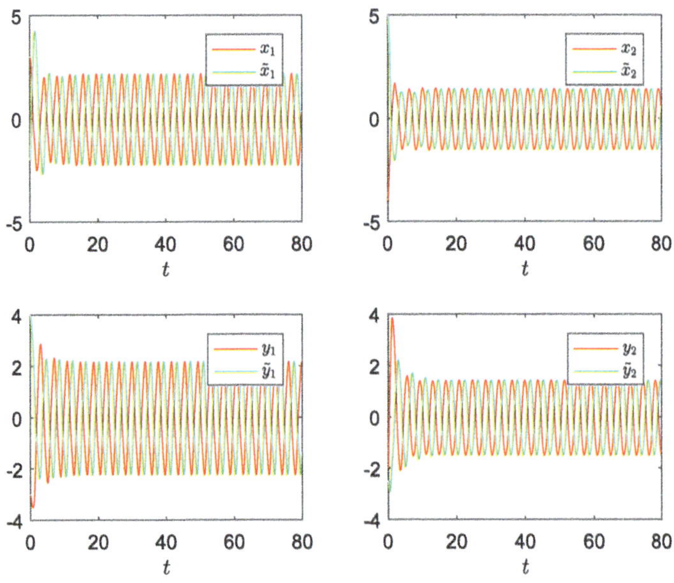

Fig. 8.1 The curves of the variables x_1 and $\tilde{x}_1$, x_2 and $\tilde{x}_2$, y_1 and $\tilde{y}_1$, y_2 and $\tilde{y}_2$ without the controller in Example 8.1

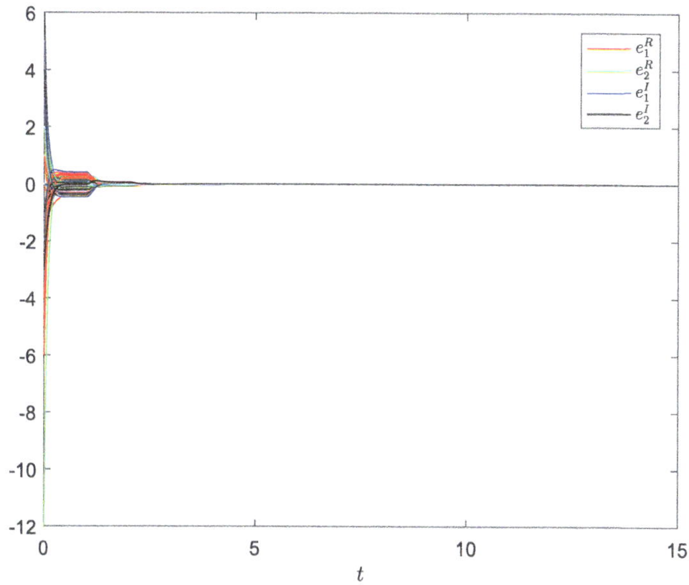

Fig. 8.2 Under 15 arbitrary initial conditions, the curves of errors e_1^R, e_1^I, e_2^R and e_2^I in Example 8.1

For simulations, see Figs. 8.3 and 8.4. Thereinto, Fig. 8.3 shows the curves of the variables z_1, z_2, $\tilde{z}_1$ and $\tilde{z}_2$ without controller and Fig. 8.4 displays the curves of errors e_1 and e_2 under the controller, which indicates that systems (8.1) and (8.3) can be anti-synchronized under the controller (8.20). Figures 8.3 and 8.4 further verify the effectiveness of the obtained result in Theorem 8.2.

8.6 Conclusion

In this chapter, we have analyzed the anti-synchronization control for CVNNs with leakage delay and time-varying delays. Based on the different assumption conditions for different types of activation functions, we have obtained the corresponding sufficient conditions for ensuring anti-synchronization by using some inequalities techniques. In addition, when the leakage delay is not considered, similar results can also be obtained. The availability of the proposed results has been showed by two numerical examples.

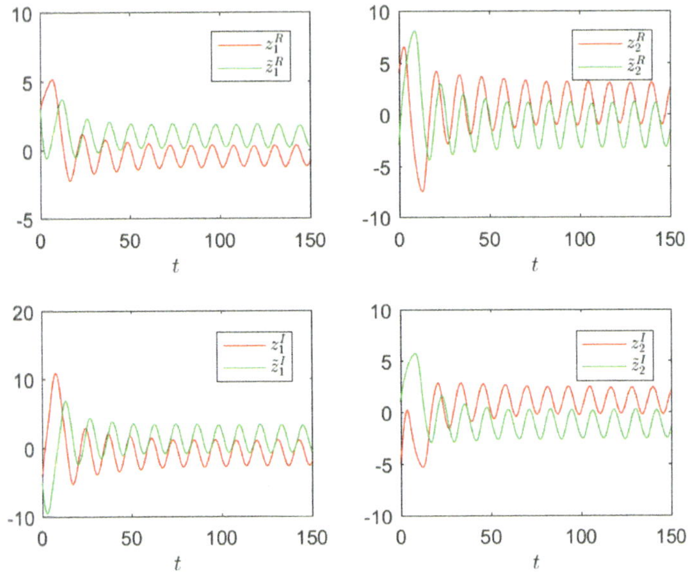

Fig. 8.3 The curves of the variables z_1, z_2, $\tilde{z}_1$ and $\tilde{z}_2$ without the controller in Example 8.2

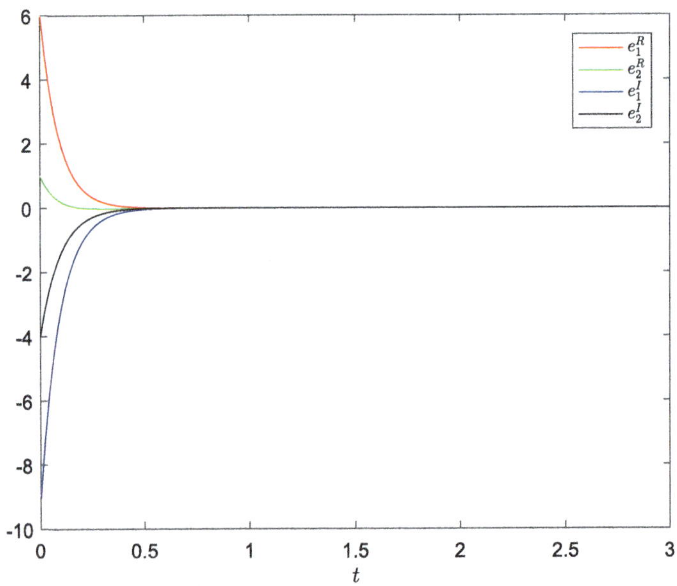

Fig. 8.4 The curves of errors e_1 and e_2 under the controller in Example 8.2

References

Wang S, Cao Y, Huang T, Wen S (2019) Passivity and passification of memristive neural networks with leakage term and time-varying delays. Appl Math Comput 361:294–310

Suresh R, Sugumaran G, Raja R, Zhu Q (2016) New stability criterion of neural networks with leakage delays and impulses: a piecewise delay method. Cogn Neurodyn 10:85–98

Balasubramaniam P, Vembarasan V, Rakkiyappan R (2011) Leakage delays in T-S fuzzy cellular neural networks. Neural Process Lett 33:111–136

Maharajan C, Raja R, Cao J, Rajchakit G, Tu Z, Alsaedi A (2018) LMI-based results on exponential stability of BAM-type neural networks with leakage and both time-varying delays: A non-fragile state estimation approach. Appl Math Comput 326:33–55

Zhu H, Rakkiyappan R, Li X (2018) Delayed state-feedback control for stabilization of neural networks with leakage delay. Neural Netw 105:248–255

Aouiti C, Assali EA (2019) Stability analysis for a class of impulsive high-order Hopfield neural networks with leakage time-varying delays. Neural Comput Appl 31:7781–7803

Ali MS, Narayanan G, Shekher V, Alsulami H, Saeed T (2020) Dynamic stability analysis of stochastic fractional-order memristor fuzzy BAM neural networks with delay and leakage terms. Appl Math Comput 369:124896

Chen X, Song Q (2013) Global stability of complex-valued neural networks with both leakage time delay and discrete time delay on time scales. Neurocomputing 121:254–264

Song Q, Zhao Z (2016) Stability criterion of complex-valued neural networks with both leakage delay and time-varying delays on time scales. Neurocomputing 171:179–184

Samidurai R, Sriraman R, Zhu S (2019) Leakage delay-dependent stability analysis for complex-valued neural networks with discrete and distributed time-varying delays. Neurocomputing 338:262–273

Gong W, Liang J, Cao J (2015) Global μ-stability of complex-valued delayed neural networks with leakage delay. Neurocomputing 168:135–144

Chen X, Song Q, Zhao Z, Liu Y (2016) Global μ-stability analysis of discrete-time complex-valued neural networks with leakage delay and mixed delays. Neurocomputing 175:723–735

Song Q, Shu H, Zhao Z, Liu Y, Alsaadi FE (2017) Lagrange stability analysis for complex-valued neural networks with leakage delay and mixed time-varying delays. Neurocomputing 244:33–41

Ali MS, Hymavathi M, Senan S, Shekher V, Arik S (2019) Global asymptotic synchronization of impulsive fractional-order complex-valued memristor-based neural networks with time varying delays. Commun Nonlinear Sci Numer Simulat 78:104869

Kan Y, Lu J, Qiu J, Kurths J (2019) Exponential synchronization of time-varying delayed complex-valued neural networks under hybrid impulsive controllers. Neural Netw 114:157–163

Wang Z, Cao J, Cai Z, Huang L (2019) Periodicity and finite-time periodic synchronization of discontinuous complex-valued neural networks. Neural Netw 119:249–260

Liu Y, Qin Y, Huang J, Huang T, Yang X (2019) Finite-time synchronization of complex-valued neural networks with multiple time-varying delays and infinite distributed delays. Neural Process Lett 50:1773–1787

Xu W, Zhu S, Fang X, Wang W (2019) Adaptive anti-synchronization of memristor-based complex-valued neural networks with time delays. Phys A 535:22427

Liu D, Zhu S, Sun K (2019) Global anti-synchronization of complex-valued memristive neural networks with time delays. IEEE Trans Cybern 49(5):1735–1747

Hu B, Song Q, Li K, Zhao Z, Liu Y, Alsaadi FE (2018) Global μ-synchronization of impulsive complex-valued neural networks with leakage delay and mixed time-varying delays. Neurocomputing 307:106–116

Wang L, Song Q, Zhao Z, Liu Y (2019) Synchronization of two nonidentical complex-valued neural networks with leakage delay and time-varying delays. Neurocomputing 332:149–158

Zhang Z, Guo R, Liu X, Lin C (2020) Lagrange exponential stability of complex-valued BAM neural networks with time-varying delays. IEEE Trans Syst Man Cybern Syst 50(8):3072–3085

Chapter 9
(Anti)-Synchronization for CVINNs with Time-Varying Delays

Abstract This chapter studies the synchronization and anti-synchronization problems of complex-valued inertial neural networks (CVINNs) with time-varying delays. Via the method of variable substitution, the considered model is expressed as the first-order complex-valued differential system. We will derive some sufficient criteria in terms of complex-valued LMIs to ensure the synchronization and anti-synchronization of CVINNs systems by constructing the appropriate Lyapunov functional.

9.1 Introduction

Neural networks model, which has been widely used in pattern recognition, signal processing, associative memory, and other fields, makes more and more scholars interested in the study of its dynamic behaviors, and has made remarkable achievements (Yang et al. 2017; Wang et al. 2018; Neyir 2019). However, it is worth mentioning that neural networks with first derivative state is the focus of previous research. Then, the dynamic characteristics of the produced second-order neural networks system by introducing the inertia term, which was named as inertial neural networks, have gradually come into the sight of research of scholars (Babcock and Westervelt 1986). Inertial term is produced by inductance and owns strong biological backgrounds. For instance, squid prominence can be achieved by designing an inductor and the semicircular canals of pigeon hair cells can also be simulated by circuits containing inductors (Mauro et al. 1970; Angelaki and Correia 1991). Obviously, different from the resistor-capacitor first-order differential equations for neural networks, inertial neural networks model not only shows more complex dynamics but also produces more complicated bifurcation behaviors and chaos. Therefore, it is highly important to think the dynamical behaviors and synthesis problems for inertial neural networks (Liu et al. 2009; Li et al. 2017; Xiao et al. 2019; Zhang et al. 2017; Tu et al. 2017; Ru et al. 2020).

As stated in the previous Chaps. 7 and 8, synchronization and anti-synchronization have been intensively studied since they possess potential applications in various fields such as secure communication, image processing, information science, and

Z. Zhang et al., *Complex-Valued Neural Networks Systems with Time Delay*, Intelligent Control and Learning Systems 4,
https://doi.org/10.1007/978-981-19-5450-4_9

optics. In fact, synchronization means that the state values of drive-response system are the same when they reach synchronization while anti-synchronization means that when the drive-response system reach anti-synchronization, their absolute values of states are the same and their symbols are opposite. Hitherto, for this issue, various types of forms containing exponential synchronization , dissipative synchronization and finite/fixed-time synchronization about multiple networks models such as markovian jumping neural networks and switched inertial neural networks, have appeared over that last decades and many achievements have been made, see Velmurugan et al. (2016), Wan et al. (2016), Shi et al. (2016), Wu et al. (2020), Liu et al. (2020), Li and Zheng (2018), Chen and Li (2019), Gong et al. (2018), Prakash et al. (2016), Zhang et al. (2020), Zhang and Cao (2019), Zhang and Ren (2019). Among them, Li and Zheng (2018), Chen and Li (2019), Gong et al. (2018), Prakash et al. (2016), Zhang et al. (2020), Zhang and Cao (2019), Zhang and Ren (2019) studies the synchronization of the inertial neural networks.

With the involvement of complex signal into inertial neural networks model, complex-valued inertial neural networks (CVINNs) attract the attention from some scholars. For example, for impulse CVINNs, the exponential stabilization and global exponential convergence have been explained by matrix measure and LMIs method in Tang and Jian (2018) and Tang and Jian (2019), respectively. For CVINNs, exponential and adaptive synchronization problems are discussed by using of non-reduced order and non-separation approach in Yu et al. (2020). To our knowledge, there has been fewer developed achievements for CVINNs with time-varying delays and the result of anti-synchronization problem has not been reported.

Inspired by the foregoing analysis, this chapter is dedicated to solving the synchronization and anti-synchronization problems of CVINNs with time-varying delays. The main contributions in our work can be presented as follows: (1) Considering rich biological backgrounds of inertial term, CVINNs system with time-varying delays expressed by second-order differential equation is regarded as the study object and the anti-synchronization problem for this system is investigated for the first time; (2) By way of the reduced order, the addressed system is translated into an equivalent first-order complex-valued one. Then, the corresponding criteria in forms of tractable LMIs for the synchronization and anti-synchronization control are established via constructing the appropriate Lyapunov functional; (3) In order to discuss this model more uniformly, it is taken as a whole to develop theoretical results in complex domain instead of separating real-imaginary parts.

9.2 Problem Formulation and Preliminaries

Consider the following CVINNs model:

$$\frac{d^2z(t)}{dt^2} = -\alpha\frac{dz(t)}{dt} - Dz(t) + Af(z(t)) + Bf(z(t-\mu(t))), \tag{9.1}$$

where $z(t) \in \mathbb{C}^n$ denotes the complex-valued state variable; the inertial term of system (9.1) refers to its second derivative term; $D > 0$ and $\alpha > 0$ are real-valued diagonal matrices; A and B, $f(z(t)) \in \mathbb{C}^n$ and $f(z(t-\mu(t))) \in \mathbb{C}^n$ denote complex-valued connection weight matrices, activation functions, respectively; $\mu(t)$ denotes the time-varying delays with $0 \leq \mu(t) \leq \mu$, $\dot{\mu}(t) \leq \delta < 1$, where $\mu > 0$ and $\delta > 0$ are constants.

The initial conditions of system (9.1) is given as

$$z(s) = \varrho(s), \quad \frac{dz(s)}{dt} = \varphi(s), \tag{9.2}$$

where $\varrho(s)$ and $\varphi(s)(s \in [-\mu, 0])$ are continuous.

Define $g(t) = \frac{dz(t)}{dt} + z(t)$, then the drive system (9.1) can be represented as

$$\begin{cases} \frac{dz(t)}{dt} = -z(t) + g(t), \\ \frac{dg(t)}{dt} = -\Delta z(t) - \Lambda g(t) + Af(z(t)) + Bf(z(t-\mu(t))), \end{cases} \tag{9.3}$$

where $\Delta = D - \alpha + I$, $\Lambda = \alpha - I$.

Remark 9.1 The inertial term, an important tool generating bifurcation and chaos, has been introduced into the artificial neural networks by scholars in recent years. For this model, many achievements have been made Li and Zheng (2018), Chen and Li (2019), Gong et al. (2018), Prakash et al. (2016), Zhang et al. (2020), Zhang and Cao (2019), Zhang and Ren (2019) via selecting the appropriate variable substitution and converting the inertial system into the first-order one. But, these achievements are based on real-valued inertial neural networks. Afterward, the references (Tang and Jian 2018, 2019; Yu et al. 2020) expand this research area and introduce the inertial term into complex-valued neural networks stated in (9.1). We will carry out our work on the basis of system (9.1) in this chapter.

Remark 9.2 In this chapter, by means of variable substitution $g(t) = \frac{dz(t)}{dt} + z(t)$, the original second-order system (9.1) is transformed into a general first-order system, which reduces the order of the original system and makes it easy to study the dynamic behaviors of this system by utilizing the existing approaches about first-order system. Moreover, except $g(t) = \frac{dz(t)}{dt} + z(t)$, there are suitable variable substitution methods, such as $g(t) = \beta\frac{dz(t)}{dt} + z(t)$ in Ru et al. (2020) and $g(t) = \frac{dz(t)}{dt} + \xi z(t)$ in Tang and Jian (2018). Based on these variable substitutions, our purpose can all be achieved.

The response system is defined as follows:

$$\begin{cases} \frac{d\breve{z}(t)}{dt} = -\breve{z}(t) + \breve{g}(t) + u_1(t), \\ \frac{d\breve{g}(t)}{dt} = -\Delta\breve{z}(t) - \Lambda\breve{g}(t) + Af(\breve{z}(t)) + Bf(\breve{z}(t-\mu(t))) + u_2(t), \end{cases} \tag{9.4}$$

where $u_1(t)$ and $u_2(t)$ are the controllers that to be designed.

Let $z(t) = \mathrm{col}\{z_1(t), z_2(t), ..., z_n(t)\}$, $f(z(t)) = \mathrm{col}\{f_1(z_1(t)), f_2(z_2(t)), ..., f_n(z_n(t))\}$, and they satisfy the following Lipschitz condition.

Assumption 9.1 For $z, z' \in \mathbb{C}$, there exist constants $\lambda_j > 0$ $(j = 1, 2, ..., n)$, such that

$$|f_j(z) - f_j(z')| \leq \lambda_j |z - z'|.$$

Lemma 9.1 (Popa 2020) *For positive definite Hermitian matrix $P \in \mathbb{C}^{n\times n}$, vector function $z(s)$: $[c, d] \to \mathbb{C}^n$ with scalars $c < d$, then*

$$\Big(\int_c^d z(s)ds\Big)^* P\Big(\int_c^d z(s)ds\Big) \leq (d-c)\int_c^d z^*(s)Pz(s)ds.$$

Lemma 9.2 (Popa 2020) *For any vector $\zeta_1, \zeta_2 \in \mathbb{C}^n$, any scalar $0 < \alpha < 1$, any positive definite Hermitian matrix $P \in \mathbb{C}^{n\times n}$ and any matrix $N \in \mathbb{C}^{n\times n}$ such that $\begin{bmatrix} P & N \\ N^* & P \end{bmatrix} \geq 0$, the following inequality holds:*

$$\frac{1}{\alpha}\zeta_1^* P\zeta_1 + \frac{1}{1-\alpha}\zeta_2^* P\zeta_2 \geq \begin{bmatrix} \zeta_1 \\ \zeta_2 \end{bmatrix}^* \begin{bmatrix} P & N \\ N^* & P \end{bmatrix} \begin{bmatrix} \zeta_1 \\ \zeta_2 \end{bmatrix}.$$

9.3 Synchronization Control Criterion

Let $e(t) = \breve{z}(t) - z(t)$, $v(t) = \breve{g}(t) - g(t)$, $\mathcal{H}(e(t)) = f(\breve{z}(t)) - f(z(t))$, and $\mathcal{H}(e(t-\mu(t))) = f(\breve{z}(t-\mu(t))) - f(z(t-\mu(t)))$, the synchronization error system from (9.3) and (9.4) is given as follows:

$$\begin{cases} \frac{de(t)}{dt} = -e(t) + v(t) + u_1(t), \\ \frac{dv(t)}{dt} = -\Delta e(t) - \Lambda v(t) + A\mathcal{H}(e(t)) + B\mathcal{H}(e(t-\mu(t))) + u_2(t). \end{cases} \tag{9.5}$$

Then, let

$$w(t) = \begin{bmatrix} e(t) \\ v(t) \end{bmatrix}, \ \widetilde{\mathcal{D}} = \begin{bmatrix} I & -I \\ \Delta & \Lambda \end{bmatrix}, \ \widetilde{\mathcal{A}} = \begin{bmatrix} 0 & 0 \\ A & 0 \end{bmatrix}, \ \widetilde{\mathcal{B}} = \begin{bmatrix} 0 & 0 \\ B & 0 \end{bmatrix},$$
$$\widetilde{\mathcal{H}}(w(t)) = \begin{bmatrix} \mathcal{H}(e(t)) \\ 0 \end{bmatrix}, \ \widetilde{\mathcal{H}}(w(t-\mu(t))) = \begin{bmatrix} \mathcal{H}(e(t-\mu(t))) \\ 0 \end{bmatrix}, \ u(t) = \begin{bmatrix} u_1(t) \\ u_2(t) \end{bmatrix}.$$

we have

$$\frac{dw(t)}{dt} = -\widetilde{\mathcal{D}}w(t) + \widetilde{\mathcal{A}}\widetilde{\mathcal{H}}(w(t)) + \widetilde{\mathcal{B}}\widetilde{\mathcal{H}}(w(t-\mu(t))) + u(t). \tag{9.6}$$

The controller $u(t)$ is designed as $u(t) = K_1 w(t)$. Moreover, $w(t)$, $\dot{w}(t)$, $w(t-\mu)$, $w(t-\mu(t))$, $\widetilde{\mathcal{H}}(w(t))$, and $\widetilde{\mathcal{H}}(w(t-\mu(t)))$ are simplified as w, $\dot{w}$, w^{μ}, $w^{\bar{\mu}}$, $\widetilde{\mathcal{H}}(w)$, and $\widetilde{\mathcal{H}}(w^{\bar{\mu}})$, respectively. Then, the error system (9.6) becomes

$$\dot{w} = (K_1 - \widetilde{\mathcal{D}})w + \widetilde{\mathcal{A}}\widetilde{\mathcal{H}}(w) + \widetilde{\mathcal{B}}\widetilde{\mathcal{H}}(w^{\bar{\mu}}). \tag{9.7}$$

Definition 9.1 Under the suitable controllers, if the error system (9.5) or (9.6) is globally asymptotically stable, then systems (9.3) and (9.4) can be globally asymptotically synchronized under this case.

Remark 9.3 So as to achieve the desired targets more easily, an equivalent system (9.6) is generated by reassembling (9.5), which is different from the handling methods without producing equivalent system (Tang and Jian 2018, 2019). This is a feature of our work. Then, based on system (9.6), the expected criteria will be established in the following content.

Theorem 9.1 *Suppose Assumption 9.1 holds, for given positive constants μ and δ, if there exist Hermitian matrices $P_1 > 0$, $P_2 > 0$, $P_3 > 0$, and $P_4 > 0$, real diagonal matrices $M_1 > 0$ and $M_2 > 0$, complex-valued matrices Q, Y, and N such that*

$$\begin{bmatrix} P_4 & N \\ * & P_4 \end{bmatrix} \geq 0, \tag{9.8}$$

$$\Omega = \begin{bmatrix} \Omega_{11} & \Omega_{12} & 0 & 0 & Q\widetilde{\mathcal{A}} & Q\widetilde{\mathcal{B}} \\ * & \Omega_{22} & P_4 - N^* & N & Q\widetilde{\mathcal{A}} & Q\widetilde{\mathcal{B}} \\ * & * & \Omega_{33} & P_4 - N^* & 0 & 0 \\ * & * & * & \Omega_{44} & 0 & 0 \\ * & * & * & * & \Omega_{55} & 0 \\ * & * & * & * & * & \Omega_{66} \end{bmatrix} < 0, \tag{9.9}$$

where $\Omega_{11} = -Q - Q^* + \mu^2 P_4$, $\Omega_{12} = P_1 - Q^* + Y - Q\widetilde{\mathcal{D}}$, $\Omega_{22} = P_2 + \Gamma M_1 \Gamma + P_3 - P_4 + Y + Y^* - Q\widetilde{\mathcal{D}} - \widetilde{\mathcal{D}}Q^*$, $\Omega_{33} = \Gamma M_2 \Gamma - (1-\delta)P_2 - 2P_4 + N + N^*$, $\Omega_{44} = -P_3 - P_4$, $\Omega_{55} = -M_1$, $\Omega_{66} = -M_2$, *then the drive system (9.3) and response system (9.4) can achieve the synchronization, and the controller gain matrix K_1 is given as $K_1 = Q^{-1}Y$.*

Proof Construct the following Lyapunov functional candidate:

$$\begin{aligned} V(t) &= w^*(t)P_1 w(t) + \int_{t-\mu(t)}^{t} w^*(s)P_2 w(s)ds \\ &+ \int_{t-\mu}^{t} w^*(s)P_3 w(s)ds + \mu \int_{-\mu}^{0}\int_{t+\theta}^{t} \dot{w}^*(s)P_4 \dot{w}(s)dsd\theta. \end{aligned} \tag{9.10}$$

Taking the derivative of $V(t)$, we have

$$
\begin{aligned}
\dot{V}(t) &= \dot{w}^* P_1 w + w^* P_1 \dot{w} + w^* P_2 w + w^* P_3 w - (1-\dot{\mu}(t))(w^{\tilde{\mu}})^* P_2 w^{\tilde{\mu}} \\
&\quad - (w^{\mu})^* P_3 w^{\mu} + \mu^2 \dot{w}^* P_4 \dot{w} - \mu \int_{t-\mu}^{t} \dot{w}^*(s) P_4 \dot{w}(s) ds \\
&\leq \dot{w}^* P_1 w + w^* P_1 \dot{w} + w^* P_2 w - (1-\delta)(w^{\tilde{\mu}})^* P_2 w^{\tilde{\mu}} + w^* P_3 w \\
&\quad - (w^{\mu})^* P_3 w^{\mu} + \mu^2 \dot{w}^* P_4 \dot{w} - \mu \int_{t-\mu}^{t} \dot{w}^*(s) P_4 \dot{w}(s) ds. \qquad (9.11)
\end{aligned}
$$

Based on Lemmas 9.1 and 9.2, we have

$$
\begin{aligned}
& -\mu \int_{t-\mu}^{t} \dot{w}^*(s) P_4 \dot{w}(s) ds \\
&= -\mu \int_{t-\mu}^{t-\mu(t)} \dot{w}(s)^* P_4 \dot{w}(s) ds - \mu \int_{t-\mu(t)}^{t} \dot{w}(s)^* P_4 \dot{w}(s) ds \\
&\leq -\frac{\mu}{\mu-\mu(t)} \Big(\int_{t-\mu}^{t-\mu(t)} \dot{w}(s) ds \Big)^* P_4 \Big(\int_{t-\mu}^{t-\mu(t)} \dot{w}(s) ds \Big) \\
&\quad - \frac{\mu}{\mu(t)} \Big(\int_{t-\mu(t)}^{t} \dot{w}(s) ds \Big)^* P_4 \Big(\int_{t-\mu(t)}^{t} \dot{w}(s) ds \Big) \\
&\leq \begin{bmatrix} w \\ w^{\mu} \\ w^{\tilde{\mu}} \end{bmatrix}^* \begin{bmatrix} -P_4 & N & P_4 - N^* \\ * & -P_4 & P_4 - N \\ * & * & -2P_4 + N + N^* \end{bmatrix} \begin{bmatrix} w \\ w^{\mu} \\ w^{\tilde{\mu}} \end{bmatrix}. \qquad (9.12)
\end{aligned}
$$

From Assumption 9.1, it follows that

$$
\begin{aligned}
& \widetilde{\mathcal{H}}^*(w) M_1 \widetilde{\mathcal{H}}(w) - w^* \Gamma M_1 \Gamma w \leq 0, \\
& \widetilde{\mathcal{H}}^*(w^{\tilde{\mu}}) M_2 \widetilde{\mathcal{H}}(w^{\tilde{\mu}}) - (w^{\tilde{\mu}})^* \Gamma M_2 \Gamma w^{\tilde{\mu}} \leq 0, \qquad (9.13)
\end{aligned}
$$

where $\Gamma = \begin{bmatrix} L & 0 \\ 0 & 0 \end{bmatrix}$, $L = \mathrm{diag}\{\lambda_1, \lambda_2, ..., \lambda_n\}$.

On the other hand, from (9.7), the following equality holds:

$$
\begin{aligned}
0 &= (\dot{w} + w)^* Q(-\dot{w} + (K_1 - \widetilde{\mathcal{D}})w + \widetilde{\mathcal{A}}\widetilde{\mathcal{H}}(w) + \widetilde{\mathcal{B}}\widetilde{\mathcal{H}}(w^{\tilde{\mu}})) \\
&\quad + (-\dot{w} + (K_1 - \widetilde{\mathcal{D}})w + \widetilde{\mathcal{A}}\widetilde{\mathcal{H}}(w) + \widetilde{\mathcal{B}}\widetilde{\mathcal{H}}(w^{\tilde{\mu}}))^* Q^*(\dot{w} + w). \qquad (9.14)
\end{aligned}
$$

Based on (9.11)–(9.14) and by defining $QK_1 = Y$, we can get that

$$
\dot{V} \leq \eta^* \Omega \eta,
$$

where $\eta = \mathrm{col}\{\dot{w}, w, w^{\tilde{\mu}}, w^{\mu}, \mathcal{H}(w), \mathcal{H}(w^{\tilde{\mu}})\}$. Then, according to (9.9), we have

$$
\dot{V}(t) \leq 0. \qquad (9.15)
$$

Thus, the error system (9.7) is stable, i.e., the drive-response systems (9.3) and (9.4) can achieve the synchronization. □

9.4 Anti-Synchronization Control Criterion

Let $\bar{e}(t) = \breve{z}(t) + z(t)$, $\bar{v}(t) = \breve{g}(t) + g(t)$, the anti-synchronization error system from (9.3) and (9.4) is given as follows:

$$\begin{cases} \frac{d\bar{e}(t)}{dt} = -\bar{e}(t) + \bar{v}(t) + u_1(t), \\ \frac{d\bar{v}(t)}{dt} = -\Delta\bar{e}(t) - \Lambda\bar{v}(t) + A\mathcal{G}(\bar{e}(t)) + B\mathcal{G}(\bar{e}(t-\mu(t))) + u_2(t), \end{cases} \tag{9.16}$$

where $\mathcal{G}(\bar{e}(t)) = f(\breve{z}(t)) + f(z(t))$ and $\mathcal{G}(\bar{e}(t-\mu(t))) = f(\breve{z}(t-\mu(t))) + f(z(t-\mu(t)))$. Then, let

$$\vartheta(t) = \begin{bmatrix} \bar{e}(t) \\ \bar{v}(t) \end{bmatrix}, \ \widetilde{\mathcal{G}}(\vartheta(t)) = \begin{bmatrix} \mathcal{G}(\bar{e}(t)) \\ 0 \end{bmatrix}, \ \widetilde{\mathcal{G}}(\vartheta(t-\mu(t))) = \begin{bmatrix} \mathcal{G}(\bar{e}(t-\mu(t))) \\ 0 \end{bmatrix},$$

we have

$$\frac{d\vartheta(t)}{dt} = -\widetilde{D}\vartheta(t) + \widetilde{A}\widetilde{\mathcal{G}}(\vartheta(t)) + \widetilde{B}\widetilde{\mathcal{G}}(\vartheta(t-\mu(t))) + u(t). \tag{9.17}$$

The controller $u(t)$ is designed as $u(t) = K_2\vartheta(t)$. Moreover, $\vartheta(t)$, $\dot{\vartheta}(t)$, $\vartheta(t-\mu)$, $\vartheta(t-\mu(t))$, $\widetilde{\mathcal{G}}(\vartheta(t))$, and $\widetilde{\mathcal{G}}(\vartheta(t-\mu(t)))$ are simplified as ϑ, $\dot{\vartheta}$, ϑ^{μ}, $\vartheta^{\bar{\mu}}$, $\widetilde{\mathcal{G}}(\vartheta)$, and $\widetilde{\mathcal{G}}(\vartheta^{\bar{\mu}})$, respectively, then the error system (9.17) becomes

$$\dot{\vartheta} = (K_2 - \widetilde{D})\vartheta(t) + \widetilde{A}\widetilde{\mathcal{G}}(\vartheta(t)) + \widetilde{B}\widetilde{\mathcal{G}}(\vartheta(t-\mu(t))). \tag{9.18}$$

Definition 9.2 Under the suitable controllers, if the anti-synchronization error system (9.16) or (9.17) is globally asymptotically stable, then systems (9.3) and (9.4) can be globally asymptotically anti-synchronized under this case.

Theorem 9.2 *Suppose component functions $f_j(\cdot)$ $(j = 1, 2, ..., n)$ are odd functions and Assumption 9.1 holds, for given positive constants μ and δ, if there exist Hermitian matrices $R_i > 0 (i = 1, 2, 3, 4)$, real diagonal matrices $N_i > 0 (i = 1, 2)$, complex-valued matrices E, S, and M, such that*

$$\begin{bmatrix} R_4 & M \\ * & R_4 \end{bmatrix} \geq 0, \tag{9.19}$$

$$\Xi = \begin{bmatrix} \Xi_{11} & \Xi_{12} & 0 & 0 & S\widetilde{A} & S\widetilde{B} \\ * & \Xi_{22} & R_4 - M^* & M & S\widetilde{A} & S\widetilde{B} \\ * & * & \Xi_{33} & R_4 - M^* & 0 & 0 \\ * & * & * & \Xi_{44} & 0 & 0 \\ * & * & * & * & \Xi_{55} & 0 \\ * & * & * & * & * & \Xi_{66} \end{bmatrix} < 0, \tag{9.20}$$

where $\Xi_{11} = -S - S^* + \mu^2 R_4$, $\Xi_{12} = R_1 - S^* + Y - S\widetilde{D}$, $\Xi_{22} = R_2 + \Gamma N_1 \Gamma + R_3 - R_4 + E + E^* - S\widetilde{D} - \widetilde{D}S^*$, $\Xi_{33} = \Gamma N_2 \Gamma - (1-\delta)R_2 - 2R_4 + M + M^*$, $\Xi_{44} = -R_3 - R_4$, $\Xi_{55} = -N_1$, $\Xi_{66} = -N_2$, *then the drive-response systems (9.3) and (9.4) are anti-synchronized, and the controller gain matrix* K_2 *is designed as* $K_2 = S^{-1}E$.

Proof Construct the following Lyapunov functional candidate:

$$\begin{aligned} V(t) = &\ \vartheta^*(t) R_1 \vartheta(t) + \int_{t-\mu(t)}^{t} \vartheta^*(s) R_2 \vartheta(s) ds \\ &+ \int_{t-\mu}^{t} \vartheta^*(s) R_3 \vartheta(s) ds + \mu \int_{-\mu}^{0} \int_{t+\theta}^{t} \dot{\vartheta}^*(s) R_4 \dot{\vartheta}(s) ds d\theta. \end{aligned} \tag{9.21}$$

Taking the derivative of $V(t)$, we have

$$\begin{aligned} \dot{V}(t) = &\ \dot{\vartheta}^* R_1 \vartheta + \vartheta^* R_1 \dot{\vartheta} + \vartheta^* R_2 \vartheta + \vartheta^* R_3 \vartheta - (1 - \dot{\mu}(t))(\vartheta^{\tilde{\mu}})^* R_2 \vartheta^{\tilde{\mu}} \\ &- (\vartheta^{\mu})^* R_3 \vartheta^{\mu} + \mu^2 \dot{\vartheta}^* R_4 \dot{\vartheta} - \mu \int_{t-\mu}^{t} \dot{\vartheta}^*(s) R_4 \dot{\vartheta}(s) ds \\ \le &\ \dot{\vartheta}^* R_1 \vartheta + \vartheta^* R_1 \dot{\vartheta} + \vartheta^* R_2 \vartheta - (1-\delta)(\vartheta^{\tilde{\mu}})^* R_2 \vartheta^{\tilde{\mu}} + \vartheta^* R_3 \vartheta \\ &- (\vartheta^{\mu})^* R_3 \vartheta^{\mu} + \mu^2 \dot{\vartheta}^* R_4 \dot{\vartheta} - \mu \int_{t-\mu}^{t} \dot{\vartheta}^*(s) R_4 \dot{\vartheta}(s) ds. \end{aligned} \tag{9.22}$$

Based on Lemmas 9.1 and 9.2, we have

$$\begin{aligned} &-\mu \int_{t-\mu}^{t} \dot{\vartheta}^*(s) R_4 \dot{\vartheta}(s) ds \\ &= -\mu \int_{t-\mu}^{t-\mu(t)} \dot{\vartheta}(s)^* R_4 \dot{\vartheta}(s) ds - \mu \int_{t-\mu(t)}^{t} \dot{\vartheta}(s)^* R_4 \dot{\vartheta}(s) ds \\ &\le -\frac{\mu}{\mu - \mu(t)} \Big(\int_{t-\mu}^{t-\mu(t)} \dot{\vartheta}(s) ds\Big)^* R_4 \Big(\int_{t-\mu}^{t-\mu(t)} \dot{\vartheta}(s) ds\Big) \\ &\quad - \frac{\mu}{\mu(t)} \Big(\int_{t-\mu(t)}^{t} \dot{\vartheta}(s) ds\Big)^* R_4 \Big(\int_{t-\mu(t)}^{t} \dot{\vartheta}(s) ds\Big) \\ &\le \begin{bmatrix} \vartheta \\ \vartheta^{\mu} \\ \vartheta^{\tilde{\mu}} \end{bmatrix}^* \begin{bmatrix} -R_4 & M & R_4 - M^* \\ * & -R_4 & R_4 - M \\ * & * & -2R_4 + M + M^* \end{bmatrix} \begin{bmatrix} \vartheta \\ \vartheta^{\mu} \\ \vartheta^{\tilde{\mu}} \end{bmatrix}. \end{aligned} \tag{9.23}$$

By means of odd and even properties of component functions $f_j(\cdot)$ and Assumption 9.1, we have

$$\begin{aligned}&\widetilde{\mathcal{G}}^*(\vartheta)N_1\widetilde{\mathcal{G}}(\vartheta)-\vartheta^*\Gamma N_1\Gamma\vartheta\leq 0,\\&\widetilde{\mathcal{G}}^*(\vartheta^{\tilde{\mu}})N_2\widetilde{\mathcal{G}}(\vartheta^{\tilde{\mu}})-(\vartheta^{\tilde{\mu}})^*\Gamma N_2\Gamma\vartheta^{\tilde{\mu}}\leq 0.\end{aligned}\tag{9.24}$$

Moreover, from (9.18), the following equality holds

$$\begin{aligned}0=&(\dot{\vartheta}+\vartheta)^*S(-\dot{\vartheta}+(K_2-\widetilde{D})\vartheta+A\widetilde{\mathcal{G}}(\vartheta)+B\widetilde{\mathcal{G}}(\vartheta^{\tilde{\mu}}))\\&+(-\dot{\vartheta}+(K_2-\widetilde{D})\vartheta+A\widetilde{\mathcal{G}}(\vartheta)+B\widetilde{\mathcal{G}}(\vartheta^{\tilde{\mu}}))^*S^*(\dot{\vartheta}+\vartheta).\end{aligned}\tag{9.25}$$

Based on (9.22)–(9.25) and by defining $SK_2=E$, we can obtain that

$$\dot{V}\leq\zeta^*\Xi\zeta,$$

where $\zeta=\mathrm{col}\{\dot{\vartheta},\vartheta,\vartheta^{\tilde{\mu}},\vartheta^{\mu},\widetilde{\mathcal{G}}(\vartheta),\widetilde{\mathcal{G}}(\vartheta^{\tilde{\mu}})\}$. Then, according to (9.20), we have

$$\dot{V}(t)\leq 0.\tag{9.26}$$

Thus, the error system (9.18) is stable, i.e., the drive-response systems (9.3) and (9.4) can achieve the anti-synchronization. □

Remark 9.4 Hitherto, few references have involved the CVINNs (Tang and Jian 2018, 2019; Yu et al. 2020). Further, for this system, no result about anti-synchronization issues has been reported. Here, we aim at this model with time-varying delays, deeply investigate it and proposes the corresponding criteria as stated in Theorems 9.1 and 9.2, which complement the existing achievements.

Remark 9.5 In Tang and Jian (2018), Tang and Jian (2019), for delayed impulsive CVINNs, the exponential stabilization and exponential convergence subjects are discussed by separating real and imaginary parts of the considered systems. Here, we treat system (9.1) as an entirety and derive our results in complex-valued domain.

Remark 9.6 Sufficient conditions guaranteeing the synchronization and anti-synchronization of CVINNs in Theorems 9.1 and 9.2 are established in the form of complex-valued LMIs that cannot be handled directly with the LMI toolbox. So, we need to change the complex-valued LMIs into real-valued ones. Then, by solving the real-valued LMIs, we can obtain the desired results. Here, we omit them in order to save the space.

9.5 Illustrative Examples

In this section, we will give two examples to illustrate the effectiveness of the proposed results in Theorems 9.1 and 9.2.

Example 9.1 Consider a two-dimension system (9.1) with

$$\alpha = \begin{bmatrix} 3 & 0 \\ 0 & 3 \end{bmatrix}, \; D = \begin{bmatrix} 2.5 & 0 \\ 0 & 2.5 \end{bmatrix}, \; f_j(z_j) = \frac{1 - e^{-\bar{z}_j}}{1 + e^{-\bar{z}_j}}, \; j = 1, 2,$$

$$A = \begin{bmatrix} -1.2 + 2.5i & 2.5 - 2.3i \\ -3.7 + 2.5i & -1.8 - 4i \end{bmatrix}, \; B = \begin{bmatrix} -2 - 3i & 2.2 - 1.3i \\ -3.5 + 3i & 2.1 - 2.9i \end{bmatrix},$$

and $\mu(t) = 0.2\cos^2(t)$ represents the time-varying delays, which means that $\mu = 0.2$ and $\delta = 0.2$. Then, we can obtain the corresponding matrices of system (9.7) as follows:

$$\widetilde{\mathcal{D}} = \begin{bmatrix} 1 & 0 & -1 & 0 \\ 0 & 1 & 0 & -1 \\ 0.5 & 0 & 2 & 0 \\ 0 & 0.5 & 0 & 2 \end{bmatrix}, \; \widetilde{\mathcal{A}} = \begin{bmatrix} 0 & 0 & 0 & 0 \\ 0 & 0 & 0 & 0 \\ -1.2 + 2.5i & 2.5 - 2.3i & 0 & 0 \\ -3.7 + 2.5i & -1.8 - 4i & 0 & 0 \end{bmatrix},$$

$$\Gamma = \begin{bmatrix} \frac{2}{3} & 0 & 0 & 0 \\ 0 & \frac{2}{3} & 0 & 0 \\ 0 & 0 & 0 & 0 \\ 0 & 0 & 0 & 0 \end{bmatrix}, \; \widetilde{\mathcal{B}} = \begin{bmatrix} 0 & 0 & 0 & 0 \\ 0 & 0 & 0 & 0 \\ -2 - 3i & -2.2 - 1.3i & 0 & 0 \\ -3.5 - 3i & -2.1 - 2.9i & 0 & 0 \end{bmatrix}.$$

As the controller $u(t) = 0$, the curves of systems (9.3) and (9.4) with the above parameters are drawn in Figs. 9.1 and 9.2.

Then, by means of LMI toolbox, the feasible solutions of (9.8) and (9.9) can be obtained, such as

$$Q = \begin{bmatrix} Q_1 & Q_2 \end{bmatrix}, \quad Y = \begin{bmatrix} Y_1 & Y_2 \end{bmatrix},$$

$$Q_1 = \begin{bmatrix} 0.7523 + 0.0000i & 54.8969 - 0.0035i \\ -54.9028 + 0.0064i & 0.7602 - 0.0003i \\ 0.0158 + 0.0016i & -0.0047 - 0.0098i \\ -0.0051 + 0.0156i & 0.0072 + 0.0000i \end{bmatrix},$$

$$Q_2 = \begin{bmatrix} -0.0141 + 0.0004i & 0.0047 + 0.0147i \\ 0.0144 - 0.0472i & -0.0226 - 0.0023i \\ 0.0814 - 0.0016i & -0.0230 - 0.0257i \\ -0.0230 + 0.0290i & 0.0436 - 0.0001i \end{bmatrix},$$

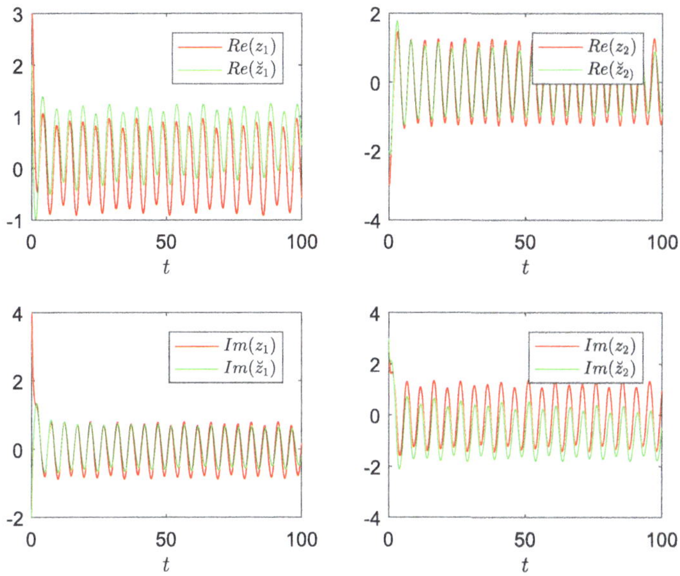

Fig. 9.1 The curves of state variables z_1, z_2 and $\breve{z}_1$, $\breve{z}_2$ when the controller $u(t) = 0$ in Example 9.1

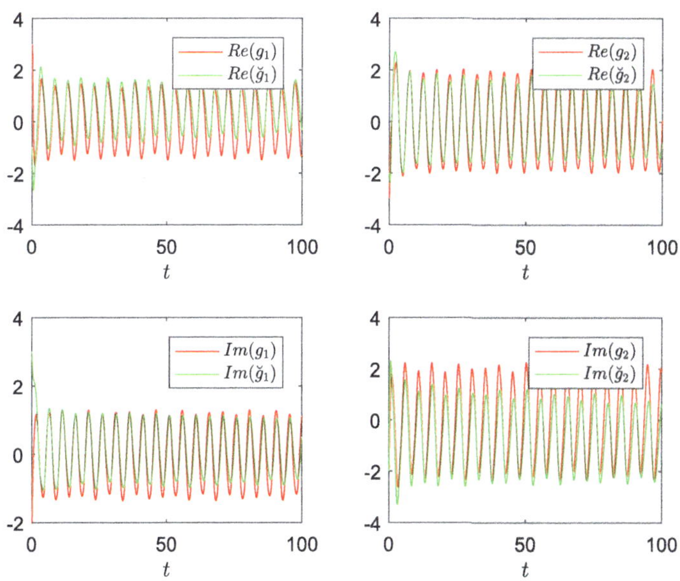

Fig. 9.2 The curves of state variables g_1, g_2 and $\breve{g}_1$, $\breve{g}_2$ when the controller $u(t) = 0$ in Example 9.1

$$Y_1 = \begin{bmatrix} -2.2419 - 0.0001i & 0.0249 + 0.0297i \\ 0.0084 + 0.0144i & -2.3697 - 0.0055i \\ 0.0419 + 0.0020i & -0.1130 - 0.1105i \\ -0.0134 + 0.0195i & 0.2191 - 0.0020i \end{bmatrix},$$

$$Y_2 = \begin{bmatrix} -0.7604 - 0.0003i & -54.8945 + 0.0034i \\ 54.9192 - 0.0274i & -0.7830 - 0.0020i \\ -2.2570 - 0.0041i & -0.0311 - 0.0530i \\ -0.0304 + 0.0357i & -2.2936 + 0.0002i \end{bmatrix}.$$

The other feasible solutions of matrices are not listed here due to the limited space. So,

$$K_1 = Q^{-1}Y = \begin{bmatrix} K_{11} & K_{12} \end{bmatrix},$$

$$K_{11} = \begin{bmatrix} -0.0005 - 0.0007i & 0.0413 - 0.0001i \\ -0.0407 + 0.0000i & -0.0005 - 0.0008i \\ 0.4460 + 0.0776i & 0.0683 + 0.3328i \\ -0.0153 + 0.1906i & 5.2870 + 0.0823i \end{bmatrix},$$

$$K_{12} = \begin{bmatrix} -1.0005 + 0.0266i & 0.0053 + 0.0165i \\ -0.0012 + 0.0038i & -0.9988 + 0.0151i \\ -43.8777 + 0.1420i & -23.8223 - 26.9892i \\ -23.9444 + 30.3672i & -82.9412 + 1.3855i \end{bmatrix}.$$

According to Theorem 9.1, the error system (9.7) based on the controller matrix K_1 is easily verified to be stable. The simulation results are given in Figs. 9.3, 9.4 and 9.5.

Example 9.2 Consider a two-dimension system (9.1) with

$$\alpha = \begin{bmatrix} 0.1 & 0 \\ 0 & 0.1 \end{bmatrix}, \; D = \begin{bmatrix} 2.5 & 0 \\ 0 & 2.5 \end{bmatrix}, \; f_j(z_j) = -\bar{z}_j, \; j = 1, 2,$$

$$A = \begin{bmatrix} -0.4 - 2i & -0.79 - 1.8i \\ -1.2 + 2i & 1.89 - 2i \end{bmatrix}, \; B = \begin{bmatrix} -1 - 1.5i & -0.8 + 1.3i \\ 0.66 - 0.58i & 1.9 - 1.25i \end{bmatrix},$$

and the same $\mu(t)$ as stated in Example 9.1. Then, we can obtain the corresponding matrices of system (9.18) as follows:

$$\widetilde{\mathcal{D}} = \begin{bmatrix} 1 & 0 & -1 & 0 \\ 0 & 1 & 0 & -1 \\ 3.4 & 0 & -0.9 & 0 \\ 0 & 3.4 & 0 & -0.9 \end{bmatrix}, \; \widetilde{\mathcal{A}} = \begin{bmatrix} 0 & 0 & 0 & 0 \\ 0 & 0 & 0 & 0 \\ -0.4 - 2i & -0.79 - 1.8i & 0 & 0 \\ -1.2 + 2i & 1.89 - 2i & 0 & 0 \end{bmatrix},$$

$$\Gamma = \begin{bmatrix} 1 & 0 & 0 & 0 \\ 0 & 1 & 0 & 0 \\ 0 & 0 & 0 & 0 \\ 0 & 0 & 0 & 0 \end{bmatrix}, \; \widetilde{\mathcal{B}} = \begin{bmatrix} 0 & 0 & 0 & 0 \\ 0 & 0 & 0 & 0 \\ -1 - 1.5i & -0.8 + 1.3i & 0 & 0 \\ 0.66 - 0.58i & 1.9 - 1.25i & 0 & 0 \end{bmatrix}.$$

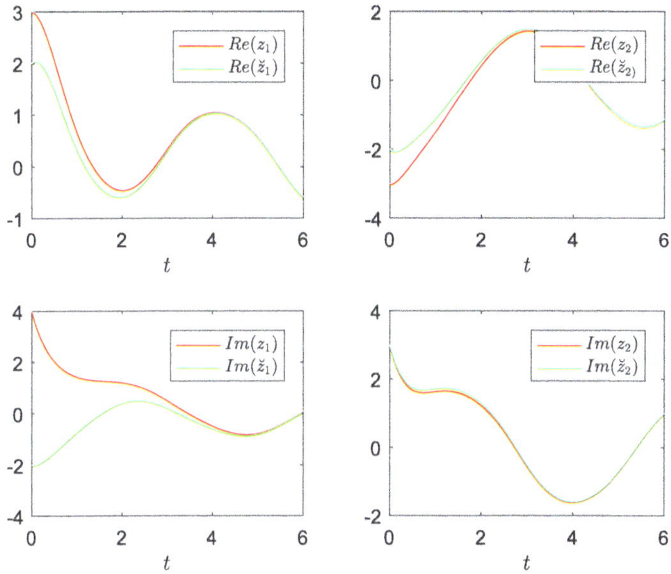

Fig. 9.3 The curves of state variables z_1, z_2 and $\breve{z}_1$, $\breve{z}_2$ under the controller in Example 9.1

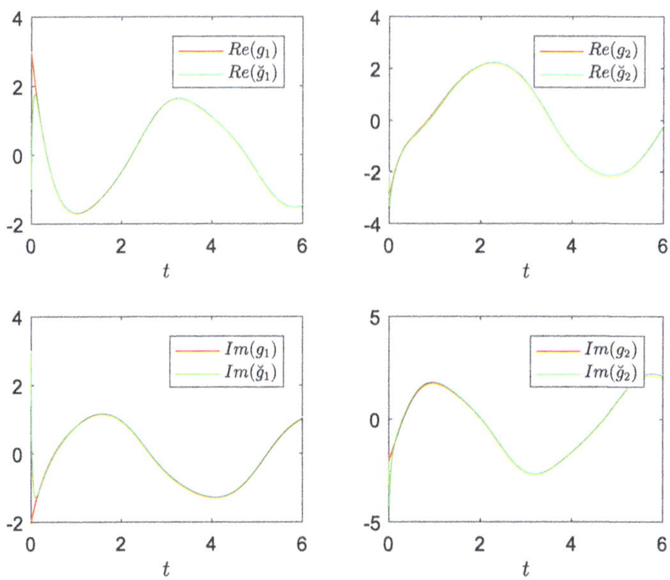

Fig. 9.4 The curves of state variables g_1, g_2 and $\breve{g}_1$, $\breve{g}_2$ under the controller in Example 9.1

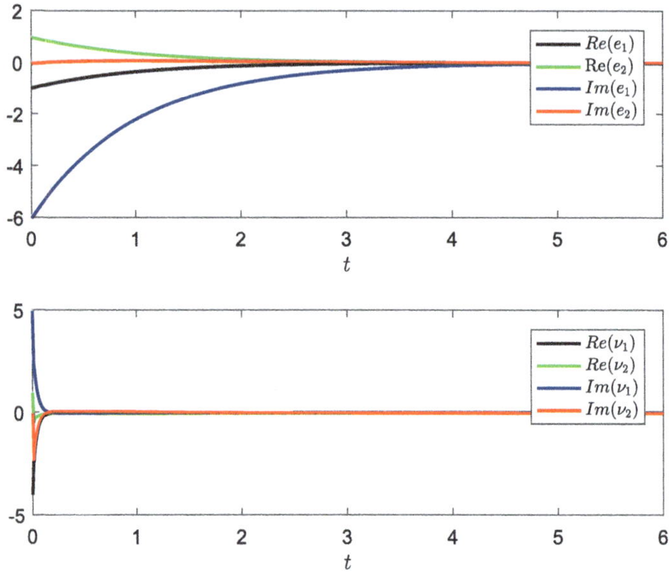

Fig. 9.5 The curves of the error system (9.7) in Example 9.1

Figures 9.6 and 9.7 depict the curves of state variables of systems (9.3) and (9.4) with the above parameters under the controller $u(t) = 0$.

Then, by means of LMI toolbox, the feasible solutions of (9.19) and (9.20) can be obtained, such as

$$S = \begin{bmatrix} S_1 \ S_2 \end{bmatrix}, \quad E = \begin{bmatrix} E_1 \ E_2 \end{bmatrix},$$

$$S_1 = 1.0e+02 \begin{bmatrix} 0.0010 - 0.0000i & -1.9894 + 0.0000i \\ 1.9894 + 0.0000i & 0.0010 - 0.0000i \\ 0.0000 - 0.0000i & 0.0000 - 0.0000i \\ 0.0000 + 0.0000i & 0.0000 + 0.0000i \end{bmatrix},$$

$$S_2 = 1.0e+02 \begin{bmatrix} -0.0000 - 0.0000i & -0.0000 + 0.0000i \\ -0.0000 - 0.0000i & -0.0000 + 0.0000i \\ 0.0001 + 0.0000i & 0.0000 - 0.0000i \\ 0.0000 + 0.0000i & 0.0001 - 0.0000i \end{bmatrix},$$

$$E_1 = 1.0e+02 \begin{bmatrix} -0.0034 + 0.0000i & -0.0000 + 0.0000i \\ -0.0000 + 0.0000i & -0.0034 - 0.0000i \\ 0.0005 - 0.0001i & 0.0000 + 0.0000i \\ 0.0005 + 0.0001i & -0.0000 + 0.0000i \end{bmatrix},$$

$$E_2 = 1.0e+02 \begin{bmatrix} -0.0010 + 0.0000i & 1.9894 - 0.0001i \\ -1.9894 + 0.0000i & -0.0010 + 0.0000i \\ -0.0034 - 0.0000i & -0.0000 + 0.0000i \\ -0.0000 - 0.0000i & -0.0034 + 0.0000i \end{bmatrix}.$$

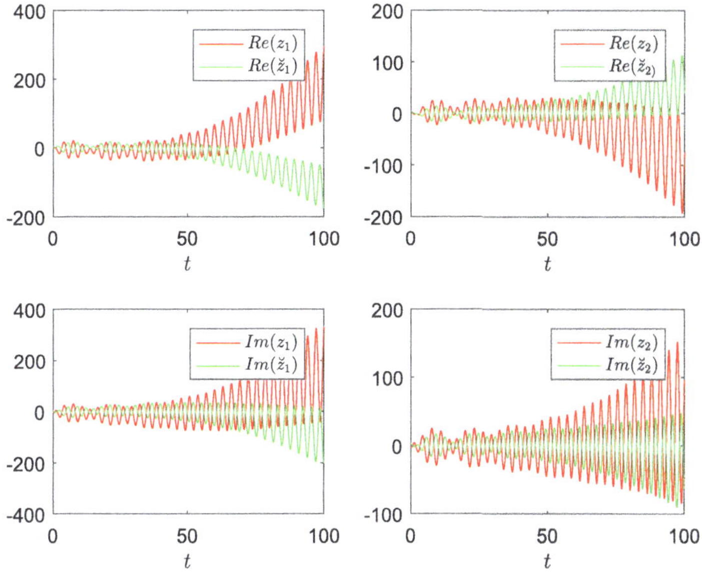

Fig. 9.6 The curves of state variables z_1, z_2 and $\breve{z}_1$, $\breve{z}_2$ without the controller in Example 9.2

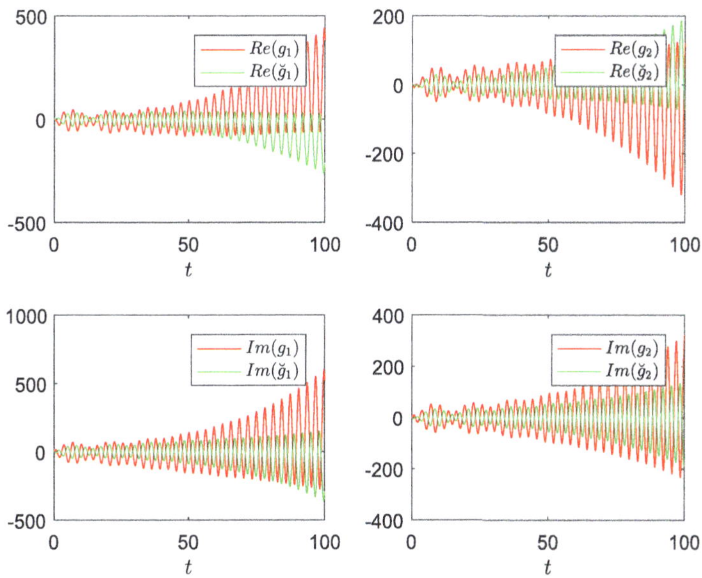

Fig. 9.7 The curves of state variables g_1, g_2 and $\breve{g}_1$, $\breve{g}_2$ without the controller in Example 9.2

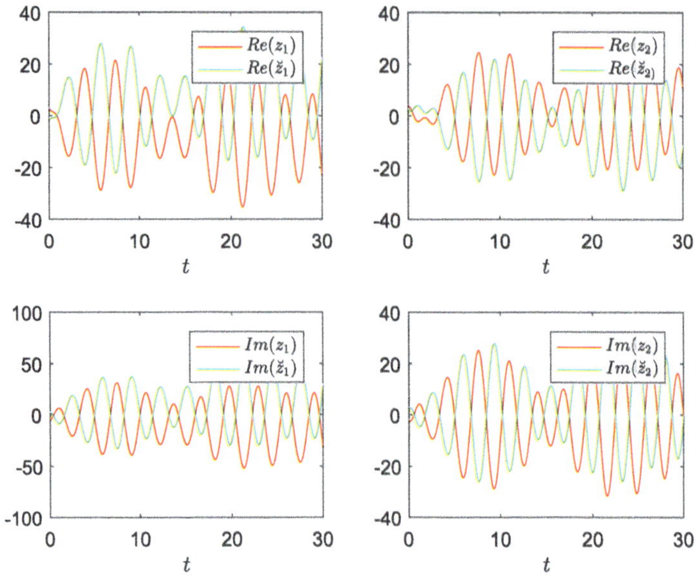

Fig. 9.8 The curves of state variables z_1, z_2 and $\check{z}_1$, $\check{z}_2$ under the controller in Example 9.2

The other feasible solutions of matrices are not listed here due to save space. So,

$$K_2 = S^{-1}E = \begin{bmatrix} K_{21} & K_{22} \end{bmatrix},$$

$$K_{21} = \begin{bmatrix} -0.0000 + 0.0000i & -0.0017 - 0.0000i \\ 0.0017 + 0.0000i & -0.0000 - 0.0000i \\ 3.5330 + 0.3280i & -0.0029 + 0.0862i \\ 3.5093 - 0.3805i & 0.0206 + 0.0288i \end{bmatrix},$$

$$K_{22} = \begin{bmatrix} -1.0000 - 0.0000i & -0.0000 + 0.0000i \\ 0.0000 + 0.0001i & -1.0000 - 0.0002i \\ -28.2873 + 1.3776i & 2.0530 - 8.1660i \\ 2.0717 + 7.9287i & -25.6925 - 1.1927i \end{bmatrix}.$$

According to Theorem 9.2, the error system (9.18) based on the controller matrix K_2 is easily verified to be stable. The simulation results are given in Figs. 9.8, 9.9 and 9.10.

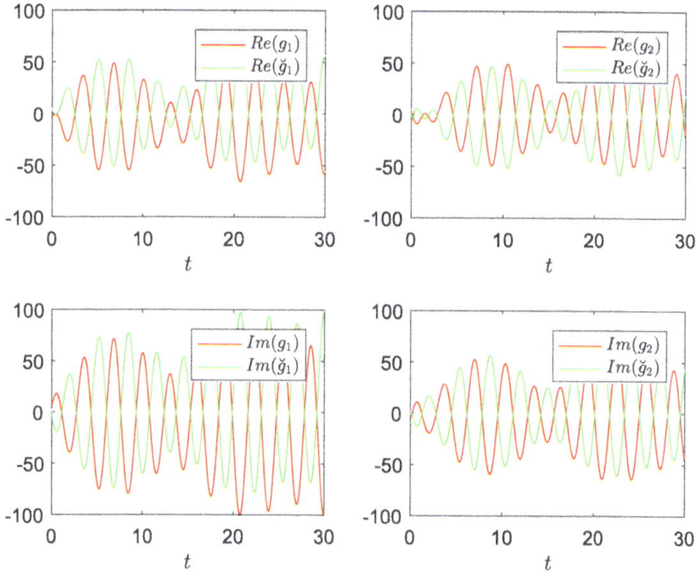

Fig. 9.9 The curves of state variables g_1, g_2 and $\breve{g}_1$, $\breve{g}_2$ under the controller in Example 9.2

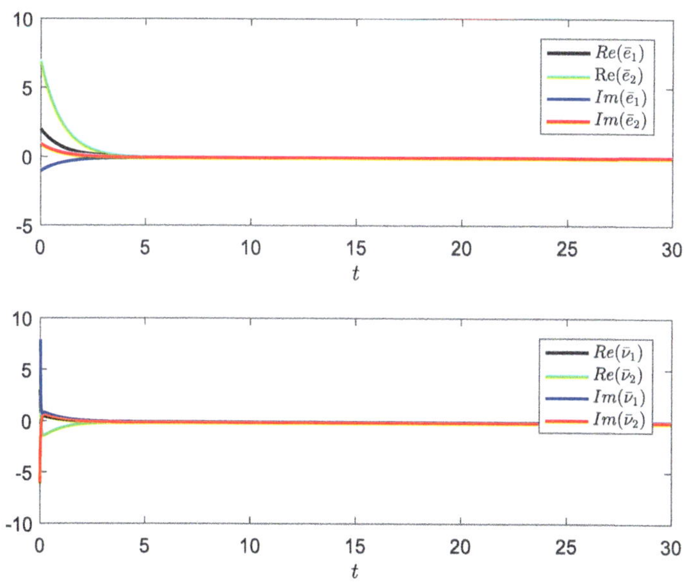

Fig. 9.10 The curves of the error system (9.18) in Example 9.2

9.6 Conclusion

In this chapter, we have studied the synchronization and anti-synchronization problems of CVINNs with time-varying delays. Via representing the inertial neural networks model as the first derivative form, we have established some corresponding sufficient conditions in terms of LMIs with regard to the synchronization and anti-synchronization control by using Lyapunov function method. In the following chapters, we will focus on the synchronization problems of complex-valued systems based on the finite/fixed-time theory.

References

Angelaki D, Correia M (1991) Models of membrane resonance in pigeon semicircular canal type II hair cells. Biol Cybern 65(1):1–10

Babcock K, Westervelt R (1986) Stability and dynamics of simple electronic neural networks with added inertia. Phys D 23(1–3):464–469

Chen C, Li L (2019) Fixed-time synchronization of inertial memristor-based neural networks with discrete delay. Neural Netw 109:81–89

Gong S, Yang S, Guo Z, Huang T (2018) Global exponential synchronization of inertial memristive neural networks with time-varying delay via nonlinear controller. Neural Netw 102:138–148

Li N, Zheng W (2018) Synchronization criteria for inertial memristor-based neural networks with linear coupling. Neural Netw 106:260–270

Li XY, Li XT, Hu C (2017) Some new results on stability and synchronization for delayed inertial neural networks based on non-reduced order method. Neural Netw 96:91–100

Liu Q, Liao X, Liu Y, Zhou S, Guo S (2009) Dynamics of an inertial two-neuron system with time delay. Nonlinear Dyn 58:573–609

Liu Y, Xia J, Meng B, Song X, Shen H (2020) Extended dissipative synchronization for semi-Markov jump complex dynamic networks via memory sampled-data control scheme. J Franklin Inst 357:10900–10920

Mauro A, Conti F, Dodge F, Schor R (1970) Subthreshold behavior and phenomenological impedance of the squid giant axon. J Gen Physiol 55(4):497–523

Neyir O (2019) Stability analysis of Cohen-Grossberg neural networks of neutral-type: multiple delays case. Neural Netw 113:20–27

Popa CA (2020) Global μ-stability of neutral-type impulsive complex-valued BAM neural networks with leakage delay and unbounded time-varying delays. Neurocomputing 376:73–94

Prakash M, Balasubramaniam P, Lakshmanan S (2016) Synchronization of Markovian jumping inertial neural networks and its applications in image encryption. Neural Netw 83:86–93

Ru T, Xia J, Huang X, Chen X, Wang J (2020) Reachable set estimation of delayed fuzzy inertial neural networks with Markov jumping parameters. J Franklin Inst 357:6882–6898

Shi L, Yang X, Li Y, Feng Z (2016) Finite-time synchronization of nonidentical chaotic systems with multiple time-varying delays and bounded perturbations. Nonlinear Dyn 83:75–87

Tang Q, Jian J (2018) Matrix measure based exponential stabilization for complex-valued inertial neural networks with time-varying delays using impulsive control. Neurocomputing 273:251–259

Tang Q, Jian J (2019) Global exponential convergence for impulsive inertial complex-valued neural networks with time-varying delays. Math Comput Simulat 159:39–56

Tu Z, Cao J, Alsaedi A (2017) Global dissipativity of memristor-based neutral type inertial neural networks. Neural Netw 88:125–133

Velmurugan G, Rakkiyappan R, Cao J (2016) Finite-time synchronization of fractional-order memristor-based neural networks with time delays. Neural Netw 73:36–46

Wan Y, Cao J, Wen G, Yu W (2016) Robust fixed-time synchronization of delayed Cohen-Grossberg neural networks. Neural Netw 73:86–94

Wang Z, Li L, Li Y, Cheng Z (2018) Stability and Hopf bifurcation of a three-neuron network with multiple discrete and distributed delays. Neural Process Lett 48:1481–1502

Wu T, Huang X, Chen X, Wang J (2020) Sampled-data H_∞ exponential synchronization for delayed semi-Markov jump CDNs: A looped-functional approach. Appl Math Comput 377:125156

Xiao Q, Huang Z, Zeng Z (2019) Passivity analysis for memristor-based inertial neural networks with discrete and distributed delays. IEEE Trans Syst Man Cybern Syst 49:375–385

Yang B, Wang J, Wang J (2017) Stability analysis of delayed neural networks via a new integral inequality. Neural Netw 88:49–57

Yu J, Hu C, Jiang H, Wang L (2020) Exponential and adaptive synchronization of inertial complex-valued neural networks: A non-reduced order and non-separation approach. Neural Netw 124:50–59

Zhang Z, Cao J (2019) Novel finite-time synchronization criteria for inertial neural networks with time delays via integral inequality method. IEEE Trans Neural Netw Learn Syst 30:1476–1485

Zhang Z, Ren L (2019) New sufficient conditions on global asymptotic synchronization of inertial delayed neural networks by using integrating inequality techniques. Nonlinear Dyn 95:905–917

Zhang W, Huang T, He X, Li C (2017) Global exponential stability of inertial memristor-based neural networks with time-varying delays and impulses. Neural Netw 95:102–109

Zhang G, Zeng Z, Ning D (2020) Novel results on synchronization for a class of switched inertial neural networks with distributed delays. Inf Sci 511:114–126

Chapter 10
Fixed-Time Synchronization for CVBAMNNs with Time Delays

Abstract This chapter focuses on the fixed-time synchronization problem for complex-valued bidirectional associative memory neural networks (CVBAMNNs) with time delays. We will design a new nonlinear delayed controller different from the existing ones. Then, we will present a new criterion to guarantee that the addressed systems achieve the synchronization in fixed time and give a more accurate estimation independent on the initial conditions for the settling time by using on the fixed-time stability, the Lyapunov function method and some inequalities techniques.

10.1 Introduction

Because synchronization has a fundamental role in perceiving unknown dynamical system via another known system, it exists in many practical systems involving chemical reactions, power converters, and biological systems. Besides, it has been widely used in various engineering fields, such as human heartbeat regulation, secure communication, and information science. Nowadays, lots of achievements for asymptotic and exponential synchronization have been obtained via various methods. Moreover, to meet the requirement of practical problems, finite-time synchronization is concerned and has drawn wide attentions.

Different from the classical synchronization analysis, finite-time synchronization means that the dynamical behaviors of coupled systems achieve the same time spatial state in finite settling time. So far, a lot of relevant articles have been reported (Selvaraj et al. 2018; Jiang et al. 2015; Abdurahman et al. 2016; Du et al. 2014; Velmurugan et al. 2016; Yang et al. 2016). However, there exists a drawback in these results, i.e., the estimation of the settling time is heavily dependent on the initial conditions of the considered systems. This dependence leads to serious inconvenience in practical applications especially when the information about initial conditions is unavailable beforehand. In addition, in practical engineering situations, it is more desirable that the synchronization of the systems can be accomplished within a fixed time interval for any initial conditions. In 2012, fixed-time stability was first proposed in (Polyakov, 2012), in which the settling time of the finite-time stable system can

Z. Zhang et al., *Complex-Valued Neural Networks Systems with Time Delay*,
Intelligent Control and Learning Systems 4,
https://doi.org/10.1007/978-981-19-5450-4_10

be bounded by a fixed constant resting upon initial values no longer. Afterward, the research results for fixed-time theory appear in various fields, such as power system (Ni et al. 2017), traffic (Muralidharan et al. 2015), rigid spacecraft (Jiang et al. 2016), consensus problems (Zuo and Tie 2016; Fu and Wang 2016), stabilization problems (Polyakov et al. 2015; Hua et al. 2017; Zhang et al. 2018), fixed-time synchronization (Ding et al. 2017), and so on.

Compared with the finite-time synchronization, fixed-time synchronization not only induces faster convergence and higher precision control performance but also makes the derived settling time irrelevant to any initial values of the considered systems. Hence, it is meaningful to study the fixed-time synchronization of various systems. For neural networks model, Liu and Chen (2018) discusses fixed-time cluster synchronization problem for complex networks with or without pinning control. Hu et al. (2017), Wang et al. (2017), Wan et al. (2016), Cao and Li (2017) investigate fixed-time synchronization for neural networks with discontinuous activation functions, Cohen-Grossberg neural networks, and coupled memristive neural networks, respectively.

For bidirectional associative memory neural networks (BAMNNs), some achievements on the finite-time analysis have arisen (Liu et al. 2013; Xiao et al. 2017). However, up to present, there has no result on the fixed-time analysis for this kind of model. Later, for complex-valued neural networks, only Wu et al. (2018), Zhang et al. (2018), Zhou et al. (2017), Ding et al. (2017), Zhang et al. (2018) are concerned about the finite/fixed-time synchronization and stabilizability. As complex-valued bidirectional associative memory neural networks (CVBAMNNs) enters scholars' vision, although there are available references with regard to dynamical behaviors (Wang and Huang 2016; Subramanian and Muthukumar 2018; Zhang et al. 2018; Guo et al. 2017, 2018; Zhang et al. 2020), it is still very few, let alone fixed-time synchronization research.

Motivated by the above analysis, this chapter aims to investigate the fixed-time synchronization problem for a class of CVBAMNNs with time delays. The main contributions can be summarized as follows: (1) The fixed-time synchronization of CVBAMNNs with time delays is studied for the first time; (2) The proposed nonlinear delayed controller with independent parameters extends the designed ones in the existing results and includes them as special cases. So, our work can serve a broader field; (3) Based on fixed-time stability and some inequalities technique, a novel synchronization criterion via the new designed controller is developed and a more accurate estimation independent on the initial conditions is given for the settling time; and (4) When CVBAMNNs model is reduced to real-valued one, similar criterion can be derived and corresponding numerical simulation is provided to verify its effectiveness.

10.2 Problem Formulation and Preliminaries

Consider the following CVBAMNNs with time delays as the drive system:

$$\begin{cases} \dot{u}_j(t) = -d_{1j}u_j(t) + \sum_{k=1}^{m} a_{jk} f_k(v_k(t)) + \sum_{k=1}^{m} b_{jk} f_k(v_k(t-\tau_{2k})) + I_{1j}(t), \\ \dot{v}_k(t) = -d_{2k}v_k(t) + \sum_{j=1}^{n} c_{kj} g_j(u_j(t)) + \sum_{j=1}^{n} e_{kj} g_j(u_j(t-\tau_{1j})) + I_{2k}(t), \end{cases} \tag{10.1}$$

where $j = 1, 2, ..., n$, $k = 1, 2, ..., m$, d_{1j} and d_{2k} are positive constants; $u_j(t)$ and $v_k(t)$ are states of the neuron j and k, respectively; a_{jk}, b_{jk}, c_{kj}, and e_{kj} are complex-valued connection weights; f_k and g_j denote the complex-valued activation functions in U-layer and V-layer, respectively; τ_{1j} and τ_{2k} correspond to the constant transmission delays, $\tau_1 = \max_{1\le j\le n}\{\tau_{1j}\}$ and $\tau_2 = \max_{1\le k\le m}\{\tau_{2k}\}$; $I_{1j}(t)$ and $I_{2k}(t)$ are the external inputs of the jth neuron and kth neuron at time t.

The corresponding response system is given by

$$\begin{cases} \dot{\hat{u}}_j(t) = -d_{1j}\hat{u}_j(t) + \sum_{k=1}^{m} a_{jk} f_k(\hat{v}_k(t)) \\ \qquad + \sum_{k=1}^{m} b_{jk} f_k(\hat{v}_k(t-\tau_{2k})) + I_{1j}(t) + U_j(t), \\ \dot{\hat{v}}_k(t) = -d_{2k}\hat{v}_k(t) + \sum_{j=1}^{n} c_{kj} g_j(\hat{u}_j(t)) \\ \qquad + \sum_{j=1}^{n} e_{kj} g_j(\hat{u}_j(t-\tau_{1j})) + I_{2k}(t) + V_k(t), \end{cases} \tag{10.2}$$

where $\hat{u}_j(t)$ and $\hat{v}_k(t)$ are neural states of the response system; $U_j(t)$ and $V_k(t)$ are the suitable control inputs that will be designed. The activation functions satisfy the assumptions listed below.

Assumption 10.1 For $u_j = x'_j + iy'_j$ and $v_k = x''_k + iy''_k$, $f_k(.)$ and $g_j(.)$ can be separated into the real and imaginary parts as

$$f_k(v_k) = f_k^R(x''_k, y''_k) + i f_k^I(x''_k, y''_k),$$
$$g_j(u_j) = g_j^R(x'_j, y'_j) + i g_j^I(x'_j, y'_j).$$

Moreover, for any x'_j, y'_j, x''_k, y''_k, $\bar{x}'_j$, $\bar{y}'_j$, $\bar{x}''_k$, $\bar{y}''_k$, there exist positive constants l_k^{RR}, l_k^{RI}, l_k^{IR}, l_k^{II} and m_j^{RR}, m_j^{RI}, m_j^{IR}, m_j^{II} such that

$$\begin{cases} |f_k^R(x_k'', y_k'') - f_k^R(\hat{x}_k'', \hat{y}_k'')| \le l_k^{RR}|x_k'' - \hat{x}_k''| + l_k^{RI}|y_k'' - \hat{y}_k''|, \\ |f_k^I(x_k'', y_k'') - f_k^I(\hat{x}_k'', \hat{y}_k'')| \le l_k^{IR}|x_k'' - \hat{x}_k''| + l_k^{II}|y_k'' - \hat{y}_k''|, \\ |g_j^R(x_j', y_j') - g_j^R(\hat{x}_j', \hat{y}_j')| \le m_j^{RR}|x_j' - \hat{x}_j'| + m_j^{RI}|y_j' - \hat{y}_j'|, \\ |g_j^I(x_j', y_j') - g_j^I(\hat{x}_j', \hat{y}_j')| \le m_j^{IR}|x_j' - \hat{x}_j'| + m_j^{II}|y_j' - \hat{y}_j'|. \end{cases}$$

By defining $z_j(t) = \hat{u}_j(t) - u_j(t)$ and $e_k(t) = \hat{v}_k(t) - v_k(t)$ as the error state, the error system between (10.1) and (10.2) can be described by

$$\begin{cases} \dot{z}_j(t) = -d_{1j}z_j(t) + \sum_{k=1}^{m} a_{jk}f_k(e_k(t)) + \sum_{k=1}^{m} b_{jk}f_k(e_k(t-\tau_{2k})) + U_j(t), \\ \dot{e}_k(t) = -d_{2k}e_k(t) + \sum_{j=1}^{n} c_{kj}g_j(z_j(t)) + \sum_{j=1}^{n} e_{kj}g_j(z_j(t-\tau_{1j})) + V_k(t), \end{cases} \tag{10.3}$$

where $f_k(e_k(t)) = f_k(\hat{v}_k(t)) - f_k(v_k(t))$, $g_j(z_j(t)) = g_j(\hat{u}_j(t)) - g_j(u_j(t))$, $f_k(e_k(t-\tau_{2k})) = f_k(\hat{v}_k(t-\tau_{2k})) - f_k(v_k(t-\tau_{2k}))$, and $g_j(z_j(t-\tau_{1j})) = g_j(\hat{u}_j(t-\tau_{1j})) - g_j(u_j(t-\tau_{1j}))$.

Let $z_j(t) = (\hat{x}_j'(t) - x_j'(t)) + i(\hat{y}_j'(t) - y_j'(t)) = x_j(t) + iy_j(t)$, $e_k(t) = (\hat{x}_k''(t) - x_k''(t)) + i(\hat{y}_k''(t) - y_k''(t)) = \tilde{x}_k(t) + i\tilde{y}_k(t)$, $U_i(t) = U_i^R(t) + iU_i^I(t)$, and $V_k(t) = V_k^R(t) + iV_k^I(t)$. For convenience, we abbreviate $x_j(t)$, $y_j(t)$, $\tilde{x}_k(t)$, $\tilde{y}_k(t)$, $x_j'(t)$, $y_j'(t)$, $\hat{x}_j'(t)$, $\hat{y}_j'(t)$, $x_k''(t)$, $y_k''(t)$, $\hat{x}_k''(t)$, $\hat{y}_k''(t)$, $x_j'(t-\tau_{1j})$, $y_j'(t-\tau_{1j})$, $\hat{x}_j'(t-\tau_{1j})$, $\hat{y}_j'(t-\tau_{1j})$, $x_k''(t-\tau_{2k})$, $y_k''(t-\tau_{2k})$, $\hat{x}_k''(t-\tau_{2k})$, $\hat{y}_k''(t-\tau_{2k})$, $U_j^R(t)$, $U_j^I(t)$, $V_k^R(t)$, and $V_k^I(t)$ to x_j, y_j, $\tilde{x}_k$, $\tilde{y}_k$, x_j', y_j', $\hat{x}_j'$, $\hat{y}_j'$, x_k'', y_k'', $\hat{x}_k''$, $\hat{y}_k''$, $x_j^{\prime\tau_{1j}}$, $y_j^{\prime\tau_{1j}}$, $\hat{x}_j^{\prime\tau_{1j}}$, $\hat{y}_j^{\prime\tau_{1j}}$, $x_k^{\prime\prime\tau_{2k}}$, $y_k^{\prime\prime\tau_{2k}}$, $\hat{x}_k^{\prime\prime\tau_{2k}}$, $\hat{y}_k^{\prime\prime\tau_{2k}}$, U_j^R, U_j^I, V_k^R, and V_k^I, respectively. Then, the system (10.3) can be expressed by separating it into the real and imaginary parts as below

$$\begin{aligned} \dot{x}_j = & -d_{1j}x_j + \sum_{k=1}^{m} a_{jk}^R(f_k^R(\hat{x}_k'', \hat{y}_k'') - f_k^R(x_k'', y_k'')) - \sum_{k=1}^{m} a_{jk}^I(f_k^I(\hat{x}_k'', \hat{y}_k'') \\ & - f_k^I(x_k'', y_k'')) + \sum_{k=1}^{m} b_{jk}^R(f_k^R(\hat{x}_k^{\prime\prime\tau_{2k}}, \hat{y}_k^{\prime\prime\tau_{2k}}) - f_k^R(x_k^{\prime\prime\tau_{2k}}, y_k^{\prime\prime\tau_{2k}})) \\ & - \sum_{k=1}^{m} b_{jk}^I(f_k^I(\hat{x}_k^{\prime\prime\tau_{2k}}, \hat{y}_k^{\prime\prime\tau_{2k}}) - f_k^I(x_k^{\prime\prime\tau_{2k}}, y_k^{\prime\prime\tau_{2k}})) + U_j^R, \\ \dot{y}_j = & -d_{1j}y_j + \sum_{k=1}^{m} a_{jk}^I(f_k^R(\hat{x}_k'', \hat{y}_k'') - f_k^R(x_k'', y_k'')) + \sum_{k=1}^{m} a_{jk}^R(f_k^I(\hat{x}_k'', \hat{y}_k'') \\ & - f_k^I(x_k'', y_k'')) + \sum_{k=1}^{m} b_{jk}^I(f_k^R(\hat{x}_k^{\prime\prime\tau_{2k}}, \hat{y}_k^{\prime\prime\tau_{2k}}) - f_k^R(x_k^{\prime\prime\tau_{2k}}, y_k^{\prime\prime\tau_{2k}})) \\ & + \sum_{k=1}^{m} b_{jk}^R(f_k^I(\hat{x}_k^{\prime\prime\tau_{2k}}, \hat{y}_k^{\prime\prime\tau_{2k}}) - f_k^I(x_k^{\prime\prime\tau_{2k}}, y_k^{\prime\prime\tau_{2k}})) + U_j^I, \end{aligned}$$

$$
\begin{aligned}
\dot{\tilde{x}}_k &= -d_{2k}\tilde{x}_k + \sum_{j=1}^{n} c_{kj}^R (g_j^R(\hat{x}_j', \hat{y}_j') - g_j^R(x_j', y_j')) - \sum_{j=1}^{n} c_{kj}^I (g_j^I(\hat{x}_j', \hat{y}_j') \\
&\quad - g_j^I(x_j', y_j')) + \sum_{j=1}^{n} e_{kj}^R (g_j^R(\hat{x}_j^{\prime\tau_{1j}}, \hat{y}_j^{\prime\tau_{1j}}) - g_j^R(x_j^{\prime\tau_{1j}}, y_j^{\prime\tau_{1j}})) \\
&\quad - \sum_{j=1}^{n} e_{kj}^I (g_j^I(\hat{x}_j^{\prime\tau_{1j}}, \hat{y}_j^{\prime\tau_{1j}}) - g_j^I(x_j^{\prime\tau_{1j}}, y_j^{\prime\tau_{1j}})) + V_k^R, \\
\dot{\tilde{y}}_k &= -d_{2k}\tilde{y}_k + \sum_{j=1}^{n} c_{kj}^I (g_j^R(\hat{x}_j', \hat{y}_j') - g_j^R(x_j', y_j')) + \sum_{j=1}^{n} c_{kj}^R (g_j^I(\hat{x}_j', \hat{y}_j') \\
&\quad - g_j^I(x_j', y_j')) + \sum_{j=1}^{n} e_{kj}^I (g_j^R(\hat{x}_j^{\prime\tau_{1j}}, \hat{y}_j^{\prime\tau_{1j}}) - g_j^R(x_j^{\prime\tau_{1j}}, y_j^{\prime\tau_{1j}})) \\
&\quad + \sum_{j=1}^{n} e_{kj}^R (g_j^I(\hat{x}_j^{\prime\tau_{1j}}, \hat{y}_j^{\prime\tau_{1j}}) - g_j^I(x_j^{\prime\tau_{1j}}, y_j^{\prime\tau_{1j}})) + V_k^I,
\end{aligned}
\tag{10.4}
$$

or

$$
\begin{aligned}
\dot{\omega}_j &= -\begin{bmatrix} d_{1j} & 0 \\ 0 & d_{1j} \end{bmatrix} \omega_j + \sum_{k=1}^{m} \begin{bmatrix} a_{jk}^R & -a_{jk}^I \\ a_{jk}^I & a_{jk}^R \end{bmatrix} \begin{bmatrix} f_k^R(\hat{x}_k'', \hat{y}_k'') - f_k^R(x_k'', y_k'') \\ f_k^I(\hat{x}_k'', \hat{y}_k'') - f_k^I(x_k'', y_k'') \end{bmatrix} \\
&\quad + \sum_{k=1}^{m} \begin{bmatrix} b_{jk}^R & -b_{jk}^I \\ b_{jk}^I & b_{jk}^R \end{bmatrix} \begin{bmatrix} f_k^R(\hat{x}_k^{\prime\prime\tau_{2k}}, \hat{y}_k^{\prime\prime\tau_{2k}}) - f_k^R(x_k^{\prime\prime\tau_{2k}}, y_k^{\prime\prime\tau_{2k}}) \\ f_k^I(\hat{x}_k^{\prime\prime\tau_{2k}}, \hat{y}_k^{\prime\prime\tau_{2k}}) - f_k^I(x_k^{\prime\prime\tau_{2k}}, y_k^{\prime\prime\tau_{2k}}) \end{bmatrix} + \begin{bmatrix} U_j^R \\ U_j^I \end{bmatrix}, \\
\dot{\tilde{\omega}}_k &= -\begin{bmatrix} d_{2k} & 0 \\ 0 & d_{2k} \end{bmatrix} \tilde{\omega}_k + \sum_{j=1}^{n} \begin{bmatrix} c_{kj}^R & -c_{kj}^I \\ c_{kj}^I & c_{kj}^R \end{bmatrix} \begin{bmatrix} g_j^R(\hat{x}_j', \hat{y}_j') - g_j^R(x_j', y_j') \\ g_j^I(\hat{x}_j', \hat{y}_j') - g_j^I(x_j', y_j') \end{bmatrix} \\
&\quad + \sum_{j=1}^{n} \begin{bmatrix} e_{kj}^R & -e_{kj}^I \\ e_{kj}^I & e_{kj}^R \end{bmatrix} \begin{bmatrix} g_j^R(\hat{x}_j^{\prime\tau_{1j}}, \hat{y}_j^{\prime\tau_{1j}}) - g_j^R(x_j^{\prime\tau_{1j}}, y_j^{\prime\tau_{1j}}) \\ g_j^I(\hat{x}_j^{\prime\tau_{1j}}, \hat{y}_j^{\prime\tau_{1j}}) - g_j^I(x_j^{\prime\tau_{1j}}, y_j^{\prime\tau_{1j}}) \end{bmatrix} + \begin{bmatrix} V_k^R \\ V_k^I \end{bmatrix},
\end{aligned}
\tag{10.5}
$$

where $\omega_j = \begin{bmatrix} x_j \\ y_j \end{bmatrix}$, $\tilde{\omega}_k = \begin{bmatrix} \tilde{x}_k \\ \tilde{y}_k \end{bmatrix}$.

Next, we will present the following definition and lemmas, which can be used in establishing the main result.

Definition 10.1 The synchronization between the drive system (10.1) and the response system (10.2) can be achieved in a fixed time, if for any initial condition ω_0, there exists a fixed-time $T_{\max} \geq 0$ and a settling time function $T(\omega_0) \geq 0$ such that $\lim_{t \to T(\omega_0)} \|\omega(t)\| = 0$, $\omega(t) = 0$ for all $t \geq T(\omega_0)$, and $T(\omega_0) \leq T_{\max}$, where $\omega(t) = (\omega_1^T, ..., \omega_n^T, \tilde{\omega}_1^T, ..., \tilde{\omega}_m^T)^T$ and $\omega_0 = \omega(0)$.

Lemma 10.1 (Hu et al. 2017) *If there exists a regular, positive definite and radially unbounded function* $V(\omega(t)) : \mathbb{R}^n \to \mathbb{R}^+$ *such that for any solution of system (10.4) or (10.5), the following inequality holds*

$$\frac{d}{dt}V(\omega(t)) \leq -\Big(\alpha V^a(\omega(t)) + \beta V^b(\omega(t))\Big)^\rho, \tag{10.6}$$

for $\omega(t) \in \mathbb{R}^n \setminus \{0\}$, *where* α, β, b, $\rho > 0$, $a \geq 0$, $a\rho < 1$, *and* $b\rho > 1$, *then system (10.4) or (10.5) is globally fixed-time stable at the origin. Moreover, the settling time* $T(\omega_0)$ *satisfies*

$$T(\omega_0) \leq T^1_{\max} = \frac{1}{\alpha^\rho}(\frac{\alpha}{\beta})^{\frac{1-a\rho}{b-a}}(\frac{1}{1-a\rho} + \frac{1}{b\rho - 1}). \tag{10.7}$$

Especially, when $\rho = 1$,

$$T^2_{\max} = \frac{1}{\alpha}(\frac{\alpha}{\beta})^{\frac{1-a}{b-a}}(\frac{1}{1-a} + \frac{1}{b - 1}). \tag{10.8}$$

Further, if $\beta \leq \alpha e$ $(e = 2.718281828459...)$, $T^2_{\max}$ *has an infimum given by*

$$\inf T^2_{\max} = \frac{1}{\alpha(1-a)}. \tag{10.9}$$

Remark 10.1 Recently, for the fixed-time stability theory, compared with the classical result given in Polyakov (2012), a relaxed condition as stated in (10.6) was proposed in Hu et al. (2017). Moreover, for the settling time, a more accurate estimation $T^1_{\max}$ in (10.7) was given in Hu et al. (2017) than the existing ones in Polyakov (2012), Zuo and Tie (2016). In this chapter, we will derive a fixed-time synchronization criterion for systems (10.1) and (10.2) and compute a more accurate estimation of the settling time by Lemma 10.1.

Lemma 10.2 (Hua et al. 2017) *Let* $a_1, a_2, \ldots, a_n$, *and* ρ *be the positive numbers, then the following inequalities hold:*

$$(a_1 + a_2 + \ldots + a_n)^\rho \leq \max\{n^{\rho-1}, 1\}(a_1^\rho + a_2^\rho + \ldots + a_n^\rho). \tag{10.10}$$

10.3 Fixed-Time Synchronization Criterion

In this section, we will present a new criterion to guarantee that systems (10.1) and (10.2) can achieve the fixed-time synchronization. First, we design the following controller

$$
\begin{aligned}
U_j^R(t) &= -\xi_{1j}x_j - (\delta_{1j}|x_j|^{\mu_1} + \alpha_{1j}|x_j|^{\beta_1})\mathrm{sgn}(x_j) - \chi_{1j}\sum_{k=1}^{m}|\tilde{x}_k^{\tau_{2k}}|\mathrm{sgn}(x_j), \\
U_j^I(t) &= -\xi_{2j}y_j - (\delta_{2j}|y_j|^{\mu_2} + \alpha_{2j}|y_j|^{\beta_2})\mathrm{sgn}(y_j) - \chi_{2j}\sum_{k=1}^{m}|\tilde{y}_k^{\tau_{2k}}|\mathrm{sgn}(y_j), \\
V_k^R(t) &= -\xi_{3k}\tilde{x}_k - (\delta_{3k}|\tilde{x}_k|^{\mu_3} + \alpha_{3k}|\tilde{x}_k|^{\beta_3})\mathrm{sgn}(\tilde{x}_k) - \chi_{3k}\sum_{j=1}^{n}|x_j^{\tau_{1j}}|\mathrm{sgn}(\tilde{x}_k), \\
V_k^I(t) &= -\xi_{4k}\tilde{y}_k - (\delta_{4k}|\tilde{y}_k|^{\mu_4} + \alpha_{4k}|\tilde{y}_k|^{\beta_4})\mathrm{sgn}(\tilde{y}_k) - \chi_{4k}\sum_{j=1}^{n}|y_j^{\tau_{1j}}|\mathrm{sgn}(\tilde{y}_k),
\end{aligned}
\tag{10.11}
$$

where $\xi_{1j}, \xi_{2j}, \xi_{3k}, \xi_{4k}, \delta_{1j} > 0, \delta_{2j} > 0, \delta_{3k} > 0, \delta_{4k} > 0, \chi_{1j}, \chi_{2j}, \chi_{3k}, \chi_{4k}, \alpha_{1j} > 0$, $\alpha_{2j} > 0$, $\alpha_{3k} > 0$, $\alpha_{4k} > 0$, $0 < \mu_q < 1$, $\beta_q > 1$ are constants for $j = 1, 2, \cdots, n$, $k = 1, 2, \cdots, m$, $q = 1, 2, 3, 4$, and $\mathrm{sgn}(\cdot)$ is defined as

$$
\mathrm{sgn}(\varpi) = \begin{cases} 1, & |\varpi| > 0, \\ 0, & |\varpi| = 0, \\ -1, & |\varpi| < 0, \end{cases}
$$

Remark 10.2 It is worth noting that different from the controllers in Liu et al. (2013), Ding et al. (2017), parameters μ_q and β_q $(q = 1, 2, 3, 4)$ are independent of each other in (10.11), which can bring great difficulties in deriving our main result. This point can be seen in the proof of Theorem 10.1 below. Moreover, the proposed controller (10.11) in this chapter includes the one in Liu et al. (2013), Ding et al. (2017) as a special case. Thus, our work can serve a broader field.

Next, for convenience, we denote

$$
\mathcal{D}_1 = \begin{bmatrix} D_1 & 0 \\ 0 & D_1 \end{bmatrix}, \mathcal{D}_2 = \begin{bmatrix} D_2 & 0 \\ 0 & D_2 \end{bmatrix}, \Upsilon = \begin{bmatrix} \Upsilon_1 & 0 \\ 0 & \Upsilon_2 \end{bmatrix}, \bar{\Upsilon} = \begin{bmatrix} \bar{\Upsilon}_1 & 0 \\ 0 & \bar{\Upsilon}_2 \end{bmatrix},
$$

$$
L = \begin{bmatrix} L^{RR} & L^{RI} \\ L^{IR} & L^{II} \end{bmatrix}, M = \begin{bmatrix} M^{RR} & M^{RI} \\ M^{IR} & M^{II} \end{bmatrix}, \mathcal{A} = \begin{bmatrix} A^R & A^I \\ A^I & A^R \end{bmatrix}, \mathcal{B} = \begin{bmatrix} B^R & B^I \\ B^I & B^R \end{bmatrix},
$$

$$
\mathcal{D} = \begin{bmatrix} \mathcal{D}_1 + \Upsilon & 0 \\ 0 & \mathcal{D}_2 + \bar{\Upsilon} \end{bmatrix}, \mathcal{C} = \begin{bmatrix} C^R & C^I \\ C^I & C^R \end{bmatrix}, \mathcal{E} = \begin{bmatrix} E^R & E^I \\ E^I & E^R \end{bmatrix}, \mathcal{M} = \begin{bmatrix} M & 0 \\ 0 & L \end{bmatrix},
$$

$$
\mathcal{L} = \begin{bmatrix} L & 0 \\ 0 & M \end{bmatrix}, \Gamma = \begin{bmatrix} \Gamma_1 & 0 \\ 0 & \Gamma_2 \end{bmatrix}, \tilde{\Gamma} = \begin{bmatrix} \Gamma & 0 \\ 0 & \bar{\Gamma} \end{bmatrix}, \bar{\Gamma} = \begin{bmatrix} \bar{\Gamma}_1 & 0 \\ 0 & \bar{\Gamma}_2 \end{bmatrix}, \Lambda_1 = \begin{bmatrix} \mathcal{C} & 0 \\ 0 & \mathcal{A} \end{bmatrix},
$$

$$
\Lambda_2 = \begin{bmatrix} \mathcal{B} & 0 \\ 0 & \mathcal{E} \end{bmatrix}, A^R = (|a_{jk}^R|)_{n\times m}, A^I = (|a_{jk}^I|)_{n\times m}, B^R = (|b_{jk}^R|)_{n\times m},
$$

$$
B^I = (|b_{jk}^I|)_{n\times m}, C^R = (|c_{kj}^R|)_{m\times n}, C^I = (|c_{kj}^I|)_{m\times n}, E^R = (|e_{kj}^R|)_{m\times n},
$$

$$
E^I = (|e_{kj}^I|)_{m\times n}, D_1 = \mathrm{diag}\{d_{11}, d_{12}, ..., d_{1n}\}, D_2 = \mathrm{diag}\{d_{21}, d_{22}, ..., d_{2m}\},
$$

$$
\Upsilon_1 = \mathrm{diag}\{\xi_{11}, \xi_{12}, \ldots, \xi_{1n}\}, \Upsilon_2 = \mathrm{diag}\{\xi_{21}, \xi_{22}, \ldots, \xi_{2n}\}, \Gamma_1 = \mathrm{tr}(\Gamma_1')E_n,
$$

$$\bar{\Upsilon}_1 = \text{diag}\{\xi_{31}, \xi_{32}, \ldots, \xi_{3m}\},\ \bar{\Upsilon}_2 = \text{diag}\{\xi_{41}, \xi_{42}, \ldots, \xi_{4m}\},\ \Gamma_2 = \text{tr}(\Gamma_2')E_n,$$
$$L^{RR} = \text{diag}\{l_1^{RR}, l_2^{RR}, \ldots, l_m^{RR}\},\ L^{RI} = \text{diag}\{l_1^{RI}, l_2^{RI}, \ldots, l_m^{RI}\},$$
$$L^{IR} = \text{diag}\{l_1^{IR}, l_2^{IR}, \ldots, l_m^{IR}\},\ L^{II} = \text{diag}\{l_1^{II}, l_2^{II}, \ldots, l_m^{II}\},$$
$$M^{RR} = \text{diag}\{m_1^{RR}, m_2^{RR}, \ldots, m_n^{RR}\},\ M^{RI} = \text{diag}\{m_1^{RI}, m_2^{RI}, \ldots, m_n^{RI}\},$$
$$M^{IR} = \text{diag}\{m_1^{IR}, m_2^{IR}, \ldots, m_n^{IR}\},\ M^{II} = \text{diag}\{m_1^{II}, m_2^{II}, \ldots, m_n^{II}\},$$
$$\Gamma_1' = \text{diag}\{\chi_{11}, \chi_{12}, \ldots, \chi_{1n}\},\ \Gamma_2' = \text{diag}\{\chi_{21}, \chi_{22}, \ldots, \chi_{2n}\},\ \bar{\Gamma}_1 = \text{tr}(\bar{\Gamma}_1')E_m,$$
$$\bar{\Gamma}_1' = \text{diag}\{\chi_{31}, \chi_{32}, \ldots, \chi_{3m}\},\ \bar{\Gamma}_2' = \text{diag}\{\chi_{41}, \chi_{42}, \ldots, \chi_{4m}\},\ \bar{\Gamma}_2 = \text{tr}(\bar{\Gamma}_2')E_m,$$
$$E_n = \text{diag}\{1, 1, ..., 1\},\ I_{2n+2m} = [1, 1, ..., 1]^T,\ E_m = \text{diag}\{1, 1, ..., 1\}. \tag{10.12}$$

Theorem 10.1 *Suppose that Assumption 10.1 holds, systems (10.1) and (10.2) can be synchronized in fixed time under the controller (10.11) if the following inequalities hold*

$$\begin{aligned}(\mathcal{D} - (\Lambda_1\mathcal{M})^T)I_{2n+2m} > 0,\\ (\tilde{\Gamma} - (\Lambda_2\mathcal{L})^T)I_{2n+2m} > 0.\end{aligned} \tag{10.13}$$

Moreover, the settling time T is estimated by

$$T \leq T_{max} = \frac{1}{\bar{\rho}}\Big(2^{\frac{3\bar{\beta}-2\bar{\beta}\bar{\mu}+\bar{\mu}-2}{\bar{\beta}-\bar{\mu}}}\Big)\Big(\frac{1}{1-\bar{\mu}} + \frac{1}{\bar{\beta}-1}\Big), \tag{10.14}$$

where $\bar{\rho} = \min\{\delta, \alpha n^{1-\beta_1}, \alpha n^{1-\beta_2}, \alpha m^{1-\beta_3}, \alpha m^{1-\beta_4}\}$, $\bar{\mu} = \max_{1\leq q\leq 4}\{\mu_q\}$, $\delta = \min_{1\leq j\leq n, 1\leq k\leq m}\{\delta_{1j}, \delta_{2j}, \delta_{3k}, \delta_{4k}\}$, $\alpha = \min_{1\leq j\leq n, 1\leq k\leq m}\{\alpha_{1j}, \alpha_{2j}, \alpha_{3k}, \alpha_{4k}\}$, *and* $\bar{\beta} = \min_{1\leq q\leq 4}\{\beta_q\}$.

Proof Consider the candidate Lyapunov function as follows:

$$\begin{aligned}V(t) &\triangleq V_1 + V_2 + V_3 + V_4\\ &= \sum_{j=1}^{n}|x_j| + \sum_{j=1}^{n}|y_j| + \sum_{k=1}^{m}|\tilde{x}_k| + \sum_{k=1}^{m}|\tilde{y}_k|.\end{aligned} \tag{10.15}$$

First, based on Assumption 10.1, it is easy to get the following inequalities

$$\begin{aligned}&|f_k^\nu(x_k'', y_k'') - f_k^\nu(\hat{x}_k'', \hat{y}_k'')| \leq l_k^{\nu R}|\tilde{x}_k| + l_k^{\nu I}|\tilde{y}_k|,\\ &|f_k^\nu(x_k''^{\tau_{2k}}, y_k''^{\tau_{2k}}) - f_k^\nu(\hat{x}_k''^{\tau_{2k}}, \hat{y}_k''^{\tau_{2k}})| \leq l_k^{\nu R}|\tilde{x}_k^{\tau_{2k}}| + l_k^{\nu I}|\tilde{y}_k^{\tau_{2k}}|,\\ &|g_j^\nu(x_j', y_j') - g_j^\nu(\hat{x}_j', \hat{y}_j')| \leq m_j^{\nu R}|x_j| + m_j^{\nu I}|y_j|,\\ &|g_j^\nu(x_j'^{\tau_{1j}}, y_j'^{\tau_{1j}}) - g_j^\nu(\hat{x}_j'^{\tau_{1j}}, \hat{y}_j'^{\tau_{1j}})| \leq m_j^{\nu R}|x_j^{\tau_{1j}}| + m_j^{\nu I}|y_j^{\tau_{1j}}|,\end{aligned} \tag{10.16}$$

for $\nu = R, I$.

Next, along the solution trajectories of the error system (10.4) and by (10.16) we have

$$
\begin{aligned}
\dot{V}_1 &= \sum_{j=1}^{n} \operatorname{sgn}(x_j)\dot{x}_j \le \sum_{j=1}^{n}\Big[-(d_{1j}+\xi_{1j})|x_j| + \sum_{k=1}^{m}|a_{jk}^{R}|(l_k^{RR}|\tilde{x}_k| + l_k^{RI}|\tilde{y}_k|)\\
&\quad + \sum_{k=1}^{m}|a_{jk}^{I}|(l_k^{IR}|\tilde{x}_k| + l_k^{II}|\tilde{y}_k|) + \sum_{k=1}^{m}|b_{jk}^{R}|(l_k^{RR}|\tilde{x}_k^{\tau_{2k}}| + l_k^{RI}|\tilde{y}_k^{\tau_{2k}}|)\\
&\quad + \sum_{k=1}^{m}|b_{jk}^{I}|(l_k^{IR}|\tilde{x}_k^{\tau_{2k}}| + l_k^{II}|\tilde{y}_k^{\tau_{2k}}|) - \delta_{1j}|x_j|^{\mu_1} - \alpha_{1j}|x_j|^{\beta_1} - \chi_{1j}\sum_{k=1}^{m}|\tilde{x}_k^{\tau_{2k}}|\Big]\\
&= \sum_{j=1}^{n}\Big[-(d_{1j}+\xi_{1j})|x_j| + \sum_{k=1}^{m}(|a_{jk}^{R}|l_k^{RR} + |a_{jk}^{I}|l_k^{IR})|\tilde{x}_k|\\
&\quad + \sum_{k=1}^{m}(|a_{jk}^{R}|l_k^{RI} + |a_{jk}^{I}|l_k^{II})|\tilde{y}_k| + \sum_{k=1}^{m}(|b_{jk}^{R}|l_k^{RI} + |b_{jk}^{I}|l_k^{II})|\tilde{y}_k^{\tau_{2k}}|\\
&\quad - \delta_{1j}|x_j|^{\mu_1} - \alpha_{1j}|x_j|^{\beta_1} + \sum_{k=1}^{m}(|b_{jk}^{R}|l_k^{RR} + |b_{jk}^{I}|l_k^{IR} - \chi_{1j})|\tilde{x}_k^{\tau_{2k}}|\Big]. \qquad (10.17)
\end{aligned}
$$

Similarly,

$$
\begin{aligned}
\dot{V}_2 &= \sum_{j=1}^{n} \operatorname{sgn}(y_j)\dot{y}_j \le \sum_{j=1}^{n}\Big[-(d_{1j}+\xi_{2j})|y_j| + \sum_{k=1}^{m}(|a_{jk}^{I}|l_k^{RR} + |a_{jk}^{R}|l_k^{IR})|\tilde{x}_k|\\
&\quad + \sum_{k=1}^{m}(|a_{jk}^{I}|l_k^{RI} + |a_{jk}^{R}|l_k^{II})|\tilde{y}_k| + \sum_{k=1}^{m}(|b_{jk}^{I}|l_k^{RR} + |b_{jk}^{R}|l_k^{IR})|\tilde{x}_k^{\tau_{2k}}|\\
&\quad - \delta_{2j}|y_j|^{\mu_2} - \alpha_{2j}|y_j|^{\beta_2} + \sum_{k=1}^{m}(|b_{jk}^{I}|l_k^{RI} + |b_{jk}^{R}|l_k^{II} - \chi_{2j})|\tilde{y}_k^{\tau_{2k}}|\Big],\\
\dot{V}_3 &= \sum_{k=1}^{m} \operatorname{sgn}(\tilde{x}_k)\dot{\tilde{x}}_k \le \sum_{k=1}^{m}\Big[-(d_{2k}+\xi_{3k})|\tilde{x}_k| + \sum_{j=1}^{n}(|c_{kj}^{R}|m_j^{RR} + |c_{kj}^{I}|m_j^{IR})|x_j|\\
&\quad + \sum_{j=1}^{n}(|c_{kj}^{R}|m_j^{RI} + |c_{kj}^{I}|m_j^{II})|y_j| + \sum_{j=1}^{n}(|e_{kj}^{R}|m_j^{RI} + |e_{kj}^{I}|m_j^{II})|y_j^{\tau_{1j}}|\\
&\quad - \delta_{3k}|\tilde{x}_k|^{\mu_3} - \alpha_{3k}|\tilde{x}_k|^{\beta_3} + \sum_{j=1}^{n}(|e_{kj}^{R}|m_j^{RR} + |e_{kj}^{I}|m_j^{IR} - \chi_{3k})|x_j^{\tau_{1j}}|\Big],
\end{aligned}
$$

$$\begin{aligned}\dot{V}_4 = \sum_{k=1}^{m} \operatorname{sgn}(\tilde{y}_k)\dot{\tilde{y}}_k \leq \sum_{k=1}^{m} \Big[&-(d_{2k}+\xi_{4k})|\tilde{x}_k| + \sum_{j=1}^{n}(|c_{kj}^{I}|m_j^{RR} + |c_{kj}^{R}|m_j^{IR})|x_j| \\ &+ \sum_{j=1}^{n}(|c_{kj}^{I}|m_j^{RI} + |c_{kj}^{R}|m_j^{II})|y_j| + \sum_{j=1}^{n}(|e_{kj}^{I}|m_j^{RR} + |e_{kj}^{R}|m_j^{IR})|x_j^{\tau_{1j}}| \\ &- \delta_{4k}|\tilde{y}_k|^{\mu_4} - \alpha_{4k}|\tilde{y}_k|^{\beta_4} + \sum_{j=1}^{n}(|e_{kj}^{I}|m_j^{RI} + |e_{kj}^{R}|m_j^{II} - \chi_{4k})|y_j^{\tau_{1j}}|\Big]. \end{aligned} \tag{10.18}$$

By combining (10.17) with (10.18), one can obtain

$$\begin{aligned}\dot{V}(t) \leq &\sum_{j=1}^{n}|x_j|\Big[-(d_{1j}+\xi_{1j}) + \sum_{k=1}^{m}(|c_{kj}^{R}|m_j^{RR} + |c_{kj}^{I}|m_j^{IR} + |c_{kj}^{I}|m_j^{RR} + |c_{kj}^{R}|m_j^{IR})\Big] \\ &+ \sum_{j=1}^{n}|y_j|\Big[-(d_{1j}+\xi_{2j}) + \sum_{k=1}^{m}(|c_{kj}^{R}|m_j^{RI} + |c_{kj}^{I}|m_j^{II} + |c_{kj}^{I}|m_j^{RI} + |c_{kj}^{R}|m_j^{II})\Big] \\ &+ \sum_{k=1}^{m}|\tilde{x}_k|\Big[-(d_{2k}+\xi_{3k}) + \sum_{j=1}^{n}(|a_{jk}^{R}|l_k^{RR} + |a_{jk}^{I}|l_k^{IR} + |a_{jk}^{I}|l_k^{RR} + |a_{jk}^{R}|l_k^{IR})\Big] \\ &+ \sum_{k=1}^{m}|\tilde{y}_k|\Big[-(d_{2k}+\xi_{4k}) + \sum_{j=1}^{n}(|a_{jk}^{R}|l_k^{RI} + |a_{jk}^{I}|l_k^{II} + |a_{jk}^{I}|l_k^{RI} + |a_{jk}^{R}|l_k^{II})\Big] \\ &+ \sum_{j=1}^{n}\sum_{k=1}^{m}(|e_{kj}^{R}|m_j^{RR} + |e_{kj}^{I}|m_j^{IR} + |e_{kj}^{I}|m_j^{RR} - \chi_{3k} + |e_{kj}^{R}|m_j^{IR})|x_j^{\tau_{1j}}| \\ &+ \sum_{j=1}^{n}\sum_{k=1}^{m}(|e_{kj}^{R}|m_j^{RI} + |e_{kj}^{I}|m_j^{II} - \chi_{4k} + |e_{kj}^{I}|m_j^{RI} + |e_{kj}^{R}|m_j^{II})|y_j^{\tau_{1j}}| \end{aligned}$$

$$\begin{aligned} &+ \sum_{k=1}^{m}\sum_{j=1}^{n}(|b_{jk}^{R}|l_k^{RR} + |b_{jk}^{I}|l_k^{IR} + |b_{jk}^{I}|l_k^{RR} - \chi_{1j} + |b_{jk}^{R}|l_k^{IR})|\tilde{x}_k^{\tau_{2k}}| \\ &+ \sum_{k=1}^{m}\sum_{j=1}^{n}(|b_{jk}^{R}|l_k^{RI} + |b_{jk}^{I}|l_k^{II} - \chi_{2j} + b_{jk}^{I}|l_k^{RI} + |b_{jk}^{R}|l_k^{II})|\tilde{y}_k^{\tau_{2k}}| \\ &- \sum_{j=1}^{n}\delta_{1j}|x_j|^{\mu_1} - \sum_{j=1}^{n}\delta_{2j}|y_j|^{\mu_2} - \sum_{k=1}^{m}\delta_{3k}|\tilde{x}_k|^{\mu_3} - \sum_{k=1}^{m}\delta_{4k}|\tilde{y}_k|^{\mu_4} \\ &- \sum_{j=1}^{n}\alpha_{1j}|x_j|^{\beta_1} - \sum_{j=1}^{n}\alpha_{2j}|y_j|^{\beta_2} - \sum_{k=1}^{m}\alpha_{3k}|\tilde{x}_k|^{\beta_3} - \sum_{k=1}^{m}\alpha_{4k}|\tilde{y}_k|^{\beta_4}. \end{aligned} \tag{10.19}$$

On the other hand, from (10.13), we can get

$$
\begin{aligned}
& d_{1j} + \xi_{1j} - \sum_{k=1}^{m}(|c_{kj}^{R}|m_{j}^{RR} + |c_{kj}^{I}|m_{j}^{IR} + |c_{kj}^{I}|m_{j}^{RR} + |c_{kj}^{R}|m_{j}^{IR}) > 0, \\
& d_{1j} + \xi_{2j} - \sum_{k=1}^{m}(|c_{kj}^{R}|m_{j}^{RI} + |c_{kj}^{I}|m_{j}^{II} + |c_{kj}^{I}|m_{j}^{RI} + |c_{kj}^{R}|m_{j}^{II}) > 0, \\
& d_{2k} + \xi_{3k} - \sum_{j=1}^{n}(|a_{jk}^{R}|l_{k}^{RR} + |a_{jk}^{I}|l_{k}^{IR} + |a_{jk}^{I}|l_{k}^{RR} + |a_{jk}^{R}|l_{k}^{IR})] > 0, \\
& d_{2k} + \xi_{4k} - \sum_{j=1}^{n}(|a_{jk}^{R}|l_{k}^{RI} + |a_{jk}^{I}|l_{k}^{II} + |a_{jk}^{I}|l_{k}^{RI} + |a_{jk}^{R}|l_{k}^{II}) > 0
\end{aligned}
\tag{10.20}
$$

and

$$
\begin{aligned}
& \sum_{k=1}^{m}\chi_{3k} - \sum_{k=1}^{m}(|e_{kj}^{R}|m_{j}^{RR} + |e_{kj}^{I}|m_{j}^{IR} + e_{kj}^{I}|m_{j}^{RR} + |e_{kj}^{R}|m_{j}^{IR}) > 0, \\
& \sum_{k=1}^{m}\chi_{4k} - \sum_{k=1}^{m}(|e_{kj}^{R}|m_{j}^{RI} + |e_{kj}^{I}|m_{j}^{II} + e_{kj}^{I}|m_{j}^{RI} + |e_{kj}^{R}|m_{j}^{II}) > 0, \\
& \sum_{j=1}^{n}\chi_{1j} - \sum_{j=1}^{n}(|b_{jk}^{R}|l_{k}^{RR} + |b_{jk}^{I}|l_{k}^{IR} + b_{jk}^{I}|l_{k}^{RR} + |b_{jk}^{R}|l_{k}^{IR}) > 0, \\
& \sum_{j=1}^{n}\chi_{2j} - \sum_{j=1}^{n}(|b_{jk}^{R}|l_{k}^{RI} + |b_{jk}^{I}|l_{k}^{II} + b_{jk}^{I}|l_{k}^{RI} + |b_{jk}^{R}|l_{k}^{II}) > 0.
\end{aligned}
\tag{10.21}
$$

Then, combining (10.19)-(10.21), one has

$$
\begin{aligned}
\dot{V}(t) \leq & -\sum_{j=1}^{n}\delta_{1j}|x_j|^{\mu_1} - \sum_{j=1}^{n}\delta_{2j}|y_j|^{\mu_2} - \sum_{k=1}^{m}\delta_{3k}|\tilde{x}_k|^{\mu_3} - \sum_{k=1}^{m}\delta_{4k}|\tilde{y}_k|^{\mu_4} \\
& - \sum_{j=1}^{n}\alpha_{1j}|x_j|^{\beta_1} - \sum_{j=1}^{n}\alpha_{2j}|y_j|^{\beta_2} - \sum_{k=1}^{m}\alpha_{3k}|\tilde{x}_k|^{\beta_3} - \sum_{k=1}^{m}\alpha_{4k}|\tilde{y}_k|^{\beta_4} \\
\leq & -\delta(\sum_{j=1}^{n}|x_j|^{\mu_1} + \sum_{j=1}^{n}|y_j|^{\mu_2} + \sum_{k=1}^{m}|\tilde{x}_k|^{\mu_3} + \sum_{k=1}^{m}|\tilde{y}_k|^{\mu_4}) \\
& - \alpha(\sum_{j=1}^{n}|x_j|^{\beta_1} + \sum_{j=1}^{n}|y_j|^{\beta_2} + \sum_{k=1}^{m}|\tilde{x}_k|^{\beta_3} + \sum_{k=1}^{m}|\tilde{y}_k|^{\beta_4}),
\end{aligned}
\tag{10.22}
$$

where $\delta = \min_{1\leq j\leq n, 1\leq k\leq m}\{\delta_{1j}, \delta_{2j}, \delta_{3k}, \delta_{4k}\}$, $\alpha = \min_{1\leq j\leq n, 1\leq k\leq m}\{\alpha_{1j}, \alpha_{2j}, \alpha_{3k}, \alpha_{4k}\}$. Moreover, due to $0 < \mu_q < 1$ and $\beta_q > 1$ $(q = 1, 2, 3, 4)$, by means of Lemma 10.2, we have

$$\begin{aligned}
&\sum_{j=1}^{n}|x_j|^{\mu_1} + \sum_{j=1}^{n}|y_j|^{\mu_2} + \sum_{k=1}^{m}|\tilde{x}_k|^{\mu_3} + \sum_{k=1}^{m}|\tilde{y}_k|^{\mu_4} \\
&\geq V_1^{\mu_1} + V_2^{\mu_2} + V_3^{\mu_3} + V_4^{\mu_4}, \\
&\sum_{j=1}^{n}|x_j|^{\beta_1} + \sum_{j=1}^{n}|y_j|^{\beta_2} + \sum_{k=1}^{m}|\tilde{x}_k|^{\beta_3} + \sum_{k=1}^{m}|\tilde{y}_k|^{\beta_4} \\
&\geq n^{1-\beta_1}V_1^{\beta_1} + n^{1-\beta_2}V_2^{\beta_2} + m^{1-\beta_3}V_3^{\beta_3} + m^{1-\beta_4}V_4^{\beta_4}.
\end{aligned} \tag{10.23}$$

Let $\bar{\rho} = \min\{\delta, \alpha n^{1-\beta_1}, \alpha n^{1-\beta_2}, \alpha m^{1-\beta_3}, \alpha m^{1-\beta_4}\}$, substituting (10.23) into (10.22), we have

$$\dot{V}(t) \leq -\bar{\rho}\Big(V_1^{\mu_1} + V_2^{\mu_2} + V_3^{\mu_3} + V_4^{\mu_4} + V_1^{\beta_1} + V_2^{\beta_2} + V_3^{\beta_3} + V_4^{\beta_4}\Big). \tag{10.24}$$

Now, let $0 < \bar{\mu} = \max_{1\leq q\leq 4}\{\mu_q\} < 1$, $\bar{\beta} = \min_{1\leq q\leq 4}\{\beta_q\} > 1$, $(q = 1, 2, 3, 4)$, obviously, $\beta_q \geq \bar{\beta} > 1 > \bar{\mu} \geq \mu_q$. Then, we can get

$$V_1^{\bar{\mu}} \leq V_1^{\mu_1} + V_1^{\beta_1}, \quad V_1^{\bar{\beta}} \leq V_1^{\mu_1} + V_1^{\beta_1},$$

i.e.,

$$V_1^{\mu_1} + V_1^{\beta_1} \geq \frac{1}{2}(V_1^{\bar{\mu}} + V_1^{\bar{\beta}}). \tag{10.25}$$

Similarly,

$$\begin{aligned}
V_2^{\mu_2} + V_2^{\beta_2} &\geq \frac{1}{2}(V_2^{\bar{\mu}} + V_2^{\bar{\beta}}), \\
V_3^{\mu_3} + V_3^{\beta_3} &\geq \frac{1}{2}(V_3^{\bar{\mu}} + V_3^{\bar{\beta}}), \\
V_4^{\mu_4} + V_4^{\beta_4} &\geq \frac{1}{2}(V_4^{\bar{\mu}} + V_4^{\bar{\beta}}).
\end{aligned} \tag{10.26}$$

Then, by substituting (10.25) and (10.26) into (10.24) and applying Lemma 10.2, one has

$$\begin{aligned}\dot{V}(t) &\le -\frac{1}{2}\bar{\rho}\Big(V_1^{\bar{\mu}} + V_2^{\bar{\mu}} + V_3^{\bar{\mu}} + V_4^{\bar{\mu}} + V_1^{\bar{\beta}} + V_2^{\bar{\beta}} + V_3^{\bar{\beta}} + V_4^{\bar{\beta}}\Big)\\ &\le -\frac{1}{2}\bar{\rho}\Big((V_1 + V_2 + V_3 + V_4)^{\bar{\mu}} + 4^{1-\bar{\beta}}(V_1 + V_2 + V_3 + V_4)^{\bar{\beta}}\Big)\\ &= -\frac{1}{2}\bar{\rho}(V^{\bar{\mu}}(t) + 4^{1-\bar{\beta}}V^{\bar{\beta}}(t)). \end{aligned} \tag{10.27}$$

Therefore, according to Lemma 10.1, the error system (10.4) or (10.5) is fixed-time stable, that is, the fixed-time synchronization between the drive system (10.1) and response system (10.2) under the designed controller (10.11) can be achieved, and the bound of settling time T can be given as stated in (10.14). □

Remark 10.3 Since the proposed controller (10.11) contains independent parameters, it becomes very difficult to derive $\dot{V}(t) \le -(\alpha V^a(t) + \beta V^b(t))^\rho$. To solve this obstacle, the inequalities (10.25) and (10.26) in the proof of Theorem 10.1 are utilized fully. Then, the fixed-time synchronization can be achieved.

Remark 10.4 For real-valued neural networks or BAMNNs, many results about finite-time synchronization have been developed, see the references and therein (Selvaraj et al. 2018; Jiang et al. 2015; Abdurahman et al. 2016; Velmurugan et al. 2016; Yang et al. 2016; Xiao et al. 2017). In comparison, less references (Liu and Chen 2018; Hu et al. 2017; Wang et al. 2017; Wan et al. 2016; Cao and Li 2017) devote to the fixed-time synchronization of neural networks. On the other hand, in complex domain, only (Wu et al. 2018; Zhang et al. 2018; Zhou et al. 2017; Ding et al. 2017; Zhang et al. 2018) are concerned about the finite/fixed-time synchronization and stabilizability for complex-valued neural networks. Until now, there has no information involving the fixed-time stability of CVBAMNNs. Theorem 10.1 provides the fixed-time synchronization criterion for this kind of system with time delays. It is the first time that this problem is investigated. Thus, our work is very valuable in complex domain and supplement the previous results.

Remark 10.5 If systems (10.1) and (10.2) are reduced to real-valued ones, then in the designed controller (10.11), the corresponding parameters should be $\xi_{2j} = \xi_{4k} = \chi_{2j} = \chi_{4k} = \delta_{2j} = \delta_{4k} = \alpha_{2j} = \alpha_{4k} = 0$ $(j = 1, 2, k = 1, 2)$ and similar result for the fixed-time synchronization of BAMNNs can be obtained. Here, we omit it to save space. But, the effectiveness of this result can be shown by a numerical simulation. This point can be seen from Example 10.2 in the next section.

Remark 10.6 So far, there are available references with regard to CVBAMNNs (Wang and Huang 2016; Subramanian and Muthukumar 2018; Zhang et al. 2018; Guo et al. 2017, 2018; Zhang et al. 2020) which mainly involve exponential input-to-state stability, asymptotic stability and exponential stability analysis. In this chapter, the fixed-time synchronization problem for this kind of system with time delays is studied, and the novel fixed-time synchronization criterion for the addressed model is obtained for the first time based on fixed-time stability and some inequalities technique. Meanwhile, the nonlinear delayed controller with independent parameters is proposed different from the existing results, and a more accurate estimation independent on the initial conditions is given for the settling time, this point will be reflected in the examples section.

10.4 Illustrative Examples

In this section, we will give two examples to illustrate the effectiveness of the proposed result in Theorem 10.1.

Example 10.1 Consider a two-dimension system (10.1) with

$$\begin{aligned}
&d_{11}=0.2,\ d_{12}=0.3,\ d_{21}=0.1,\ d_{22}=0.2,\ a_{11}=1.5+2i,\ \tau_{11}=0.2,\\
&a_{12}=-2-i, a_{21}=-3-i,\ a_{22}=-1-i,\ b_{11}=2-i,\ b_{12}=-1+i,\\
&b_{21}=-1+2i,\ b_{22}=2+i,\ c_{11}=-1+3i,\ c_{12}=2-i,\ c_{21}=1.5+i,\\
&c_{22}=-1-1.5i,\ e_{11}=2+i,\ e_{12}=-1-i,\ e_{21}=-1+2i,\ e_{22}=2-1.5i,\\
&I_{11}(t)=7\sin(t-1)-i5\cos(t-1),\ \tau_{12}=0.1,\ \tau_{21}=0.1,\ \tau_{22}=0.2,\\
&I_{12}(t)=6\cos(t-1)-i8\sin(t),\ I_{21}(t)=4\cos(t)+i7\sin(t-1),\\
&I_{22}(t)=5\sin(t+1)+i6\cos(t+1).
\end{aligned}$$

In addition, take the activation functions as follows:

$$f_k(v_k)=\frac{1-e^{-x_k''}}{1+e^{-x_k''}}+i\frac{1}{1+e^{-y_k''}},\ g_j(u_j)=\frac{1-e^{-y_j'}}{1+e^{-y_j'}}+i\frac{1}{1+e^{-x_j'}},$$

which imply that $l_k^{RR}=0.5, l_k^{II}=0.25, l_k^{RI}=l_k^{IR}=0$ and $m_j^{RR}=m_j^{II}=0, m_j^{RI}=0.5, m_j^{IR}=0.25$.

We choose five different sets of initial values:

(1) $u_1(s)=2-2.3i, u_2(s)=-1.2+i, v_1(s)=0.5+0.3i,$
$v_2(s)=2.2-2i, \hat{u}_1(s)=-1.2+1.2i, \hat{u}_2(s)=2-i,$
$\hat{v}_1(s)=-2-1.5i, \hat{v}_2(s)=1+0.2i,$

(2) $u_1(s)=1-2.3i, u_2(s)=-2.2+i, v_1(s)=1.5+1.3i,$
$v_2(s)=2.2-i, \hat{u}_1(s)=-2.2+3.2i, \hat{u}_2(s)=2-i,$
$\hat{v}_1(s)=-2-1.5i, \hat{v}_2(s)=1+2.1i,$

(3) $u_1(s)=1-2i, u_2(s)=-1.2+i, v_1(s)=0.5+1.3i,$
$v_2(s)=2.2-2i, \hat{u}_1(s)=-1.5+1.2i, \hat{u}_2(s)=1-i,$
$\hat{v}_1(s)=-2-1.5i, \hat{v}_2(s)=1+0.5i,$

(4) $u_1(s)=-1-i, u_2(s)=-1.2+i, v_1(s)=-1.5+1.3i,$
$v_2(s)=2.2+2i, \hat{u}_1(s)=2.5+3.2i, \hat{u}_2(s)=1-2i,$
$\hat{v}_1(s)=2-1.5i, \hat{v}_2(s)=1-1.5i,$

(5) $u_1(s) = -1 + i$, $u_2(s) = -1.2 + i$, $v_1(s) = -2.5 + 1.3i$,
$v_2(s) = -2.2 + i$, $\hat{u}_1(s) = 1.5 + 1.2i$, $\hat{u}_2(s) = 1 - 2i$,
$\hat{v}_1(s) = 2 + 1.5i$, $\hat{v}_2(s) = 1 - 1.5i$, $\forall s \in [-1, 0]$,

under which the drive system (10.1) does not synchronize with the response system (10.2) without controller. For instance, when the initial conditions are chosen as the first set, the trajectories of systems (10.1) and (10.2) without controller are depicted in Figs. 10.1 and 10.2. It can be seen from these figures that the drive system (10.1) does not synchronize with the response system (10.2).

By Theorem 10.1, we can obtain that when $\xi_{11} > 1.425$, $\xi_{12} > 1.075$, $\xi_{21} > 3.05$, $\xi_{22} > 2.45$, $\chi_{11} + \chi_{12} > 3$, $\chi_{21} + \chi_{22} > 1.5$, $\xi_{31} > 3.65$, $\xi_{32} > 2.3$, $\xi_{41} > 1.775$, $\xi_{42} > 1.05$, $\chi_{31} + \chi_{32} > 1.5$, $\chi_{41} + \chi_{42} > 3$, systems (10.1) and (10.2) can achieve the synchronization in fixed time. Now, we take two types of parameters:

(1) $\xi_{11} = 1.6$, $\xi_{12} = 1.1$, $\xi_{21} = 3.1$, $\xi_{22} = 2.5$, $\chi_{11} = 1.5$, $\chi_{12} = 1.55$, $\chi_{21} = 0.5$, $\chi_{22} = 1.5$, $\xi_{31} = 3.7$, $\xi_{32} = 2.35$, $\xi_{41} = 1.8$, $\xi_{42} = 1.1$, $\chi_{31} = 1.1$, $\chi_{32} = 0.5$, $\chi_{41} = 1.6$, $\chi_{42} = 1.5$, $\delta_{11} = 1.4$, $\delta_{12} = 1.3$, $\delta_{21} = 1.5$, $\delta_{22} = 1.3$, $\delta_{31} = 1.4$, $\delta_{32} = 1.6$, $\delta_{41} = 1.6$, $\delta_{42} = 1.5$, $\alpha_{11} = 1.6$, $\alpha_{12} = 1.8$, $\alpha_{21} = 1.6$, $\alpha_{22} = 1.7$, $\alpha_{31} = 1.6$, $\alpha_{32} = 1.7$, $\alpha_{41} = 1.8$, $\alpha_{42} = 2$, $\mu_1 = 0.4$, $\mu_2 = 0.5$, $\mu_3 = 0.3$, $\mu_4 = 0.2$, $\beta_1 = 1.1$, $\beta_2 = 1.4$, $\beta_3 = 1.3$, $\beta_4 = 1.2$,

(2) $\mu_1 = 0.6$, $\mu_2 = 0.2$, $\mu_3 = 0.3$, $\mu_4 = 0.1$, $\beta_1 = 1.4$, $\beta_2 = 1.8$, $\beta_3 = 1.9$, $\beta_4 = 2$, the other parameters are same to the first type.

Then, by Theorem 10.1 and Lemma 10.1, under the first type of parameters, $\bar{\mu} = 0.5$, $\bar{\beta} = 1.1$, and the settling time T can be bounded by $T \leq 3.2987$ second. But, by the proposed result of Polyakov (2012), the upper bound of the settling time T can be computed as 22.2447 s second. On the other hand, for the second type of parameters, $\bar{\mu} = 0.6$, $\bar{\beta} = 1.4$, and the settling time T satisfies $T \leq 6.25$ second via Theorem 10.1 and Lemma 10.1 whereas the upper bound of T is 17.1319 s second on the basis of the result of Zuo and Tie (2016). Obviously, Lemma 10.1 can provide a more accurate estimation of the settling time.

Moreover, under five sets of initial values, the time revolutions of real-imaginary parts of systems (10.1) and (10.2) with the first type of parameters under controller are depicted in Figs. 10.3 and 10.4. The time responses of synchronization errors between them with the two types of parameters are depicted in Figs. 10.5 and 10.6, respectively. The upper bound of the settling time T has been marked with "o". It can be seen from Figs. 10.3, 10.4, 10.5 and 10.6 that systems (10.1) and (10.2) can realize the synchronization in fixed time under the controller (10.11) with the above parameters and the upper bound of the settling time is independent on the initial conditions.

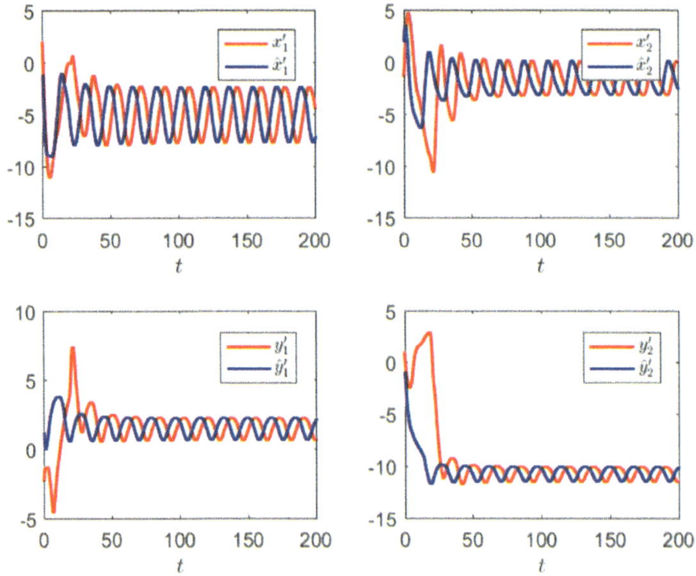

Fig. 10.1 The curves of real and imaginary parts of variables u_1 and $\hat{u}_1$, u_2 and $\hat{u}_2$ without controller for Example 10.1

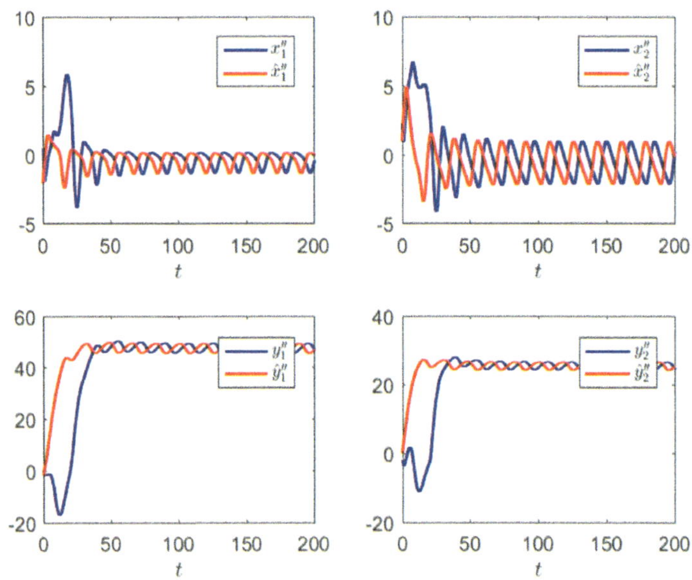

Fig. 10.2 The curves of real and imaginary parts of variables v_1 and $\hat{v}_1$, v_2 and $\hat{v}_2$ without controller for Example 10.1

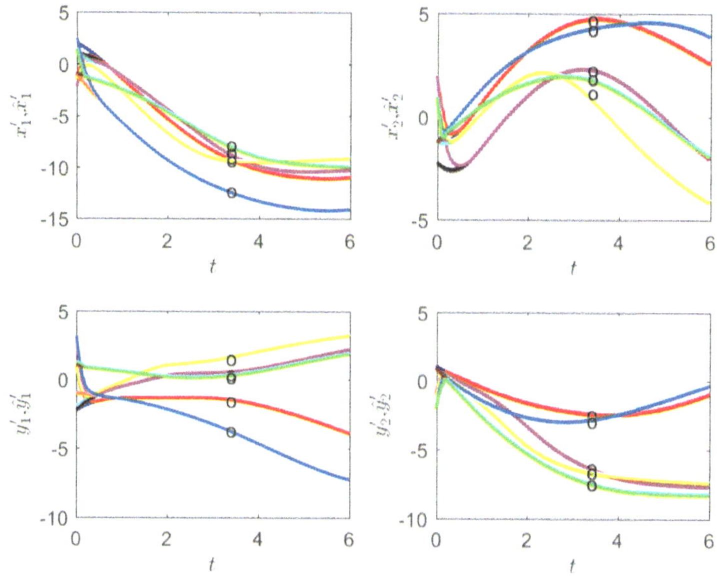

Fig. 10.3 Time revolutions of real and imaginary parts of variables u_1 and $\hat{u}_1$, u_2 and $\hat{u}_2$ under controller for Example 10.1

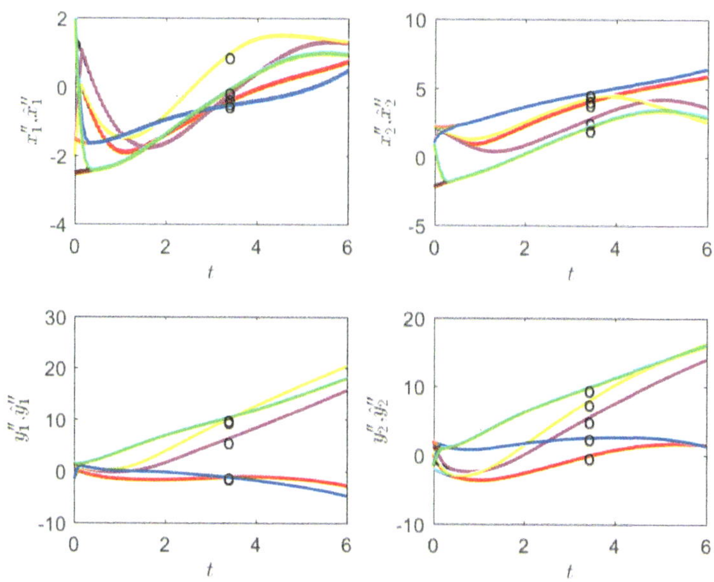

Fig. 10.4 Time revolutions of real and imaginary parts of variables v_1 and $\hat{v}_1$, v_2 and $\hat{v}_2$ under controller for Example 10.1

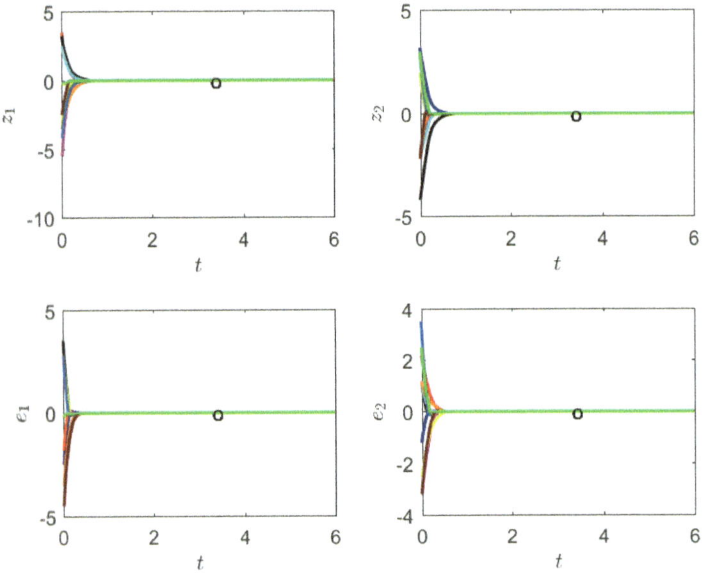

Fig. 10.5 The curves of synchronization errors variables z_1 and z_2, e_1 and e_2 under the first type of parameters for Example 10.1

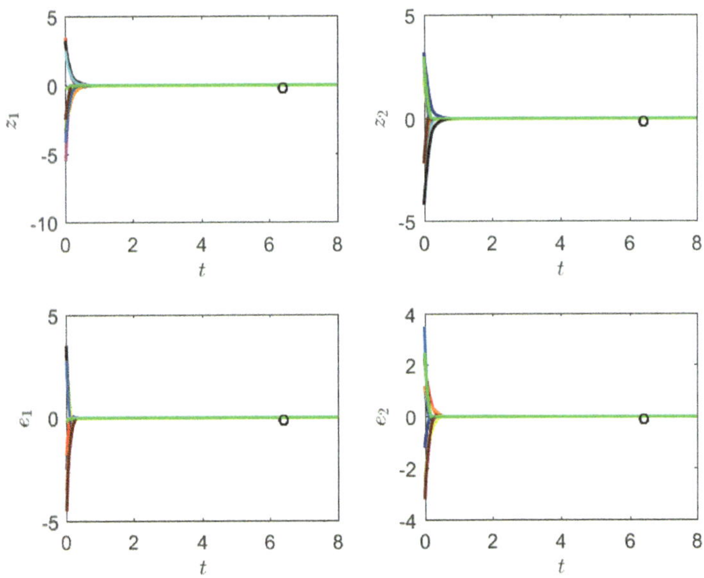

Fig. 10.6 The curves of synchronization errors variables z_1 and z_2, e_1 and e_2 under the second type of parameters for Example 10.1

Example 10.2 Consider a real-valued two-dimension system (10.1) with

$$d_{11} = 3,\ d_{12} = 5,\ d_{21} = 4,\ d_{22} = 2,\ a_{11} = 3,\ a_{12} = 0.5,\ a_{21} = -3,$$
$$a_{22} = 2,\ b_{11} = 2,\ b_{12} = -2,\ b_{21} = 2,\ b_{22} = 1,\ c_{11} = -2,$$
$$c_{12} = -2.5,\ c_{21} = -1.5,\ c_{22} = 4,\ e_{11} = 2.5,\ e_{12} = -1,\ e_{21} = 3,\ e_{22} = -1,$$
$$I_{11}(t) = 6\sin(t+1),\ I_{12}(t) = -4\cos(t),\ \tau_{11} = \tau_{22} = 0.1,$$

$$I_{21}(t) = 9\cos(t),\ I_{22}(t) = -6\sin(t-1),\ \tau_{12} = \tau_{21} = 0.2.$$

In addition, take the activation functions as follows:

$$f_k(v_k) = 0.5(|v_k + 1| - |v_k - 1|),\ g_j(u_j) = 0.5(|u_j + 1| - |u_j - 1|),$$

which imply that $l_k = m_j = 1$ for $k = 1, 2,\ j = 1, 2$.

We choose five different sets of initial values:

(1) $u_1(s) = -0.2,\ u_2(s) = 0.2,\ \hat{u}_1(s) = -1,\ \hat{u}_2(s) = -0.6,\ v_1(s) = 1,$
$v_2(s) = -0.6,\ \hat{v}_1(s) = 0.5,\ \hat{v}_2(s) = 0.2,$

(2) $u_1(s) = 0.5,\ u_2(s) = 0.2,\ \hat{u}_1(s) = -2,\ \hat{u}_2(s) = -0.6,\ v_1(s) = -0.2,$
$v_2(s) = -0.6,\ \hat{v}_1(s) = -0.5,\ \hat{v}_2(s) = 0.2,$

(3) $u_1(s) = -0.5,\ u_2(s) = 0.2,\ \hat{u}_1(s) = 1,\ \hat{u}_2(s) = -0.7,\ v_1(s) = 1,$
$v_2(s) = -0.3,\ \hat{v}_1(s) = -0.2,\ \hat{v}_2(s) = 0.4,$

(4) $u_1(s) = 1,\ u_2(s) = 0.3,\ \hat{u}_1(s) = -2,\ \hat{u}_2(s) = -0.4,\ v_1(s) = -0.2,$
$v_2(s) = 0.1,\ \hat{v}_1(s) = 0.3,\ \hat{v}_2(s) = 0.4,$

(5) $u_1(s) = -0.4,\ u_2(s) = 1.1,\ \hat{u}_1(s) = 0.8,\ \hat{u}_2(s) = -0.8,\ v_1(s) = 0.2,$
$v_2(s) = -0.1,\ \hat{v}_1(s) = -0.4,\ \hat{v}_2(s) = 0.2,\ \forall s \in [-0.2, 0],$

under which the drive system (10.1) does not synchronize with the response system (10.2) without controller. For instance, when the initial conditions are chosen as the first set, the trajectories of systems (10.1) and (10.2) without controller are depicted in Fig. 10.7. It can be seen from them the drive system (10.1) does not synchronize with the response system (10.2).

By Theorem 10.1 and Remark 10.5, we can obtain that when $\xi_{11} > 0.5$, $\xi_{12} > 1.5$, $\xi_{31} > 2$, $\xi_{32} > 0.5$, $\chi_{11} + \chi_{12} > 4$, $\chi_{31} + \chi_{32} > 5.5$, systems (10.1) and (10.2) can achieve the synchronization in fixed time. Meanwhile, we take two types of parameters:

(1) $\xi_{11} = 0.55,\ \xi_{12} = 1.6,\ \xi_{31} = 2.1,\ \xi_{32} = 0.6,\ \chi_{11} = 2.3,\ \chi_{12} = 1.75,$
$\chi_{31} = 2.82,\ \chi_{32} = 2.7,\ \delta_{11} = 2.1,\ \delta_{12} = 1,\ \delta_{31} = 1.7,\ \delta_{32} = 1.3,\ \alpha_{11} = 2,$
$\alpha_{12} = 1,\ \alpha_{31} = 2.6,\ \alpha_{32} = 1.2,\ \mu_1 = 0.6,\ \mu_3 = 0.5,\ \beta_1 = 2.2,\ \beta_3 = 2,$

(2) $\mu_1 = 0.2,\ \mu_3 = 0.1,\ \beta_1 = 1.4,\ \beta_3 = 1.8$, the other parameters are same to the first type.

Then, by Theorem 10.1 and Lemma 10.1, under the first type of parameters, $\bar{\mu} = 0.6$, $\bar{\beta} = 2$, and the settling time T can be bounded by $T \le 11.4863$ second. But, by the proposed result of Polyakov (2012), the upper bound of the settling time T can be

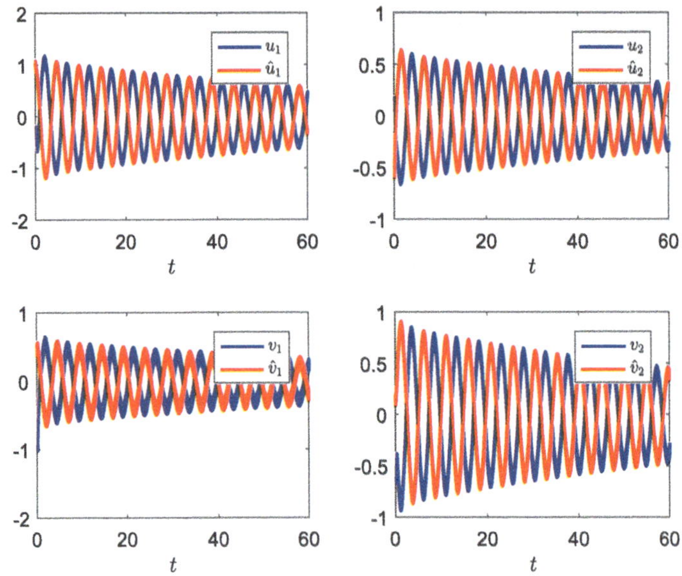

Fig. 10.7 The curves of the variables u_1 and $\hat{u}_1$, u_2 and $\hat{u}_2$, v_1 and $\hat{v}_1$, v_2 and $\hat{v}_2$ without controller for Example 10.2

computed as 29.8645 s second. On the other hand, for the second type of parameters, $\bar{\mu} = 0.2$, $\bar{\beta} = 1.4$, and the settling time T satisfies $T \leq 4.3531$ second via Theorem 10.1 and Lemma 10.1 whereas the upper bound of T is 19.5116 s second on the basis of the result of Zuo and Tie (2016). It is clear that Lemma 10.1 can provide a more accurate estimation of the settling time.

Moreover, under five sets of initial values, the time revolutions of systems (10.1) and (10.2) with the first type of parameters under controller are depicted in Figs. 10.8 and 10.9. The time responses of synchronization errors between them with the two types of parameters are depicted in Figs. 10.10 and 10.11, respectively. The upper bound of the settling time T has been marked with "o". It can be seen from Figs. 10.8, 10.9, 10.10 and 10.11 that systems (10.1) and (10.2) can realize the synchronization in fixed time under the controller (10.11) with the above parameters and the upper bound of the settling time is independent on the initial conditions.

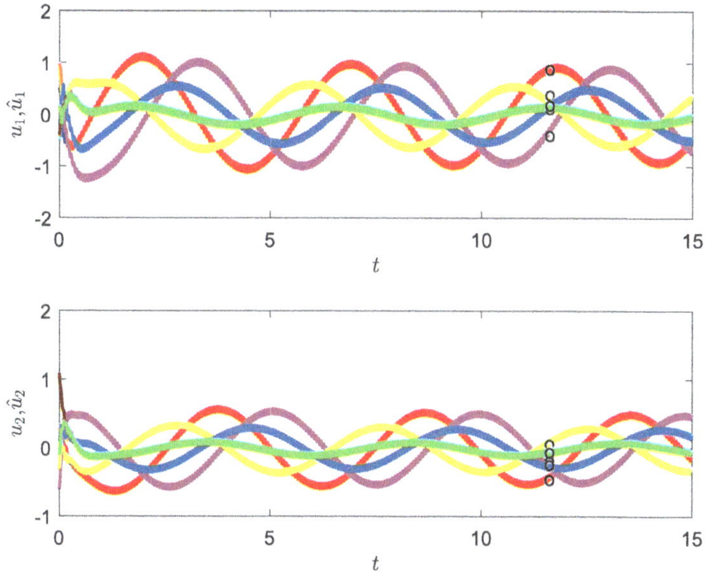

Fig. 10.8 Time revolutions of the variables u_1 and $\hat{u}_1$, u_2 and $\hat{u}_2$ under controller for Example 10.2

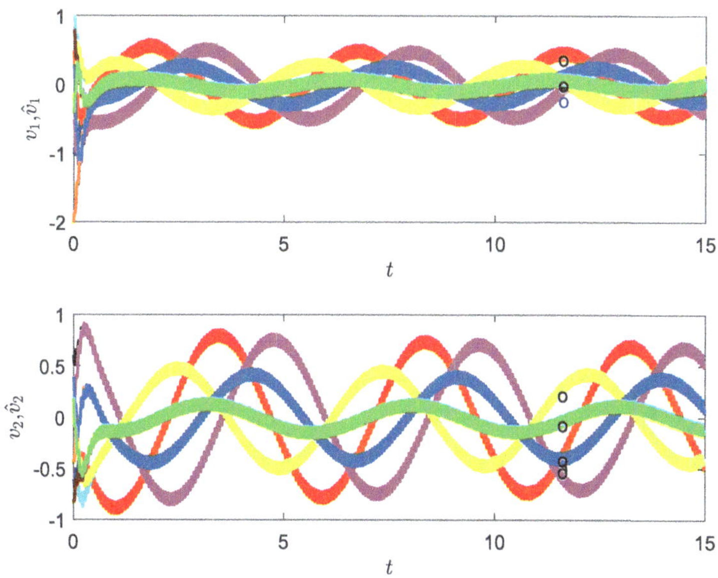

Fig. 10.9 Time revolutions of the variables v_1 and $\hat{v}_1$, v_2 and $\hat{v}_2$ under controller for Example 10.2

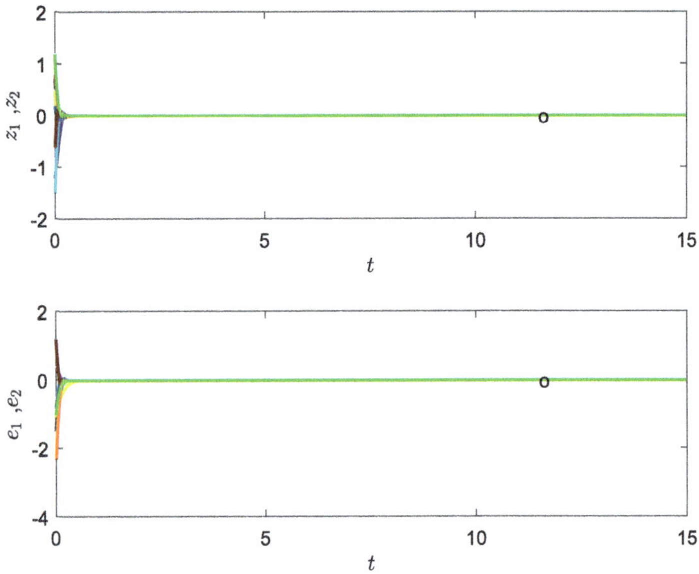

Fig. 10.10 The curves of synchronization errors variables z_1, z_2, e_1 and e_2 under the first type of parameters for Example 10.2

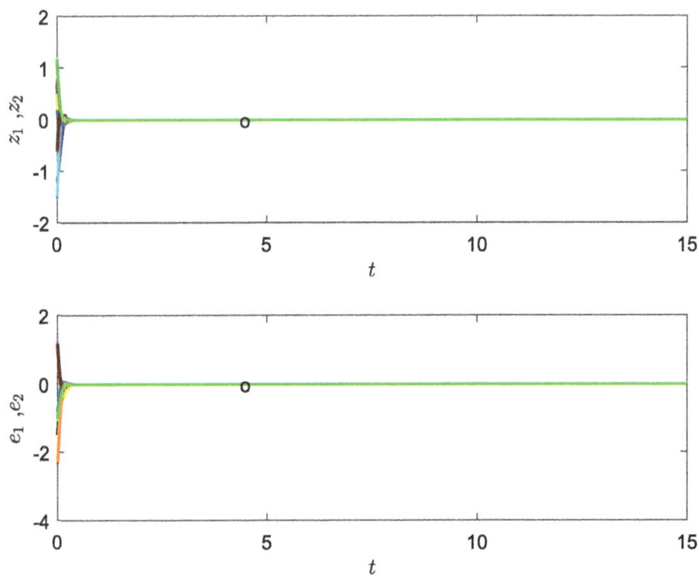

Fig. 10.11 The curves of synchronization errors variables z_1, z_2, e_1 and e_2 under the second type of parameters for Example 10.2

10.5 Conclusion

In this chapter, we have studied the fixed-time synchronization for a class of CVBAMNNs with time delays. We have presented a new nonlinear delayed controller with independent parameters different from the existing ones. By utilizing the fixed-time stability theory, the Lyapunov function method and some inequalities techniques, we have established new sufficient condition of fixed-time synchronization for the addressed drive-response system under the designed controller. Finally, we have verified the validity of the theoretical result by two numerical examples.

References

Selvaraj P, Sakthivel R, Kwon OM (2018) Finite-time synchronization of stochastic coupled neural networks subject to Markovian switching and input saturation. Neural Netw 105:154–165

Jiang M, Wang S, Mei J, Shen Y (2015) Finite-time synchronization control of a class of memristor-based recurrent neural networks. Neural Netw 63:133–140

Abdurahman A, Jiang H, Teng Z (2016) Finite-time synchronization for fuzzy cellular neural networks with time-varying delays. Fuzzy Sets Syst 297:96–111

Du H, He Y, Cheng Y (2014) Finite-time synchronization of a class of second-order nonlinear multi-agent systems using output feedback control. IEEE Trans Circuits Syst I Reg Papers 61(6):1778–1788

Velmurugan G, Rakkiyappan R, Cao J (2016) Finite-time synchronization of fractional order memristor-based neural networks with time delays. Neural Netw 73:36–46

Yang F, Mei J, Wu Z (2016) Finite-time synchronisation of neural networks with discrete and distributed delays via periodically intermittent memory feedback control. IET Control Theory A 10(14):1630–1640

Polyakov A (2012) Nonlinear feedback design for fixed-time stabilization of linear control systems. IEEE Trans Autom Control 57(8):2106–2110

Ni J, Liu L, Liu C, Hu X, Li S (2017) Fast fixed-time nonsingular terminal sliding mode control and its application to chaos suppression in power system. IEEE Trans Circuits Syst II Exp Briefs 64(2):151–155

Muralidharan A, Pedarsani R, Varaiya P (2015) Analysis of fixed-time control. Transport Res B-Meth 73:81–90

Jiang B, Hu Q, Friswell MI (2016) Fixed-time attitude control for rigid spacecraft with actuator saturation and faults. IEEE Trans Control Syst Technol 24:1892–1898

Zuo Z, Tie L (2016) Distributed robust finite-time nonlinear consensus protocols for multi-agent systems. Int J Syst Sci 47(6):1366–1375

Fu J, Wang J (2016) Fixed-time coordinated tracking for second-order multiagent systems with bounded input uncertainties. Syst Control Lett 93:1–12

Polyakov A, Efimov D, Perruquetti W (2015) Finite-time and fixed-time stabilization: implicit Lyapunov function approach. Automatica 51:332–340

Hua C, Li Y, Guan X (2017) Finite/fixed-time stabilization for nonlinear interconnected systems with dead-zone input. IEEE Trans Autom Control 62(5):2554–2560

Zhang Z, Liu X, Liu Y, Lin C, Chen B (2018) Fixed-time almost disturbance decoupling of nonlinear time-varying systems with multiple disturbances and dead-zone input. Inf Sci 450:267–283

Ding X, Cao J, Alsaedi A, Alsaadi F, Hayat T (2017) Robust fixed-time synchronization for uncertain complex-valued neural networks with discontinuous activation functions. Neural Netw 90:42–55

Liu X, Chen T (2018) Finite-time and fixed-time cluster synchronization with or without pinning control. IEEE Trans Cybern 48(1):240–252
Hu C, Yu J, Chen Z, Jiang H, Huang T (2017) Fixed-time stability of dynamical systems and fixed-time synchronization of coupled discontinuous neural networks. Neural Netw 89:74–83
Wang L, Zeng Z, Hu J, Wang X (2017) Controller design for global fixed-time synchronization of delayed neural networks with discontinuous activations. Neural Netw 87:122–131
Wan Y, Cao J, Wen G, Yu W (2016) Robust fixed-time synchronization of delayed Cohen-Grossberg neural networks. Neural Netw 73:86–94
Cao J, Li R (2017) Fixed-time synchronization of delayed memristor-based recurrent neural networks. Sci China Inf Sci 60(3):032201
Liu X, Jiang N, Cao J, Wang S, Wang Z (2013) Finite-time stochastic stabilization for BAM neural networks with uncertainties. J Frankl Inst 350(8):2109–2123
Xiao J, Zhong S, Li Y, Xu F (2017) Finite-time Mittag-Leffler synchronization of fractional-order memristive BAM neural networks with time delays. Neurocomputing 219:431–439
Wu E, Yang X, Xu C, Alsaadi FE, Hayat T (2018) Finite-time synchronization of complex-valued delayed neural networks with discontinuous activations. Asian J Control 20(6):2237–2247
Zhang Z, Liu X, Lin C, Chen B (2018) Finite-time synchronization for complex-valued recurrent neural networks with time delays. Complexity 8456737
Zhou C, Zhang W, Yang X, Xu C, Feng J (2017) Finite-time synchronization of complex-valued neural networks with mixed delays and uncertain perturbations. Neural Process Lett 46(1):271–291
Zhang Z, Liu X, Zhou D, Lin C, Chen J, Wang H (2018) Finite-time stabilizability and instabilizability for complex-valued memristive neural networks with time delays. IEEE Trans Syst Man Cybern Syst 48(12):2371–2382
Wang Z, Huang L (2016) Global stability analysis for delayed complex-valued BAM neural networks. Neurocomputing 173:2083–2089
Subramanian K, Muthukumar P (2018) Existence, uniqueness, and global asymptotic stability analysis for delayed complex-valued Cohen-Grossberg BAM neural networks. Neural Comput Appl 29:565–584
Zhang Z, Liu X, Guo R, Lin C (2018) Finite-time stability for delayed complex-valued BAM neural networks. Neural Process Lett 48:179–193
Guo R, Zhang Z, Liu X, Lin C (2017) Existence, uniqueness, and exponential stability analysis for complex-value d memristor-based BAM neural networks with time delays. Appl Math Comput 311:100–117
Guo R, Zhang Z, Liu X, Lin C, Wang H, Chen J (2018) Exponential input-to-state stability for complex-valued memristor-based BAM neural networks with multiple time-varying delays. Neurocomputing 275:2041–2054
Zhang Z, Guo R, Liu X, Lin C (2020) Lagrange exponential stability of complex-valued BAM neural networks with time-varying delays. IEEE Trans Syst Man Cybern Syst 50(8):3072–3085

Chapter 11
Fixed-Time Pinning Synchronization for CVINNs with Time-Varying Delays

Abstract This chapter concentrates on the fixed-time pinning synchronization and adaptive synchronization problems of complex-valued inertial neural networks (CVINNs) with time-varying delays. We will regard this model as an entirety instead of reducing it to first-order differential equation, separate the real and imaginary parts into an equivalent real-valued one, and establish a novel Lyapunov function. The fixed-time stability for the closed-loop error system is guaranteed via partial nodes controlled directly by a new pinning controller which involves the state derivatives and other proper terms. Moreover, from the point of saving cost and avoiding resources waste, we will continue to further develop a new pinning adaptive controller and derive a sufficient condition ensuring the adaptive fixed-time stability for the closed-loop error system.

11.1 Introduction

As stated in the previous Introduction and Chap. 9, Babcock and Westervelt first advanced the inertial neural networks (Babcock and Westervelt 1986, 1987), which represented the inertial characteristics by introducing inductor into the neural circuit, and this dynamic model is described by a second-order differential equation. It can simulate squid protrusion through inductive circuits (Angelaki and Correia 1991), so its biological backgrounds are broader. Compared with the standard first-order model, the dynamic behaviors of this model are more complex. As a consequence, it is of great significance to investigate the dynamic behaviors and the control methods of inertial neural networks (Lakshmanan et al. 2018; Zhang and Wang 2019; Wang and Tian 2019; Wang et al. 2021; Qin et al. 2020).

Recently, with the development of science and technology, complex signals are absorbed in many industrial production, which causes the research upsurge of complex-valued neural networks. Further, since the inertial term was introduced into complex-valued neural networks (CVINNs) model, the attentions of the dynamic behaviors of CVINNs are increasing gradually. For example, the synchronization and anti-synchronization (Wei et al. 2021), exponential stabilization (Li et al. 2020) and exponential convergence (Tang and Jian 2019) of CVINNs are reported by using

Z. Zhang et al., *Complex-Valued Neural Networks Systems with Time Delay*, Intelligent Control and Learning Systems 4,
https://doi.org/10.1007/978-981-19-5450-4_11

the reduced-order method. On the contrary, based on the non-reduced order method, some synchronization results are discussed for the second-order CVINNs model (Li and Huang 2021; Yu et al. 2020).

Synchronization plays a very important role in the dynamic behaviors of neural networks, which is widely used in bioengineering, secure communication, information processing and so on, and has become an active research issue. However, many large-scale neural networks cannot achieve the synchronization by themselves, so it is necessary to design appropriate controllers to realize that. In the past few decades, many effective synchronization criteria for neural networks have been obtained by using different control methods. For example, the event-triggered control method is applied to realize the periodic synchronization (Wang et al. 2021). The adaptive control scheme is used to analyze the asymptotic and exponential synchronizations (Li et al. 2020; Wang et al. 2015; Xu et al. 2019; Hu and Zeng 2017). Pinning control method for the synchronization problem (Jia and Zeng 2020; Wu and Huang 2019; Zhu et al. 2020; Fu et al. 2021; Li et al. 2018) is adopted. Region-partitioning-dependent intermittent control (Ding et al. 2019) and single impulsive control (Ling et al. 2021) are utilized to reach the quasi-synchronization problems. As far as we know, for one thing, in the case of unknown disturbance, it is difficult to get accurate controller parameters. But adaptive control can automatically adjust controller parameters by designing adaptive laws, so that system can reach the synchronization. Hence, adaptive control method is more effective in application areas. For another thing, pinning control method is superior to the other in attaining the purpose of reducing control costs and saving resources because it only controls the key nodes. Thus, it is essential to further explore deeply the synchronization problem via pinning control and adaptive control strategy. This is the first motivation of our work in this chapter.

It is worth mentioning that the above synchronization results are reached in infinite time. For the sake of meet actual requirement, the finite-time synchronization is developed (Jian and Duan 2020; Liu et al. 2021; Zhou et al. 2017; Zhang et al. 2018; Zheng et al. 2020), but its settling time must depend on the initial conditions, which will cause great inconvenience in practice. To compensate for this shortcoming, the researchers came up with the idea of the fixed-time synchronization (Zhang et al. 2021; Ding et al. 2017; Xiao et al. 2022; Zhang et al. 2021; Zhang and Deng 2020; Guo et al. 2020). Here, the settling time can be bounded by an upper bound $T_{\max}$ independent of initial values, which effectively improves the theoretical application. However, for inertial neural networks system in complex domain, the results of synchronization control in fixed-time interval are few Long et al. (2021), Guo et al. (2021). This is the second motivation of our work in this chapter.

Based on what has been discussed above, the main aim of this chapter is to investigate pinning synchronization and adaptive synchronization of CVINNs with time-varying delays in fixed-time interval. The main contributions are listed as follows: (1) Considering the object model as an whole instead of using the reduced order method based on variable substitution, fixed-time pinning synchronization and adaptive synchronization problems of the second-order CVINNs with time-varying delays are studied; (2) A new pinning controller which involves the state derivatives and other proper terms is introduced to only control partial nodes directly. A novel Lyapunov function containing the state derivatives is established. Based on these ideas, the stability for the closed-loop error system is guaranteed in fixed-time interval; (3) To further meet the factual requirements, a new pinning adaptive controller with adaptive control laws is developed and sufficient condition ensuring the adaptive fixed-time stability for the closed-loop error system is derived.

11.2 Problem Formulation and Preliminaries

Consider the following CVINNs with time-varying delays as the drive system:

$$\ddot{v}_l(t) = -a_l v_l(t) - b_l \dot{v}_l(t) + \sum_{q=1}^{n} c_{lq} f_q(v_q(t)) + \sum_{q=1}^{n} d_{lq} f_q(v_q(t - \tau_q(t))), \tag{11.1}$$

where $l = 1, 2, ..., n, a_l > 0$ and $b_l > 0$ are elements of feedback template; $v_l(t) \in \mathbb{C}$ is the state of the neuron l and its second derivative is called inertia term; $c_{lq} \in \mathbb{C}$ and $d_{lq} \in \mathbb{C}$ are complex-valued connection weights; $f_q(v_q(t))$ and $f_q(v_q(t - \tau_q(t))) \in \mathbb{C}$ denote the complex-valued activation functions; $\tau_q(t)$ denotes the time-varying delays with $0 \le \tau_q(t) \le \tau_q$.

The initial conditions of system (11.1) are

$$v_l(s) = \varphi_l(s), \dot{v}_l(s) = \psi_l(s), -\tau \le s \le 0, \tag{11.2}$$

where $\tau = \max\{\tau_1, \tau_2, \ldots, \tau_n\}$, $\varphi_l(s)$ and $\psi_l(s)$ are continuous.

The response system can be expressed as

$$\ddot{z}_l(t) = -a_l v_l(t) - b_l \dot{z}_l(t) + \sum_{q=1}^{n} c_{lq} f_q(z_q(t)) + \sum_{q=1}^{n} d_{lq} f_q(z_q(t - \tau_q(t))) + u_l(t), \tag{11.3}$$

where $z_l(t) \in \mathbb{C}$ is the state of the neuron l; $u_l(t)$ is the control input that will be designed.

The initial conditions of system (11.3) are

$$z_l(s) = \tilde{\varphi}_l(s), \dot{z}_l(s) = \tilde{\psi}_l(s), -\tau \le s \le 0, \tag{11.4}$$

where $\tilde{\varphi}_l(s)$ and $\tilde{\psi}_l(s)$ are continuous.

Remark 11.1 On the one hand, the inertia term is generated by inductor, which has abundant biological backgrounds. When introduced into neural networks, it can produce more complex bifurcation behaviors and chaos. On the other hand, in industrial production, due to the involvement of complex signals, complex-valued neural networks model is superior to the real-valued one. Thus, it is worthy of further study on the dynamics for CVINNs. Hitherto, there have been few reports about this model (Wei et al. 2021; Li et al. 2020; Tang and Jian 2019; Li and Huang 2021; Yu et al. 2020). Here, we will develop our work on this system.

By defining $e_l(t) = z_l(t) - v_l(t)$ as the error state, the error system can be described as

$$\ddot{e}_l(t) = -a_l e_l(t) - b_l \dot{e}_l(t) + \sum_{q=1}^{n} c_{lq} \tilde{f}_q(e_q(t)) + \sum_{q=1}^{n} d_{lq} \tilde{f}_q(e_q(t - \tau_q(t))) + u_l(t), \tag{11.5}$$

where $\tilde{f}_q(e_q(t)) = f_q(z_q(t)) - f_q(v_q(t))$, $\tilde{f}_q(e_q(t - \tau_q(t))) = f_q(z_q(t - \tau_q(t))) - f_q(v_q(t - \tau_q(t)))$, $l = 1, 2, \ldots, n$.

Obviously, the initial conditions of (11.5) are

$$e_l(s) = \varrho_l(s) = \tilde{\varphi}_l(s) - \varphi_l(s), \dot{e}_l(s) = \zeta_l(s) = \tilde{\psi}_l(s) - \psi_l(s), \tau \leq s \leq 0, \tag{11.6}$$

where $\varrho_l(s)$ and $\zeta_l(s)$ are continuous.

Let $e_l(t) = e_l^R(t) + ie_l^I(t)$, $u_l(t) = u_l^R(t) + iu_l^I(t)$, $\tilde{f}_q(e_q(\cdot)) = \tilde{f}_q^R(e_q^R(\cdot)) + i\tilde{f}_q^I(e_q^I(\cdot))$, $(\cdot)$ denotes t or $t - \tau_q(t)$. Then, system (11.5) can be separated into

$$\begin{aligned} \ddot{e}_l^R(t) = & -a_l e_l^R(t) - b_l \dot{e}_l^R(t) + \sum_{q=1}^{n} c_{lq}^R \tilde{f}_q^R(e_q^R(t)) - \sum_{q=1}^{n} c_{lq}^I \tilde{f}_q^I(e_q^I(t)) \\ & + \sum_{q=1}^{n} d_{lq}^R \tilde{f}_q^R(e_q^R(t - \tau_q(t))) - \sum_{q=1}^{n} d_{lq}^I \tilde{f}_q^I(e_q^I(t - \tau_q(t))) + u_l^R(t) \end{aligned} \tag{11.7}$$

and

$$\begin{aligned} \ddot{e}_l^I(t) = & -a_l e_l^I(t) - b_l \dot{e}_l^I(t) + \sum_{q=1}^{n} c_{lq}^R \tilde{f}_q^I(e_q^I(t)) + \sum_{q=1}^{n} c_{lq}^I \tilde{f}_q^R(e_q^R(t)) \\ & + \sum_{q=1}^{n} d_{lq}^R \tilde{f}_q^I(e_q^I(t - \tau_q(t))) + \sum_{q=1}^{n} d_{lq}^I \tilde{f}_q^R(e_q^R(t - \tau_q(t))) + u_l^I(t). \end{aligned} \tag{11.8}$$

The initial conditions in (11.7) and (11.8) are

$$e_l^R(s) = \varrho_l^R(s), \dot{e}_l^R(s) = \zeta_l^R(s), e_l^I(s) = \varrho_l^I(s), \dot{e}_l^I(s) = \zeta_l^I(s), -\tau \leq s \leq 0. \tag{11.9}$$

Remark 11.2 In the most existing results for real-valued inertial neural networks or complex-valued ones, the main general method is to reduce the order by variable substitution (Lakshmanan et al. 2018; Zhang and Wang 2019; Wang and Tian 2019; Wang et al. 2021; Qin et al. 2020; Wei et al. 2021; Li et al. 2020; Tang and Jian 2019; Jian and Duan 2020), which increases the number of state variables and dimensions of systems. To avoid this case, we directly consider the original second-order system rather than reducing order. This maintains the features of the original systems and more conforms to practical applications.

The following preliminaries will be used in the derivation of the next section.

Assumption 11.1 For the activation functions $f_q(v_q(t)) = f_q^R(v_q^R(t)) + if_q^I(v_q^I(t))$, $v_q(t) = v_q^R(t) + iv_q^I(t)$, there exist positive real constants M_q^R and M_q^I such that $|f_q^R(x) - f_q^R(y)| \leq M_q^R|x-y|$, $|f_q^I(x) - f_q^I(y)| \leq M_q^I|x-y|$, $\forall x, y \in \mathbb{R}$.

Definition 11.1 The synchronization between the drive-response system (11.1) and (11.3) can be achieved in fixed time, if for any initial condition ϱ_0, there exists fixed-time $T_{\max} \geq 0$ and a settling time function $T(\varrho_0) \geq 0$ such that $\lim\limits_{t \to T(\varrho_0)} \|e(t)\| = 0$, $e(t) = 0$ for all $t \geq T(\varrho_0)$ and $T(\varrho_0) \leq T_{\max}$, where $e(t) = (e_1^T, e_2^T, \ldots, e_n^T)^T$ and $\varrho_0 = e(0)$.

Lemma 11.1 (Zhang et al. 2021) *If there exists a regular, positive definite and radially unbounded function $V(e(t)) : \mathbb{R}^n \to \mathbb{R}^+$ such that for any function of system (11.5), the following inequality holds*

$$\frac{d}{dt}V(e(t)) \leq -\alpha V^a(e(t)) - \beta V^b(e(t)), \tag{11.10}$$

for $e(t) \in \mathbb{R}^n \backslash \{0\}$, where $\alpha, \beta > 0$, $0 < a < 1$, $b > 1$, then system (11.5) is globally fixed-time stable at the origin. Moreover, the settling time $T(\varrho_0)$ satisfies

$$T(\varrho_0) \leq T_{\max} = \frac{1}{\alpha}\left(\frac{\alpha}{\beta}\right)^{\frac{1-a}{b-a}}\left(\frac{1}{1-a} + \frac{1}{b-1}\right). \tag{11.11}$$

Especially, when $\beta = 0$, the conclusion is reduced to the finite-time stability case, that is to say, system (11.5) is globally finite-time stable and the settling time

$$T(\varrho_0) = \frac{V^{1-a}(0)}{\alpha(1-a)}. \tag{11.12}$$

Lemma 11.2 (Zhang et al. 2021) *Let $a_1, a_2, \ldots, a_n$ and ρ be the positive numbers, then the following inequality holds:*

$$(a_1 + a_2 + \ldots + a_n)^\rho \leq \max\{n^{\rho-1}, 1\}(a_1^\rho + a_2^\rho + \ldots + a_n^\rho). \tag{11.13}$$

11.3 Main Results

This section focuses on the analysis of the synchronization problem of CVINNs (11.1) and (11.3) under pinning control and pinning adaptive control by using new Lyapunov function instead of the common reduced order techniques.

11.3.1 Fixed-Time Synchronization Criterion

We design the following pinning control input. First, let

$$\Omega_1^R(t) = \frac{\sum_{l=1}^{n} |e_l^R(t)|}{\sum_{l=1}^{m} |\dot{e}_l^R(t)|}, \quad \Omega_2^R(t) = \frac{\sum_{l=1}^{n} |\dot{e}_l^R(t)|}{\sum_{l=1}^{m} |\dot{e}_l^R(t)|}, \quad \Omega_3^R(t) = \frac{\sum_{l=1}^{n} |e_l^R(t-\tau_l(t))|}{\sum_{l=1}^{m} |\dot{e}_l^R(t)|},$$

$$\Omega_4^R(t) = \frac{\sum_{l=1}^{n} |e_l^R(t)|^{\mu_1+1}}{\sum_{l=1}^{m} |\dot{e}_l^R(t)|}, \quad \Omega_5^R(t) = \frac{\sum_{l=1}^{n} |\dot{e}_l^R(t)|^{\mu_2+1}}{\sum_{l=1}^{m} |\dot{e}_l^R(t)|}, \quad \Omega_6^R(t) = \frac{\sum_{l=1}^{n} |e_l^R(t)|^{\lambda_1+1}}{\sum_{l=1}^{m} |\dot{e}_l^R(t)|},$$

$$\Omega_7^R(t) = \frac{\sum_{l=1}^{n} |\dot{e}_l^R(t)|^{\lambda_2+1}}{\sum_{l=1}^{m} |\dot{e}_l^R(t)|}, \Omega_1^I(t) = \frac{\sum_{l=1}^{n} |e_l^I(t)|}{\sum_{l=1}^{m} |\dot{e}_l^I(t)|}, \Omega_2^I(t) = \frac{\sum_{l=1}^{n} |\dot{e}_l^I(t)|}{\sum_{l=1}^{m} |\dot{e}_l^I(t)|},$$

$$\Omega_3^I(t) = \frac{\sum_{l=1}^{n} |e_l^I(t-\tau_l(t))|}{\sum_{l=1}^{m} |\dot{e}_l^I(t)|}, \quad \Omega_4^I(t) = \frac{\sum_{l=1}^{n} |e_l^I(t)|^{\mu_3+1}}{\sum_{l=1}^{m} |\dot{e}_l^I(t)|}, \quad \Omega_5^I(t) = \frac{\sum_{l=1}^{n} |\dot{e}_l^I(t)|^{\mu_4+1}}{\sum_{l=1}^{m} |\dot{e}_l^I(t)|},$$

$$\Omega_6^I(t) = \frac{\sum_{l=1}^{n} |e_l^I(t)|^{\lambda_3+1}}{\sum_{l=1}^{m} |\dot{e}_l^I(t)|}, \Omega_7^I(t) = \frac{\sum_{l=1}^{n} |\dot{e}_l^I(t)|^{\lambda_4+1}}{\sum_{l=1}^{m} |\dot{e}_l^I(t)|},$$

$0 < \mu_p < 1, \lambda_p > 1, p \in \beth, \beth = \{1 \leq p \leq 4, p \in N^+\}$.
when $1 \leq l \leq m$,

$$u_l^R(t) = -\text{sgn}(\dot{e}_l^R(t))\Omega_1^R(t)\sum_{l=1}^{n}\theta_l^R|e_l^R(t)| - \text{sgn}(\dot{e}_l^R(t))\Omega_2^R(t)\sum_{l=1}^{n}\gamma_l^R|\dot{e}_l^R(t)|$$
$$- \xi^R\text{sgn}(\dot{e}_l^R(t))\Omega_3^R(t)\sum_{l=1}^{n}|e_l^R(t-\tau_l(t))| - \alpha_1^R\text{sgn}(\dot{e}_l^R(t))\Omega_4^R(t)$$
$$- \alpha_2^R\text{sgn}(\dot{e}_l^R(t))\Omega_5^R(t) - \beta_1^R\text{sgn}(\dot{e}_l^R(t))\Omega_6^R(t) - \beta_2^R\text{sgn}(\dot{e}_l^R(t))\Omega_7^R(t),$$
$$u_l^I(t) = -\text{sgn}(\dot{e}_l^I(t))\Omega_1^I(t)\sum_{l=1}^{n}\theta_l^I|e_l^I(t)| - \text{sgn}(\dot{e}_l^I(t))\Omega_2^I(t)\sum_{l=1}^{n}\gamma_l^I|\dot{e}_l^I(t)|$$
$$- \xi^I\text{sgn}(\dot{e}_l^I(t))\Omega_3^I(t)\sum_{l=1}^{n}|e_l^I(t-\tau_l(t))| - \alpha_3^I\text{sgn}(\dot{e}_l^I(t))\Omega_4^I(t)$$
$$- \alpha_4^I\text{sgn}(\dot{e}_l^I(t))\Omega_5^I(t) - \beta_3^I\text{sgn}(\dot{e}_l^I(t))\Omega_6^I(t) - \beta_4^I\text{sgn}(\dot{e}_l^I(t))\Omega_7^I(t), \quad (11.14)$$

otherwise

$$u_l^R(t) = 0,\ u_l^I(t) = 0, \quad (11.15)$$

where $\text{sgn}(x) = \begin{cases} 1\ , & x \geq 0, \\ -1\ , & x < 0, \end{cases}$ θ_l^R, γ_l^R, θ_l^I, γ_l^I, ξ^R, ξ^I are control gains, and α_1^R, α_2^R, β_1^R, β_2^R, α_3^I, α_4^I, β_3^I, β_4^I are positive constants.

Remark 11.3 As we know, since only controlling a small number of nodes of high-dimensional systems, pinning control method can reduce control cost. Moreover, in control theory, the finite/fixed-time control can bring that the states of systems tend to the equilibrium point in faster time. Especially, the settling time of fixed-time stability can be bounded by a boundary value independent on the initial conditions. Thus, it is very significant to study the pinning control in fixed-time interval and for CVINNs, no information about this issue has been reported. Our work will enrich this field.

Remark 11.4 In the proposed controller (11.14) and (11.15), only m nodes of neurons are controlled directly and $\sum_{l=1}^{m}|\dot{e}_l^R(t)|$ and $\sum_{l=1}^{m}|\dot{e}_l^I(t)|$ can't be zero so as to make the control approach meaning. The given third term in (11.14) and (11.15) is used to compensate the delay information. The last four terms are involved to obtain the fixed-time synchronization results.

Theorem 11.1 *Suppose that Assumption 11.1 holds, systems (11.1) and (11.3) can be synchronized in fixed time under the controller (11.14) and (11.15) if for* $l = 1, 2, ..., n$, *the following inequalities hold*

$$\theta_l^R \geq \Gamma_l^R, \theta_l^I \geq \Gamma_l^I, \gamma_l^R \geq \Delta_l^R, \gamma_l^I \geq \Delta_l^I, \xi^R \geq \Theta^R, \xi^I \geq \Theta^I, \quad (11.16)$$

where

$$\Gamma_l^R = \frac{1}{2}(1 + a_l) + \frac{1}{2}\sum_{q=1}^{n}(|c_{ql}^R|M_l^R + |c_{ql}^I|M_l^R),$$

$$\Gamma_l^I = \frac{1}{2}(1 + a_l) + \frac{1}{2}\sum_{q=1}^{n}(|c_{ql}^I|M_l^I + |c_{ql}^R|M_l^I),$$

$$\Delta_l^R = -b_l + \frac{1}{2}(1 + a_l) + \frac{1}{2}\sum_{q=1}^{n}(|c_{lq}^R|M_q^R + |c_{lq}^I|M_q^I + |d_{lq}^R|M_q^R + |d_{lq}^I|M_q^I),$$

$$\Delta_l^I = -b_l + \frac{1}{2}(1 + a_l) + \frac{1}{2}\sum_{q=1}^{n}(|c_{lq}^R|M_q^I + |c_{lq}^I|M_q^R + |d_{lq}^R|M_q^I + |d_{lq}^I|M_q^R),$$

$$\Theta_l^R = \frac{1}{2}\sum_{q=1}^{n}(|d_{ql}^R|M_l^R + |d_{ql}^I|M_l^R),\ \Theta^R = \max\{\Theta_1^R, \Theta_2^R, \Theta_3^R\},$$

$$\Theta_l^I = \frac{1}{2}\sum_{q=1}^{n}(|d_{ql}^I|M_l^I + |d_{ql}^R|M_l^I),\ \Theta^I = \max\{\Theta_1^I, \Theta_2^I, \Theta_3^I\}.$$

Moreover, the upper bound of the settling time is estimated by

$$T_{\max}^1 = \frac{1}{\bar{\rho}}\Big(2^{\frac{2\bar{\lambda}-\bar{\mu}\bar{\lambda}-1}{\bar{\lambda}-\bar{\mu}}}\Big)\Big(\frac{2}{1-\bar{\mu}} + \frac{2}{\bar{\lambda}-1}\Big), \tag{11.17}$$

where $\bar{\rho} = \min_{p\in\beth}\{\alpha 2^{\frac{\mu_p+1}{2}}, \beta 2^{\frac{\lambda_p+1}{2}} n^{\frac{1-\lambda_p}{2}}\}$, $\alpha = \min\{\alpha_1^R, \alpha_2^R, \alpha_3^I, \alpha_4^I\}$, $\beta = \min\{\beta_1^R, \beta_2^R, \beta_3^I, \beta_4^I\}$, $0 < \bar{\mu} = \max_{p\in\beth}\{\mu_p\} < 1$, $\bar{\lambda} = \min_{p\in\beth}\{\lambda_p\} > 1$.

Proof Consider the following Lyapunov function:

$$\begin{aligned} V(t) &\triangleq V_1(t) + V_2(t) + V_3(t) + V_4(t) \\ &= \frac{1}{2}\sum_{l=1}^{n}[e_l^R(t)]^2 + \frac{1}{2}\sum_{l=1}^{n}[\dot{e}_l^R(t)]^2 + \frac{1}{2}\sum_{l=1}^{n}[e_l^I(t)]^2 + \frac{1}{2}\sum_{l=1}^{n}[\dot{e}_l^I(t)]^2. \end{aligned} \tag{11.18}$$

Taking the derivative of $V(t)$, we can get

$$\dot{V}_1(t) = \sum_{l=1}^{n} e_l^R(t)\dot{e}_l^R(t) \le \frac{1}{2}\sum_{l=1}^{n}[e_l^R(t)]^2 + \frac{1}{2}\sum_{l=1}^{n}[\dot{e}_l^R(t)]^2, \tag{11.19}$$

$$\dot{V}_2(t) = \sum_{l=1}^{n} \dot{e}_l^R(t)\ddot{e}_l^R(t)$$

$$
\begin{aligned}
&= \sum_{l=1}^{n} \dot{e}_l^R(t)\Big(- a_l e_l^R(t) - b_l \dot{e}_l^R(t) + \sum_{q=1}^{n} c_{lq}^R \tilde{f}_q^R(e_q^R(t)) - \sum_{q=1}^{n} c_{lq}^I \tilde{f}_q^I(e_q^I(t)) \\
&\quad + \sum_{q=1}^{n} d_{lq}^R \tilde{f}_q^R(e_q^R(t-\tau_q(t)) - \sum_{q=1}^{n} d_{lq}^I \tilde{f}_q^I(e_q^I(t-\tau_q(t)) + u_l^R(t)\Big) \\
&\le \sum_{l=1}^{n} a_l |e_l^R(t)||\dot{e}_l^R(t)| - \sum_{l=1}^{n} b_l [\dot{e}_l^R(t)]^2 + \sum_{l=1}^{n}\sum_{q=1}^{n} |c_{lq}^R||\tilde{f}_q^R(e_q^R(t))||\dot{e}_l^R(t)| \\
&\quad + \sum_{l=1}^{n}\sum_{q=1}^{n} |c_{lq}^I||\tilde{f}_q^I(e_q^I(t))||\dot{e}_l^R(t)| + \sum_{l=1}^{n}\sum_{q=1}^{n} |d_{lq}^R||\tilde{f}_q^R(e_q^R(t-\tau_q(t))||\dot{e}_l^R(t)| \\
&\quad + \sum_{l=1}^{n}\sum_{q=1}^{n} |d_{lq}^I||\tilde{f}_q^I(e_q^I(t-\tau_q(t))||\dot{e}_l^R(t)| + \sum_{l=1}^{n} u_l^R(t)|\dot{e}_l^R(t)|. \qquad (11.20)
\end{aligned}
$$

According to Assumption 11.1, using the inequalities $\sum_{l=1}^{n}\sum_{k=1}^{n} \theta_l^R |e_l^R(t)||e_l^R(t)| \ge \sum_{l=1}^{n} \theta_l^R [e_l^R(t)]^2$ and $\sum_{l=1}^{n}\sum_{k=1}^{n} \gamma_l^R |\dot{e}_l^R(t)||\dot{e}_l^R(t)| \ge \sum_{l=1}^{n} \gamma_l^R [\dot{e}_l^R(t)]^2$, and combining (11.14) and (11.15) into (11.20), we have

$$
\begin{aligned}
\dot{V}_2(t) &\le \sum_{l=1}^{n} a_l |e_l^R(t)||\dot{e}_l^R(t)| - \sum_{l=1}^{n} b_l [\dot{e}_l^R(t)]^2 + \sum_{l=1}^{n}\sum_{q=1}^{n} |c_{lq}^R| M_q^R |e_q^R(t)||\dot{e}_l^R(t)| \\
&\quad + \sum_{l=1}^{n}\sum_{q=1}^{n} |c_{lq}^I| M_q^I |e_q^I(t)||\dot{e}_l^R(t)| + \sum_{l=1}^{n}\sum_{q=1}^{n} |d_{lq}^R| M_q^R |e_q^R(t-\tau_q(t))||\dot{e}_l^R(t)| \\
&\quad + \sum_{l=1}^{n}\sum_{q=1}^{n} |d_{lq}^I| M_q^I |e_q^I(t-\tau_q(t))||\dot{e}_l^R(t)| - \sum_{l=1}^{n} \theta_l^R [e_l^R(t)]^2 - \sum_{l=1}^{n} \gamma_l^R [\dot{e}_l{}^R(t)]^2 \\
&\quad - \xi^R \sum_{l=1}^{n} [e_l^R(t-\tau_l(t))]^2 - \alpha_1^R \sum_{l=1}^{n} |e_l^R(t)|^{\mu_1+1} - \alpha_2^R \sum_{l=1}^{n} |\dot{e}_l^R(t)|^{\mu_2+1} \\
&\quad - \beta_1^R \sum_{l=1}^{n} |e_l^R(t)|^{\lambda_1+1} - \beta_2^R \sum_{l=1}^{n} |\dot{e}_l^R(t)|^{\lambda_2+1} \\
&\le \frac{1}{2}\sum_{l=1}^{n} a_l ([e_l^R(t)]^2 + [\dot{e}_l^R(t)]^2) - \sum_{l=1}^{n} b_l [\dot{e}_l^R(t)]^2 + \frac{1}{2}\sum_{l=1}^{n}\sum_{q=1}^{n} |c_{lq}^R| M_q^R ([e_q^R(t)]^2 \\
&\quad + [\dot{e}_l^R(t)]^2) + \frac{1}{2}\sum_{l=1}^{n}\sum_{q=1}^{n} |c_{lq}^I| M_q^I ([e_q^I(t)]^2 + [\dot{e}_l^R(t)]^2) \\
&\quad + \frac{1}{2}\sum_{l=1}^{n}\sum_{q=1}^{n} |d_{lq}^R| M_q^R ([e_q^R(t-\tau_q(t))]^2 + [\dot{e}_l^R(t)]^2)
\end{aligned}
$$

$$
\begin{aligned}
&+\frac{1}{2}\sum_{l=1}^{n}\sum_{q=1}^{n}|d_{lq}^{I}|M_{q}^{I}([e_{q}^{I}(t-\tau_{q}(t))]^{2}+[\dot{e}_{l}^{R}(t)]^{2})-\sum_{l=1}^{n}\theta_{l}^{R}[e_{l}^{R}(t)]^{2}\\
&-\sum_{l=1}^{n}\gamma_{l}^{R}[\dot{e_{l}}^{R}(t)]^{2}-\xi^{R}\sum_{l=1}^{n}[e_{l}^{R}(t-\tau_{l}(t))]^{2}-\alpha_{1}^{R}\sum_{l=1}^{n}|e_{l}^{R}(t)|^{\mu_{1}+1}\\
&-\alpha_{2}^{R}\sum_{l=1}^{n}|\dot{e}_{l}^{R}(t)|^{\mu_{2}+1}-\beta_{1}^{R}\sum_{l=1}^{n}|e_{l}^{R}(t)|^{\lambda_{1}+1}-\beta_{2}^{R}\sum_{l=1}^{n}|\dot{e}_{l}^{R}(t)|^{\lambda_{2}+1}. \qquad (11.21)
\end{aligned}
$$

Similarly,

$$
\dot{V}_{3}(t)=\sum_{l=1}^{n}e_{l}^{I}(t)\dot{e}_{l}^{I}(t)\leq\frac{1}{2}\sum_{l=1}^{n}[e_{l}^{I}(t)]^{2}+\frac{1}{2}\sum_{l=1}^{n}[\dot{e}_{l}^{I}(t)]^{2}, \qquad (11.22)
$$

$$
\begin{aligned}
\dot{V}_{4}(t)\leq&\frac{1}{2}\sum_{l=1}^{n}a_{l}([e_{l}^{I}(t)]^{2}+[\dot{e}_{l}^{I}(t)]^{2})-\sum_{l=1}^{n}b_{l}[\dot{e}_{l}^{I}(t)]^{2}+\frac{1}{2}\sum_{l=1}^{n}\sum_{q=1}^{n}|c_{lq}^{R}|M_{q}^{I}([e_{q}^{I}(t)]^{2}\\
&+[\dot{e}_{l}^{I}(t)]^{2})+\frac{1}{2}\sum_{l=1}^{n}\sum_{q=1}^{n}|c_{lq}^{I}|M_{q}^{R}([e_{q}^{R}(t)]^{2}+[\dot{e}_{l}^{I}(t)]^{2})\\
&+\frac{1}{2}\sum_{l=1}^{n}\sum_{q=1}^{n}|d_{lq}^{R}|M_{q}^{I}([e_{q}^{I}(t-\tau_{q}(t))]^{2}+[\dot{e}_{l}^{I}(t)]^{2})\\
&+\frac{1}{2}\sum_{l=1}^{n}\sum_{q=1}^{n}|d_{lq}^{I}|M_{q}^{R}([e_{q}^{R}(t-\tau_{q}(t))]^{2}+[\dot{e}_{l}^{I}(t)]^{2})-\sum_{l=1}^{n}\theta_{l}^{I}[e_{l}^{I}(t)]^{2}\\
&-\sum_{l=1}^{n}\gamma_{l}^{I}[\dot{e}_{l}^{I}(t)]^{2}-\xi^{I}\sum_{l=1}^{n}[e_{l}^{I}(t-\tau_{l}(t))]^{2}-\alpha_{3}^{I}\sum_{l=1}^{n}|e_{l}^{I}(t)|^{\mu_{3}+1}\\
&-\alpha_{4}^{I}\sum_{l=1}^{n}|\dot{e}_{l}^{I}(t)|^{\mu_{4}+1}-\beta_{3}^{I}\sum_{l=1}^{n}|e_{l}^{I}(t)|^{\lambda_{3}+1}-\beta_{4}^{I}\sum_{l=1}^{n}|\dot{e}_{l}^{I}(t)|^{\lambda_{4}+1}. \qquad (11.23)
\end{aligned}
$$

Combining (11.19), (11.21), (11.22) and (11.23), we have

$$
\begin{aligned}
\dot{V}(t)\leq&\sum_{l=1}^{n}\Big\{-\theta_{l}^{R}+\frac{1}{2}(1+a_{l})+\frac{1}{2}\sum_{q=1}^{n}(|c_{ql}^{R}|M_{l}^{R}+|c_{ql}^{I}|M_{l}^{R})\Big\}[e_{l}^{R}(t)]^{2}\\
&+\sum_{l=1}^{n}\Big\{-\theta_{l}^{I}+\frac{1}{2}(1+a_{l})+\frac{1}{2}\sum_{q=1}^{n}(|c_{ql}^{I}|M_{l}^{I}+|c_{ql}^{R}|M_{l}^{I})\Big\}[e_{l}^{I}(t)]^{2}\\
&+\sum_{l=1}^{n}\Big\{-\gamma_{l}^{R}-b_{l}+\frac{1}{2}(1+a_{l})+\frac{1}{2}\sum_{q=1}^{n}(|c_{lq}^{R}|M_{q}^{R}+|c_{lq}^{I}|M_{q}^{I}\\
&+|d_{lq}^{R}|M_{q}^{R}+|d_{lq}^{I}|M_{q}^{I})\Big\}[\dot{e}_{l}^{R}(t)]^{2}+\sum_{l=1}^{n}\Big\{-\gamma_{l}^{I}-b_{l}+\frac{1}{2}(1+a_{l})
\end{aligned}
$$

$$+\frac{1}{2}\sum_{q=1}^{n}(|c_{lq}^{R}|M_q^I+|c_{lq}^{I}|M_q^R+|d_{lq}^{R}|M_q^I+|d_{lq}^{I}|M_q^R)\Big\}[\dot{e}_l^I(t)]^2$$
$$+\sum_{l=1}^{n}\Big\{-\xi^R+\frac{1}{2}\sum_{q=1}^{n}(|d_{ql}^{R}|M_l^R+|d_{ql}^{I}|M_l^R)\Big\}[e_l^R(t-\tau_l(t))]^2$$
$$+\sum_{l=1}^{n}\Big\{-\xi^I+\frac{1}{2}\sum_{q=1}^{n}(|d_{ql}^{I}|M_l^I+|d_{ql}^{R}|M_l^I)\Big\}[e_l^I(t-\tau_l(t))]^2$$
$$-\alpha_1^R\sum_{l=1}^{n}|e_l^R(t)|^{\mu_1+1}-\alpha_2^R\sum_{l=1}^{n}|\dot{e}_l^R(t)|^{\mu_2+1}-\beta_1^R\sum_{l=1}^{n}|e_l^R(t)|^{\lambda_1+1}$$
$$-\beta_2^R\sum_{l=1}^{n}|\dot{e}_l^R(t)|^{\lambda_2+1}-\alpha_3^I\sum_{l=1}^{n}|e_l^I(t)|^{\mu_3+1}-\alpha_4^I\sum_{l=1}^{n}|\dot{e}_l^I(t)|^{\mu_4+1}$$
$$-\beta_3^I\sum_{l=1}^{n}|e_l^I(t)|^{\lambda_3+1}-\beta_4^I\sum_{l=1}^{n}|\dot{e}_l^I(t)|^{\lambda_4+1}. \tag{11.24}$$

It follows from (11.24) that:

$$\dot{V}(t)\le\sum_{l=1}^{n}(-\theta_l^R+\Gamma_l^R)[e_l^R(t)]^2+(-\theta_l^I+\Gamma_l^I)[e_l^I(t)]^2+\sum_{l=1}^{n}(-\gamma_l^R+\Delta_l^R)[\dot{e}_l^R(t)]^2$$
$$+\sum_{l=1}^{n}(-\gamma_l^I+\Delta_l^I)[\dot{e}_l^I(t)]^2+\sum_{l=1}^{n}(-\xi^R+\Theta_l^R)[e_l^R(t-\tau_l(t))]^2$$
$$+\sum_{l=1}^{n}(-\xi^I+\Theta_l^I)[e_l^I(t-\tau_l(t))]^2-\alpha_1^R\sum_{l=1}^{n}|e_l^R(t)|^{\mu_1+1}-\alpha_2^R\sum_{l=1}^{n}|\dot{e}_l^R(t)|^{\mu_2+1}$$
$$-\beta_1^R\sum_{l=1}^{n}|e_l^R(t)|^{\lambda_1+1}-\beta_2^R\sum_{l=1}^{n}|\dot{e}_l^R(t)|^{\lambda_2+1}-\alpha_3^I\sum_{l=1}^{n}|e_l^I(t)|^{\mu_3+1}$$
$$-\alpha_4^I\sum_{l=1}^{n}|\dot{e}_l^I(t)|^{\mu_4+1}-\beta_3^I\sum_{l=1}^{n}|e_l^I(t)|^{\lambda_3+1}-\beta_4^I\sum_{l=1}^{n}|\dot{e}_l^I(t)|^{\lambda_4+1}. \tag{11.25}$$

From (11.16), (11.25) can be rewritten as

$$\dot{V}(t)\le-\alpha_1^R\sum_{l=1}^{n}|e_l^R(t)|^{\mu_1+1}-\alpha_2^R\sum_{l=1}^{n}|\dot{e}_l^R(t)|^{\mu_2+1}-\beta_1^R\sum_{l=1}^{n}|e_l^R(t)|^{\lambda_1+1}$$
$$-\beta_2^R\sum_{l=1}^{n}|\dot{e}_l^R(t)|^{\lambda_2+1}-\alpha_3^I\sum_{l=1}^{n}|e_l^I(t)|^{\mu_3+1}-\alpha_4^I\sum_{l=1}^{n}|\dot{e}_l^I(t)|^{\mu_4+1}$$

$$
\begin{aligned}
&- \beta_3^I \sum_{l=1}^{n} |e_l^I(t)|^{\lambda_3+1} - \beta_4^I \sum_{l=1}^{n} |\dot{e}_l^I(t)|^{\lambda_4+1} \\
&\leq -\alpha\Big(\sum_{l=1}^{n} |e_l^R(t)|^{\mu_1+1} + \sum_{l=1}^{n} |\dot{e}_l^R(t)|^{\mu_2+1} + \sum_{l=1}^{n} |e_l^I(t)|^{\mu_3+1} \\
&+ \sum_{l=1}^{n} |\dot{e}_l^I(t)|^{\mu_4+1}\Big) - \beta\Big(\sum_{l=1}^{n} |e_l^R(t)|^{\lambda_1+1} + \sum_{l=1}^{n} |\dot{e}_l^R(t)|^{\lambda_2+1} \\
&+ \sum_{l=1}^{n} |e_l^I(t)|^{\lambda_3+1} + \sum_{l=1}^{n} |\dot{e}_l^I(t)|^{\lambda_4+1}\Big),
\end{aligned}
\tag{11.26}
$$

where $\alpha = \min\{\alpha_1^R, \alpha_2^R, \alpha_3^I, \alpha_4^I\}$, $\beta = \min\{\beta_1^R, \beta_2^R, \beta_3^I, \beta_4^I\}$.

According to Lemma 11.2, let $\bar{\rho} = \min_{p\in\beth}\{\alpha 2^{\frac{\mu_p+1}{2}}, \beta 2^{\frac{\lambda_p+1}{2}} n^{\frac{1-\lambda_p}{2}}\}$, it can be obtained

$$
\begin{aligned}
\dot{V}(t) \leq -\bar{\rho}\Big[&\Big(\frac{1}{2}\sum_{l=1}^{n}[e_l^R(t)]^2\Big)^{\frac{\mu_1+1}{2}} + \Big(\frac{1}{2}\sum_{l=1}^{n}[\dot{e}_l^R(t)]^2\Big)^{\frac{\mu_2+1}{2}} + \Big(\frac{1}{2}\sum_{l=1}^{n}[e_l^I(t)]^2\Big)^{\frac{\mu_3+1}{2}} \\
&+ \Big(\frac{1}{2}\sum_{l=1}^{n}[\dot{e}_l^I(t)]^2\Big)^{\frac{\mu_4+1}{2}} + \Big(\frac{1}{2}\sum_{l=1}^{n}[e_l^R(t)]^2\Big)^{\frac{\lambda_1+1}{2}} + \Big(\frac{1}{2}\sum_{l=1}^{n}[\dot{e}_l^R(t)]^2\Big)^{\frac{\lambda_2+1}{2}} \\
&+ \Big(\frac{1}{2}\sum_{l=1}^{n}[e_l^I(t)]^2\Big)^{\frac{\lambda_3+1}{2}} + \Big(\frac{1}{2}\sum_{l=1}^{n}[\dot{e}_l^I(t)]^2\Big)^{\frac{\lambda_4+1}{2}}\Big] \\
= -\bar{\rho}\Big(&V_1^{\frac{\mu_1+1}{2}}(t) + V_2^{\frac{\mu_2+1}{2}}(t) + V_3^{\frac{\mu_3+1}{2}}(t) + V_4^{\frac{\mu_4+1}{2}}(t) \\
&+ V_1^{\frac{\lambda_1+1}{2}}(t) + V_2^{\frac{\lambda_2+1}{2}}(t) + V_3^{\frac{\lambda_3+1}{2}}(t) + V_4^{\frac{\lambda_4+1}{2}}(t)\Big).
\end{aligned}
\tag{11.27}
$$

Now, let $0 < \bar{\mu} = \max_{p\in\beth}\{\mu_p\} < 1$, $\bar{\lambda} = \min_{p\in\beth}\{\lambda_p\} > 1$, obviously, $\lambda_p \geq \bar{\lambda} > 1 > \bar{\mu} \geq \mu_p$. Then, we get

$$
V_p^{\frac{\mu_p+1}{2}} + V_p^{\frac{\lambda_p+1}{2}} \geq \frac{1}{2}(V_p^{\frac{\bar{\mu}+1}{2}} + V_p^{\frac{\bar{\lambda}+1}{2}}),\ p \in \beth.
\tag{11.28}
$$

Thus,

$$
\dot{V}(t) \leq -\frac{1}{2}\bar{\rho}V^{\frac{\bar{\mu}+1}{2}}(t) - 2^{\bar{\lambda}}\bar{\rho}V^{\frac{\bar{\lambda}+1}{2}}(t).
\tag{11.29}
$$

According to Lemma 11.1, under the controller (11.14) and (11.15), system (11.5) is globally fixed-time stable, in other words, the systems (11.1) and (11.3) can realize the synchronization in fixed-time interval. Moreover, through simple calculation, the upper bound of the settling time $T_{\max}^{1} = \frac{1}{\tilde{\rho}}\left(2^{\frac{2\tilde{\lambda}-\tilde{\mu}\tilde{\lambda}-1}{\tilde{\lambda}-\tilde{\mu}}}\right)\left(\frac{2}{1-\tilde{\mu}} + \frac{2}{\tilde{\lambda}-1}\right)$. □

Remark 11.5 Because the original second-order systems (11.1) and (11.3) are regarded as the main object without reducing order, to carry out our work, a novel Lyapunov function including the state derivatives are proposed, which is different from the existing ones in applying the reduced order method. This is a key feature in our work.

Remark 11.6 In the controller (11.14) and (11.15), when $\mu_1 = \mu_2 = \mu_3 = \mu_4$ and $\beta_1^R = \beta_2^R = \beta_3^I = \beta_4^I = 0$, by similar derivation process and Lemma 11.1, system (11.5) is finite-time stable, that is, systems (11.1) and (11.3) can achieve the finite-time synchronization. In this case, the related settling time function can be also given. Here, to save space, we will not provide the details.

11.3.2 Fixed-Time Adaptive Synchronization Criterion

To adjust automatically the control gains, we design the pinning adaptive controller as follows.

When $1 \leq l \leq m$,

$$
\begin{aligned}
u_l^R(t) = & -\mathrm{sgn}(\dot{e}_l^R(t))\frac{\sum_{l=1}^{n}\theta_l^R(t)[e_l^R(t)]^2}{\sum_{l=1}^{m}|\dot{e}_l^R(t)|} - \mathrm{sgn}(\dot{e}_l^R(t))\frac{\sum_{l=1}^{n}\gamma_l^R(t)[\dot{e}_l^R(t)]^2}{\sum_{l=1}^{m}|\dot{e}_l^R(t)|} \\
& - \xi^R\mathrm{sgn}(\dot{e}_l^R(t))\Omega_3^R(t)\sum_{l=1}^{n}|e_l^R(t-\tau_l(t))| - \alpha_1^R\mathrm{sgn}(\dot{e}_l^R(t))\Omega_4^R(t) \\
& - \alpha_2^R\mathrm{sgn}(\dot{e}_l^R(t))\Omega_5^R(t) - \beta_1^R\mathrm{sgn}(\dot{e}_l^R(t))\Omega_6^R(t) - \beta_2^R\mathrm{sgn}(\dot{e}_l^R(t))\Omega_7^R(t), \\
u_l^I(t) = & -\mathrm{sgn}(\dot{e}_l^I(t))\frac{\sum_{l=1}^{n}\theta_l^I(t)[e_l^I(t)]^2}{\sum_{l=1}^{m}|\dot{e}_l^I(t)|} - \mathrm{sgn}(\dot{e}_l^I(t))\frac{\sum_{l=1}^{n}\gamma_l^I(t)[\dot{e}_l^I(t)]^2}{\sum_{l=1}^{m}|\dot{e}_l^I(t)|} \\
& - \xi^I\mathrm{sgn}(\dot{e}_l^I(t))\Omega_3^I(t)\sum_{l=1}^{n}|e_l^I(t-\tau_l(t))| - \alpha_3^I\mathrm{sgn}(\dot{e}_l^I(t))\Omega_4^I(t) \\
& - \alpha_4^I\mathrm{sgn}(\dot{e}_l^I(t))\Omega_5^I(t) - \beta_3^I\mathrm{sgn}(\dot{e}_l^I(t))\Omega_6^I(t) - \beta_4^I\mathrm{sgn}(\dot{e}_l^I(t))\Omega_7^I(t),
\end{aligned}
\tag{11.30}
$$

otherwise

$$u_l^R(t) = 0, \ u_l^I(t) = 0, \tag{11.31}$$

and

$$\begin{aligned}
\dot{\theta}_l^R(t) &= [e_l^R(t)]^2 - \varepsilon_1^R[\theta_l^R(t) - \hat{\theta}_l^R]^{\mu_1} - \sigma_1^R[\theta_l^R(t) - \hat{\theta}_l^R]^{\lambda_1}, \\
\dot{\gamma}_l^R(t) &= [\dot{e}_l^R(t)]^2 - \varepsilon_2^R[\gamma_l^R(t) - \hat{\gamma}_l^R]^{\mu_2} - \sigma_2^R[\gamma_l^R(t) - \hat{\gamma}_l^R]^{\lambda_2}, \\
\dot{\theta}_l^I(t) &= [e_l^I(t)]^2 - \varepsilon_3^I[\theta_l^I(t) - \hat{\theta}_l^I]^{\mu_3} - \sigma_3^I[\theta_l^I(t) - \hat{\theta}_l^I]^{\lambda_3}, \\
\dot{\gamma}_l^I(t) &= [\dot{e}_l^I(t)]^2 - \varepsilon_4^I[\gamma_l^I(t) - \hat{\gamma}_l^I]^{\mu_4} - \sigma_4^I[\gamma_l^I(t) - \hat{\gamma}_l^I]^{\lambda_4},
\end{aligned}$$

where $\theta_l^R(t)$, $\gamma_l^R(t)$, $\theta_l^I(t)$, and $\gamma_l^I(t)$ are control gains; ε_1^R, ε_2^R, σ_1^R, σ_2^R, ε_3^I, ε_4^I, σ_3^I, and σ_4^I are positive constants and the others are the same as those in (11.14).

Theorem 11.2 *Suppose that Assumption 11.1 holds, systems (11.1) and (11.2) can be synchronized in fixed time under the pinning adaptive controller (11.30) and (11.31) if for $l = 1, 2, ..., n$, the following inequalities hold,*

$$\hat{\theta}_l^R \geq \Gamma_l^R, \hat{\theta}_l^I \geq \Gamma_l^I, \hat{\gamma}_l^R \geq \Delta_l^R, \hat{\gamma}_l^I \geq \Delta_l^I, \hat{\xi}^R \geq \Theta^R, \hat{\xi}^I \geq \Theta^I. \tag{11.32}$$

Moreover, the upper bound of the settling time is estimated by

$$T_{\max}^2 = \frac{1}{\hat{\rho}}\Big(2^{\frac{5\bar{\lambda}-3\bar{\mu}\bar{\lambda}+\bar{\mu}-3}{2(\bar{\lambda}-\bar{\mu})}}\Big)\Big(\frac{2}{1-\bar{\mu}} + \frac{2}{\bar{\lambda}-1}\Big), \tag{11.33}$$

where $\hat{\rho} = \min_{p\in\beth}\{\alpha 2^{\frac{\mu_p+1}{2}}, \beta 2^{\frac{\lambda_p+1}{2}} n^{\frac{1-\lambda_p}{2}}, \varepsilon 2^{\frac{\mu_p+1}{2}}, \sigma 2^{\frac{\lambda_p+1}{2}} n^{\frac{1-\lambda_p}{2}}\}$,
$\alpha = \min\{\alpha_1^R, \alpha_2^R, \alpha_3^I, \alpha_4^I\}$, $\beta = \min\{\beta_1^R, \beta_2^R, \beta_3^I, \beta_4^I\}$, $\varepsilon = \min\{\varepsilon_1^R, \varepsilon_2^R, \varepsilon_3^I, \varepsilon_4^I\}$, $\sigma = \min\{\sigma_1^R, \sigma_2^R, \sigma_3^I, \sigma_4^I\}$, $0 < \bar{\mu} = \max_{p\in\beth}\{\mu_p\} < 1$, $\bar{\lambda} = \min_{p\in\beth}\{\lambda_p\} > 1$.

Proof Consider the candidate Lyapunov function as follows:

$$\begin{aligned}
\hat{V}(t) &\triangleq V(t) + \bar{V}(t), \\
\bar{V}(t) &= V_5(t) + V_6(t) + V_7(t) + V_8(t) \\
&= \frac{1}{2}\sum_{l=1}^{n}[\theta_l^R(t) - \hat{\theta}_l^R]^2 + \frac{1}{2}\sum_{l=1}^{n}[\gamma_l^R(t) - \hat{\gamma}_l^R]^2 \\
&\quad + \frac{1}{2}\sum_{l=1}^{n}[\theta_l^I(t) - \hat{\theta}_l^I]^2 + \frac{1}{2}\sum_{l=1}^{n}[\gamma_l^I(t) - \hat{\gamma}_l^I]^2.
\end{aligned} \tag{11.34}$$

Calculate the time derivative of $\hat{V}(t)$ as follows, firstly, similar to (11.25),

$$
\begin{aligned}
\dot{V}(t) \le & \sum_{l=1}^{n}(-\theta_l^R(t) + \Gamma_l^R)[e_l^R(t)]^2 + \sum_{l=1}^{n}(-\theta_l^I(t) + \Gamma_l^I)[e_l^I(t)]^2 \\
& + \sum_{l=1}^{n}(-\gamma_l^R(t) + \Delta_l^R)[\dot{e}_l^R(t)]^2 + \sum_{l=1}^{n}(-\gamma_l^I(t) + \Delta_l^I)[\dot{e}_l^I(t)]^2 \\
& + \sum_{l=1}^{n}(-\xi^R + \Theta_l^R)[e_l^R(t-\tau_l(t))]^2 + \sum_{l=1}^{n}(-\xi^I + \Theta_l^I)[e_l^I(t-\tau_l(t))]^2 \\
& - \alpha_1^R \sum_{l=1}^{n} |e_l^R(t)|^{\mu_1+1} - \alpha_2^R \sum_{l=1}^{n} |\dot{e}_l^R(t)|^{\mu_2+1} - \beta_1^R \sum_{l=1}^{n} |e_l^R(t)|^{\lambda_1+1} \\
& - \beta_2^R \sum_{l=1}^{n} |\dot{e}_l^R(t)|^{\lambda_2+1} - \alpha_3^I \sum_{l=1}^{n} |e_l^I(t)|^{\mu_3+1} - \alpha_4^I \sum_{l=1}^{n} |\dot{e}_l^I(t)|^{\mu_4+1} \\
& - \beta_3^I \sum_{l=1}^{n} |e_l^I(t)|^{\lambda_3+1} - \beta_4^I \sum_{l=1}^{n} |\dot{e}_l^I(t)|^{\lambda_4+1},
\end{aligned} \tag{11.35}
$$

then,

$$
\begin{aligned}
\dot{V}_5(t) = & \sum_{l=1}^{n}[\theta_l^R(t) - \hat{\theta}_l^R]\dot{\theta}_l^R(t) = \sum_{l=1}^{n}[\theta_l^R(t) - \hat{\theta}_l^R][e_l^R(t)]^2 \\
& - \varepsilon_1^R \sum_{l=1}^{n}[\theta_l^R(t) - \hat{\theta}_l^R]^{\mu_1+1} - \sigma_1^R \sum_{l=1}^{n}[\theta_l^R(t) - \hat{\theta}_l^R]^{\lambda_1+1}, \\
\dot{V}_6(t) = & \sum_{l=1}^{n}[\gamma_l^R(t) - \hat{\gamma}_l^R][\dot{e}_l^R(t)]^2 \\
& - \varepsilon_2^R \sum_{l=1}^{n}[\gamma_l^R(t) - \hat{\gamma}_l^R]^{\mu_2+1} - \sigma_2^R \sum_{l=1}^{n}[\gamma_l^R(t) - \hat{\gamma}_l^R]^{\lambda_2+1}, \\
\dot{V}_7(t) = & \sum_{l=1}^{n}[\theta_l^I(t) - \hat{\theta}_l^I][e_l^I(t)]^2 \\
& - \varepsilon_3^I \sum_{l=1}^{n}[\theta_l^I(t) - \hat{\theta}_l^I]^{\mu_3+1} - \sigma_3^I \sum_{l=1}^{n}[\theta_l^I(t) - \hat{\theta}_l^I]^{\lambda_3+1}, \\
\dot{V}_8(t) = & \sum_{l=1}^{n}[\gamma_l^I(t) - \hat{\gamma}_l^I][\dot{e}_l^I(t)]^2 \\
& - \varepsilon_4^I \sum_{l=1}^{n}[\gamma_l^I(t) - \hat{\gamma}_l^I]^{\mu_4+1} - \sigma_4^I \sum_{l=1}^{n}[\gamma_l^I(t) - \hat{\gamma}_l^I]^{\lambda_4+1}.
\end{aligned} \tag{11.36}
$$

Combining (11.35) with (11.36), one has

$$
\begin{aligned}
\dot{V}(t) \leq & \sum_{l=1}^{n}(-\hat{\theta}_l^R + \Gamma_l^R)[e_l^R(t)]^2 + \sum_{l=1}^{n}(-\hat{\theta}_l^I + \Gamma_l^I)[e_l^I(t)]^2 \\
& + \sum_{l=1}^{n}(-\hat{\gamma}_l^R + \Delta_l^R)[\dot{e}_l^R(t)]^2 + \sum_{l=1}^{n}(-\hat{\gamma}_l^I + \Delta_l^R)[\dot{e}_l^I(t)]^2 \\
& + \sum_{l=1}^{n}(-\hat{\xi}^R + \Theta_l^R)[e_l^R(t - \tau_l(t))]^2 + \sum_{l=1}^{n}(-\hat{\xi}^I + \Theta_l^I)[e_l^I(t - \tau_l(t))]^2 \\
& - \varepsilon_1^R \sum_{l=1}^{n}[\theta_l^R(t) - \hat{\theta}_l^R]^{\mu_1+1} - \sigma_1^R \sum_{l=1}^{n}[\theta_l^R(t) - \hat{\theta}_l^R]^{\lambda_1+1} \\
& - \varepsilon_2^R \sum_{l=1}^{n}[\gamma_l^R(t) - \hat{\gamma}_l^R]^{\mu_2+1} - \sigma_2^R \sum_{l=1}^{n}[\gamma_l^R(t) - \hat{\gamma}_l^R]^{\lambda_2+1} \\
& - \varepsilon_3^I \sum_{l=1}^{n}[\theta_l^I(t) - \hat{\theta}_l^I]^{\mu_3+1} - \sigma_3^I \sum_{l=1}^{n}[\theta_l^I(t) - \hat{\theta}_l^I]^{\lambda_3+1} \\
& - \varepsilon_4^I \sum_{l=1}^{n}[\gamma_l^I(t) - \hat{\gamma}_l^I]^{\mu_4+1} - \sigma_4^I \sum_{l=1}^{n}[\gamma_l^I(t) - \hat{\gamma}_l^I]^{\lambda_4+1} \\
& - \alpha_1^R \sum_{l=1}^{n}|e_l^R(t)|^{\mu_1+1} - \alpha_2^R \sum_{l=1}^{n}|\dot{e}_l^R(t)|^{\mu_2+1} - \beta_1^R \sum_{l=1}^{n}|e_l^R(t)|^{\lambda_1+1} \\
& - \beta_2^R \sum_{l=1}^{n}|\dot{e}_l^R(t)|^{\lambda_2+1} - \alpha_3^I \sum_{l=1}^{n}|e_l^I(t)|^{\mu_3+1} - \alpha_4^I \sum_{l=1}^{n}|\dot{e}_l^I(t)|^{\mu_4+1} \\
& -\beta_3^I \sum_{l=1}^{n}|e_l^I(t)|^{\lambda_3+1} - \beta_4^I \sum_{l=1}^{n}|\dot{e}_l^I(t)|^{\lambda_4+1}. \qquad (11.37)
\end{aligned}
$$

From (11.32), (11.37) can be rewritten as

$$
\begin{aligned}
\dot{V}(t) \leq & -\varepsilon_1^R \sum_{l=1}^{n}[\theta_l^R(t) - \hat{\theta}_l^R]^{\mu_1+1} - \sigma_1^R \sum_{l=1}^{n}[\theta_l^R(t) - \hat{\theta}_l^R]^{\lambda_1+1} \\
& - \varepsilon_2^R \sum_{l=1}^{n}[\gamma_l^R(t) - \hat{\gamma}_l^R]^{\mu_2+1} - \sigma_2^R \sum_{l=1}^{n}[\gamma_l^R(t) - \hat{\gamma}_l^R]^{\lambda_2+1} \\
& - \varepsilon_3^I \sum_{l=1}^{n}[\theta_l^I(t) - \hat{\theta}_l^I]^{\mu_3+1} - \sigma_3^I \sum_{l=1}^{n}[\theta_l^I(t) - \hat{\theta}_l^I]^{\lambda_3+1} \\
& - \varepsilon_4^I \sum_{l=1}^{n}[\gamma_l^I(t) - \hat{\gamma}_l^I]^{\mu_4+1} - \sigma_4^I \sum_{l=1}^{n}[\gamma_l^I(t) - \hat{\gamma}_l^I]^{\lambda_4+1} \\
& - \alpha_1^R \sum_{l=1}^{n}|e_l^R(t)|^{\mu_1+1} - \alpha_2^R \sum_{l=1}^{n}|\dot{e}_l^R(t)|^{\mu_2+1} - \beta_1^R \sum_{l=1}^{n}|e_l^R(t)|^{\lambda_1+1}
\end{aligned}
$$

$$
\begin{aligned}
&- \beta_2^R \sum_{l=1}^{n} |\dot{e}_l^R(t)|^{\lambda_2+1} - \alpha_3^I \sum_{l=1}^{n} |e_l^I(t)|^{\mu_3+1} - \alpha_4^I \sum_{l=1}^{n} |\dot{e}_l^I(t)|^{\mu_4+1} \\
&- \beta_3^I \sum_{l=1}^{n} |e_l^I(t)|^{\lambda_3+1} - \beta_4^I \sum_{l=1}^{n} |\dot{e}_l^I(t)|^{\lambda_4+1}. \qquad (11.38)
\end{aligned}
$$

Similar to (11.27)–(11.29) in Theorem 11.1, we can obtain

$$
\dot{\hat{V}}(t) \leq -\frac{1}{2}\hat{\rho} V^{\frac{\bar{\mu}+1}{2}}(t) - 2^{\frac{1-3\bar{\lambda}}{2}} \hat{\rho} V^{\frac{\bar{\lambda}+1}{2}}(t), \qquad (11.39)
$$

where $\hat{\rho} = \min_{p \in \beth}\{\alpha 2^{\frac{\mu_p+1}{2}}, \beta 2^{\frac{\lambda_p+1}{2}} n^{\frac{1-\lambda_p}{2}}, \varepsilon 2^{\frac{\mu_p+1}{2}}, \sigma 2^{\frac{\lambda_p+1}{2}} n^{\frac{1-\lambda_p}{2}}\}$,
$\alpha = \min\{\alpha_1^R, \alpha_2^R, \alpha_3^I, \alpha_4^I\}$, $\beta = \min\{\beta_1^R, \beta_2^R, \beta_3^I, \beta_4^I\}$, $\varepsilon = \min\{\varepsilon_1^R, \varepsilon_2^R, \varepsilon_3^I, \varepsilon_4^I\}$,
$\sigma = \min\{\sigma_1^R, \sigma_2^R, \sigma_3^I, \sigma_4^I\}$, $\bar{\mu} = \max_{p \in \beth}\{\mu_p\}$, $\bar{\lambda} = \min_{p \in \beth}\{\lambda_p\}$.

According to Lemma 11.1, under the pinning adaptive controller (11.30) and (11.31), system (11.5) is globally fixed-time stable, that is to say, the systems (11.1) and (11.3) can realize the synchronization in fixed-time interval. The upper bound of settling time $T_{\max}^2 = \frac{1}{\hat{\rho}}\left(2^{\frac{5\bar{\lambda}-3\bar{\mu}\bar{\lambda}+\bar{\mu}-3}{2(\bar{\lambda}-\bar{\mu})}}\right)\left(\frac{2}{1-\bar{\mu}} + \frac{2}{\bar{\lambda}-1}\right)$ can be given, as stated in (11.33). □

Remark 11.7 Sufficient conditions for the fixed-time synchronization and adaptive synchronization of CVINNs with time-varying delays via pinning control are established in Theorems 11.1 and 11.2, respectively. Through constructing Lyapunov function and fully utilizing some inequalities techniques, the expected results are derived. To our knowledge, it is the first time to discuss the fixed-time synchronization and adaptive synchronization of delayed CVINNs under the framework of pinning control scheme.

11.4 Illustrative Example

In this section, we will give a numerical example to verify the effectiveness of the above theoretical results.

Example 11.1 Consider the following three-dimension drive-response CVINNs:

$$
\ddot{v}_l(t) = -a_l v_l(t) - b_l \dot{v}_l(t) + \sum_{q=1}^{3} c_{lq} f_q(v_q(t)) + \sum_{q=1}^{3} d_{lq} f_q(v_q(t - \tau_q(t))), \qquad (11.40)
$$

$$
\ddot{z}_l(t) = -a_l v_l(t) - b_l \dot{z}_l(t) + \sum_{q=1}^{3} c_{lq} f_q(z_q(t)) + \sum_{q=1}^{3} d_{lq} f_q(z_q(t - \tau_q(t))) + u_l(t) \qquad (11.41)
$$

with

$$
\begin{aligned}
&a_l = 5.5,\ b_l = 0.5, \quad \tau_l(t) = 0.2\sin^2(t), \quad (l = 1, 2, 3),\\
&c_{11} = -1.76 + 0.16i,\ c_{12} = 1.12 + 1.07i, \quad c_{13} = -0.56 - 0.16i,\\
&c_{21} = 2.11 + 0.02i, \quad c_{22} = -0.54 + 0.72i,\ c_{23} = 0.34 - 1.45i,\\
&c_{31} = -1.38 + 0.09i,\ c_{32} = 1.55 - 0.04i, \quad c_{33} = -0.11 + 1.09i,\\
&d_{11} = -1.92 + 0.13i,\ d_{12} = 0.23 - 0.28i, \quad d_{13} = -0.87 - 1.99i,\\
&d_{21} = -1.21 + 0.04i,\ d_{22} = -0.65 + 0.15i,\ d_{23} = 1.75 + 1.84i,\\
&d_{31} = -0.99 + 0.08i,\ d_{32} = 1.87 - 0.11i, \quad d_{33} = 0.03 - 2.41i.
\end{aligned}
$$

In addition, we select the activation functions as $f_q(v_q(t)) = \tanh(v_q(t)) + i\tanh(v_q(t))$. It is easy to verify that Assumption 11.1 is satisfied and $M_1^R = M_2^R = M_3^R = M_1^I = M_2^I = M_3^I = 1$. Choose the initial values as $\varphi_1(s) = -1.85 + 1.85i$, $\varphi_2(s) = -1.02 - 1.56i$, $\varphi_3(s) = 2.98 - 0.26i$, $\psi_1(s) = -1.64 + 1.07i$, $\psi_2(s) = -1.31 - 1.03i$, $\psi_3(s) = 3.22 + 1.18$, $\tilde{\varphi}_1(s) = -1.25 + 1.34i$, $\tilde{\varphi}_2(s) = 1.36 - 0.82i$, $\tilde{\varphi}_3(s) = -0.37 - 0.48i$,$\tilde{\psi}_1(s) = -1.38 - 0.68i$, $\tilde{\psi}_2(s) = -0.63 + 1.53i$, $\tilde{\psi}_3(s) = 1.03 - 1.36i$, $\forall s \in [-0.2, 0]$, their dynamic trajectories are shown in Fig. 11.1. From this simulation figure, we know that systems (11.40) and (11.41) are unsynchronized without controller. Next, we will test the proposed results in Theorems 11.1 and 11.2 on the basis of this example.

Case 1: In Theorem 11.1, by a simple calculation, we have $\Gamma_1^R = \Gamma_1^I = 6.01$, $\Gamma_2^R = \Gamma_2^I = 5.77$, $\Gamma_3^R = \Gamma_3^I = 5.11$, $\Delta_1^R = \Delta_1^I = 7.70$, $\Delta_2^R = \Delta_2^I = 6.92$, $\Delta_3^R = \Delta_3^I = 9.05$, $\Theta^R = 4.45$, and $\Theta^I = 2.82$. Then, choosing $\theta_1^R = \theta_2^R = \theta_3^R = 48$, $\gamma_1^R = \gamma_2^R = \gamma_3^R = 9.5$, $\xi^R = 25$, $\theta_1^I = \theta_2^I = \theta_3^I = 36$, $\gamma_1^I = \gamma_2^I = \gamma_3^I = 9.5$, $\xi^I = 22$, $\alpha_1^R = 25$, $\alpha_2^R = 4$, $\beta_1^R = 20$, $\beta_2^R = 4$, $\alpha_3^I = 20$, $\alpha_4^I = 5$, $\beta_3^I = 20$, $\beta_4^I = 1$, $\mu_1 = 0.9$, $\mu_2 = \mu_3 = \mu_4 = 0.8$, $\lambda_1 = \lambda_3 = 1.4$, and $\lambda_2 = \lambda_4 = 1.5$, we can get that the inequality (11.16) holds. By Theorem 11.1, systems (11.40) and (11.41) can reach the synchronization in fixed time under the pinning controller (11.14) and (11.15) with the above parameters. Moreover, we can also compute that $\bar{\mu} = 0.9$, $\bar{\lambda} = 1.4$, and $T_{\max}^1 = 29.2446$ second.

Case 2: In Theorem 11.2, choosing $\hat{\theta}_1^R = 44$, $\hat{\theta}_2^R = 47$, $\hat{\theta}_3^R = 48$, $\hat{\gamma}_1^R = 8.5$, $\hat{\gamma}_2^R = 9$, $\hat{\gamma}_3^R = 10$, $\hat{\theta}_1^I = 37$, $\hat{\theta}_2^I = 36$, $\hat{\theta}_3^I = 33$, $\hat{\gamma}_1^I = 8.5$, $\hat{\gamma}_2^I = 7$, $\hat{\gamma}_3^I = 9.5$, $\varepsilon_1^R = 2$, $\varepsilon_2^R = 2.1$, $\varepsilon_3^I = 4.2$, $\varepsilon_4^I = 2.3$, $\sigma_1^R = 1.2$, $\sigma_2^R = 2.2$, $\sigma_3^I = 2.2$, $\sigma_4^I = 2.2$, and the other parameters are the same as ones in Case 1, it is obvious that the inequality (11.32) in Theorem 11.2 is satisfied. Thus, systems (11.40) and (11.41) can reach the adaptive synchronization in fixed time under the pinning adaptive controller (11.30) and (11.31) with the above parameters. Meanwhile, via calculation, $T_{\max}^2 = 30.0667$ second.

For simulations, the trajectories of states and synchronization errors of systems (11.40) and (11.41) under the pinning controller (11.14) and (11.15) are shown in Fig. 11.2. Their synchronization response trajectories under the pinning adaptive controller (11.30) and (11.31) are shown in Fig. 11.3. From Figs. 11.2 and 11.3, it is clear that systems (11.40) and (11.41) reach the synchronization before the time $T_{\max}^1$ and $T_{\max}^2$. Obviously, Figs. 11.2 and 11.3 indicate the effectiveness of Theorems 11.1

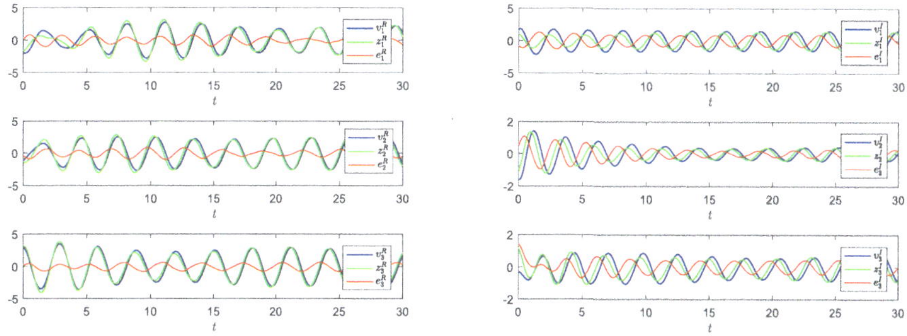

Fig. 11.1 Trajectories of states $v_l(t)$, $z_l(t)$ and synchronization errors $e_l(t)$ without the controller

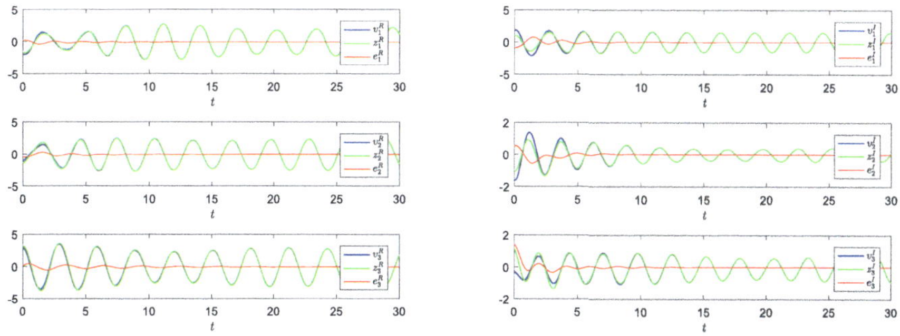

Fig. 11.2 Trajectories of states $v_l(t)$, $z_l(t)$ and synchronization errors $e_l(t)$ with the pinning controller (11.14) and (11.15)

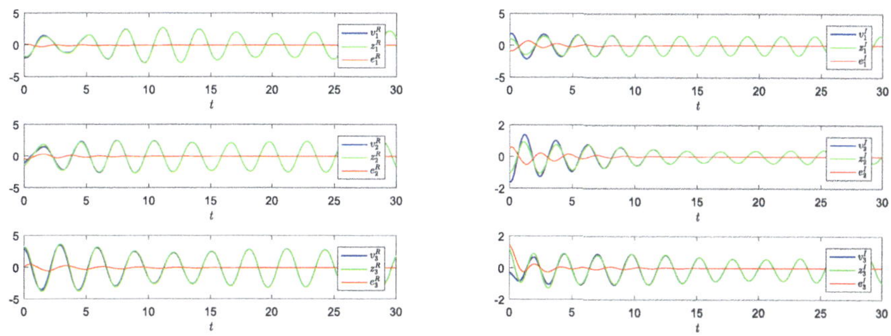

Fig. 11.3 Trajectories of states $v_l(t)$, $z_l(t)$ and synchronization errors $e_l(t)$ with the pinning adaptive controller (11.30) and (11.31)

and 11.2. Moreover, the curves of the adaptive gains $\theta_l^R(t)$, $\gamma_l^R(t)$, $\theta_l^I(t)$, and $\gamma_l^I(t)$ are depicted in Fig. 11.4, which shows that they tend to be some positive numbers finally. In fact, the ultimate limits of the adaptive gains are $\hat{\theta}_l^R$, $\hat{\gamma}_l^R$, $\hat{\theta}_l^I$, and $\hat{\gamma}_l^I$, respectively. This is consistent.

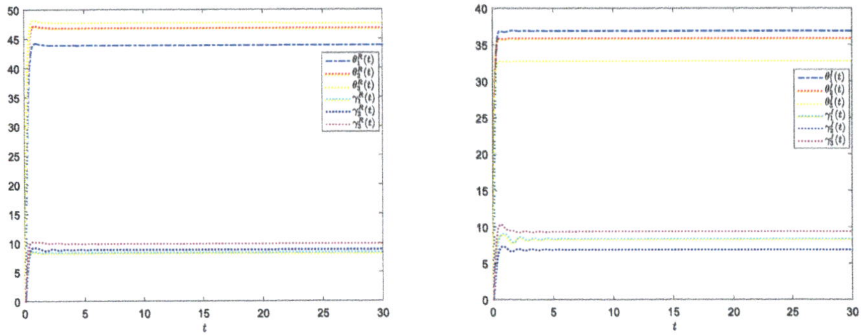

Fig. 11.4 Trajectories of control gains $\theta_l(t)$ and $\gamma_l(t)$

11.5 Conclusion

In this chapter, we have regarded CVINNs model as an entirety rather than reducing the order, designed two new pinning (adaptive) controller, established novel Lyapunov functions, investigated the pinning synchronization and adaptive synchronization problems for this model in fixed-time interval, and derived the corresponding synchronization criteria in terms of algebraic conditions. Finally, we have provided a numerical example to verify the validity of our results.

References

Babcock K, Westervelt R (1986) Stability and dynamics of simple electronic neural networks with added inertia. Phys D 23(1–3):464–469

Babcock K, Westervelt R (1987) Dynamics of simple electronic neural networks. Phys D 28(3):305–316

Angelaki D, Correia M (1991) Models of membrane resonance in pigeon semicircular canal type II hair cells. Biol Cybern 65(1):1–10

Lakshmanan S, Prakash M, Lim CP, Rakkiyappan R, Balasubramaniam P, Nahavandi S (2018) Synchronization of an inertial neural network with time-varying delays and its application to secure communication. IEEE Trans Neural Netw Learn Syst 29(1):195–207

Zhang M, Wang D (2019) Robust dissipativity analysis for delayed memristor-based inertial neural network. Neurocomputing 366:340–351

Wang J, Tian L (2019) Stability of inertial neural network with time-varying delays via sampled-data control. Neural Process Lett 50(2):1123–1138

Wang J, Wang Z, Chen X, Qiu J (2021) Synchronization criteria of delayed inertial neural networks with generally Markovian jumping. Neural Netw 139:64–76

Qin S, Gu L, Pan X (2020) Exponential stability of periodic solution for a memristor-based inertial neural network with time delays. Neural Comput Appl 32(8):3265–3281

Wei X, Zhang Z, Lin C, Chen J (2021) Synchronization and anti-synchronization for complex-valued inertial neural networks with time-varying delays. Appl Math Comput 403:126194

Li X, Fang J, Huang T (2020) Event-triggered exponential stabilization for state-based switched inertial complex-valued neural networks with multiple delays. IEEE Trans Cybern. https://doi.org/10.1109/TCYB.2020.3031379

Tang Q, Jian J (2019) Global exponential convergence for impulsive inertial complex-valued neural networks with time-varying delays. Math Comput Simulat 159:39–56

Li X, Huang T (2021) Adaptive synchronization for fuzzy inertial complex-valued neural networks with state-dependent coefficients and mixed delays. Fuzzy Sets Syst 411:174–189

Yu J, Hu C, Jiang H, Wang L (2020) Exponential and adaptive synchronization of inertial complex-valued neural networks: a non-reduced order and non-separation approach. Neural Netw 124:50–59

Wang S, Cao Y, Guo Z, Yan Z, Wen S, Huang T (2021) Periodic event-triggered synchronization of multiple memristive neural networks with switching topologies and parameter mismatch. IEEE Trans Cybern 51(1):427–437

Li Y, Luo B, Liu D, Yang Z, Zhu Y (2020) Adaptive synchronization of memristor-based neural networks with discontinuous activations. Neurocomputing 381:196–206

Wang L, Shen Y, Yin Q, Zhang G (2015) Adaptive synchronization of memristor-based neural networks with time-varying delays. IEEE Trans Neural Netw Learn Syst 26(9):2033–2042

Xu W, Zhu S, Fang X, Wang W (2019) Adaptive synchronization of memristor-based complex-valued neural networks with time delays. Neurocomputing 364:119–128

Hu J, Zeng C (2017) Adaptive exponential synchronization of complex-valued Cohen-Grossberg neural networks with known and unknown parameters. Neural Netw 86:90–101

Jia J, Zeng Z (2020) LMI-based criterion for global Mittag-Leffler lag quasi-synchronization of fractional-order memristor-based neural networks via linear feedback pinning control. Neurocomputing 412:226–243

Wu X, Huang L (2019) Pinning adaptive and exponential synchronization of fractional-order uncertain complex neural networks with time-varying delays. Neural Process Lett 50(3):2373–2388

Zhu S, Liu D, Yang C, Fu J (2020) Synchronization of memristive complex-valued neural networks with time delays via pinning control method. IEEE Trans Cybern 50(8):3806–3815

Fu Q, Zhong S, Shi K (2021) Exponential synchronization of memristive neural networks with inertial and nonlinear coupling terms: Pinning impulsive control approaches. Appl Math Comput 402:1–15

Li Y, Luo B, Liu D, Yang Z (2018) Robust synchronization of memristive neural networks with strong mismatch characteristics via pinning control. Neurocomputing 289:144–154

Ding S, Wang Z, Zhang H (2019) Quasi-synchronization of delayed memristive neural networks via region-partitioning-dependent intermittent control. IEEE Trans Cybern 49(12):4066–4077

Ling G, Ge M, Liu X, Xiao G, Fan Q (2021) Stochastic quasi-synchronization of heterogeneous delayed impulsive dynamical networks via single impulsive control. Neural Netw 139:223–236

Jian J, Duan L (2020) Finite-time synchronization for fuzzy neutral-type inertial neural networks with time-varying coefficients and proportional delays. Fuzzy Sets Syst 381:51–67

Liu C, Guo Y, Rao H, Lin M, Xu Y (2021) Finite-time synchronization for periodic T-S fuzzy master-slave neural networks with distributed delays. J Frankl Inst 358(4):2367–2381

Zhou C, Zhang W, Yang X, Xu C, Feng J (2017) Finite-time synchronization of complex-valued neural networks with mixed delays and uncertain perturbations. Neural Process Lett 46(1):271–291

Zhang Z, Li A, Yu S (2018) Finite-time synchronization for delayed complex-valued neural networks via integrating inequality method. Neurocomputing 318:248–260

Zheng B, Hu C, Yu J, Jiang H (2020) Finite-time synchronization of fully complex-valued neural networks with fractional-order. Neurocomputing 373:70–80

Zhang X, Zhou W, Karimi HR, Sun Y (2021) Finite-and fixed-time cluster synchronization of nonlinearly coupled delayed neural networks via pinning control. IEEE Trans Neural Netw Learn Syst 32(11):5222–5231

Ding X, Cao J, Alsaedi A, Alsaadi FE, Hayat T (2017) Robust fixed-time synchronization for uncertain complex-valued neural networks with discontinuous activation functions. Neural Netw 90:42–55

Xiao J, Cheng J, Shi K, Zhang R (2022) A general approach to fixed-time synchronization problem for fractional-order multidimension-valued fuzzy neural networks based on memristor. IEEE Trans Fuzzy Syst 30(4):968–977

Zhang Z, Guo R, Liu X, Zhong M, Lin C, Chen B (2021) Fixed-time synchronization for complex-valued BAM neural networks with time delays. Asian J Control 23(1):298–314

Zhang Y, Deng S (2020) Fixed-time synchronization of complex-valued memristor-based neural networks with impulsive effects. Neural Process Lett 52(2):1263–1290

Guo Y, Luo Y, Wang W, Luo X, Ge C, Kurths J, Yuan M, Gao Y (2020) Fixed-time synchronization of complex-valued memristive BAM neural network and applications in image encryption and decryption. Int J Control Autom 18(2):462–476

Long C, Zhang G, Hu J (2021) Fixed-time synchronization for delayed inertial complex-valued neural networks. Appl Math Comput 405:126272

Guo R, Lu J, Li Y, Lv W (2021) Fixed-time synchronization of inertial complex-valued neural networks with time delays. Nonlinear Dyn 105:1643–1656

Index

Z. Zhang et al., *Complex-Valued Neural Networks Systems with Time Delay*, Intelligent Control and Learning Systems 4,
https://doi.org/10.1007/978-981-19-5450-4

The manufacturer's authorised representative in the EU is Springer Nature Customer Service Centre GmbH, Europaplatz 3, 69115 Heidelberg, Germany. If you have any concerns regarding our products, please contact ProductSafety@springernature.com

Printed and bound by CPI Group (UK) Ltd, Croydon, CR0 4YY
30/09/2026
02225239-0005